The GUESTS *of* ANTS

The GUESTS
of ANTS

How Myrmecophiles
Interact with Their Hosts

**BERT HÖLLDOBLER &
CHRISTINA L. KWAPICH**

THE BELKNAP PRESS OF HARVARD UNIVERSITY PRESS

Cambridge, Massachusetts · London, England · 2022

Printed in Canada

FIRST PRINTING

Library of Congress Cataloging-in-Publication Data

Names: Hölldobler, Bert, 1936– author. | Kwapich, Christina L., 1985– author.
Title: The guests of ants : how myrmecophiles interact with their
    hosts / Bert Hölldobler and Christina L. Kwapich.
Description: Cambridge, Massachusetts : The Belknap Press of Harvard
    University Press, 2022. | Includes bibliographical references and index.
Identifiers: LCCN 2021053975 | ISBN 9780674265516 (cloth)
Subjects: LCSH: Myrmecophyes. | Host-parasite relationships. | Ants—Behavior. |
    Ant communities. | Parasites.
Classification: LCC QL523.M5 H55 2022 | DDC 595.7/54—dc23/eng/20220118
LC record available at https://lccn.loc.gov/2021053975

BOOK DESIGN BY ANNAMARIE MCMAHON WHY

Frontispiece: *Apocephalus* flies hovering over a soldier ant of *Pheidole dentata*.
(Courtesy of Alex Wild / alexanderwild.com).

In Memoriam Karl Hölldobler and Ulrich Maschwitz

Contents

Preface

AS A BOY I often accompanied my father, Karl Hölldobler, when he searched for myrmecophiles in an abandoned limestone quarry near our hometown, Ochsenfurt, in Northern Bavaria, Germany. Over decades the old quarry metamorphosed to a naturalist-myrmecologist's jewel, colonized by many species of ants, which could easily be observed and collected by turning over the many limestone rocks scattered over the grassland. A considerable number of these rocks sheltered ant colonies; a remarkable diversity of species were packed into a relatively small habitat of about 4,000–5,000 square meters. Most of these ant colonies were hosts to a variety of ant guests, the so-called myrmecophiles. Here I first encountered the myrmecophilous rove beetles *Lomechusa* (formerly called *Atemeles*) *emarginata, L. pubicollis,* and *Lomechusoides* (formerly called *Lomechusa*) *strumosus.* Here I observed the nitidulid beetle *Amphotis marginata,* the so-called highwayman beetle in the ants' world, and here I first saw rove beetle scavengers and predators of the genus *Pella.* My father introduced me to the clavigerite ant guest *Claviger testaceus* and the histerid myrmecophile *Haeterius ferrugineus.* He pointed out the strange myrmecophilous woodlouse *Platyarthrus hoffmannseggii;* a kind of silverfish, *Atelura formicaria,* which lives in ant nests; the slug-like myrmecophilous larva of the syrphid flies, *Microdon* spp.; the larvae of the chrysomelid myrmecophile *Clytra quadripunctata;* and many more. However, my father was most eager to catch the mysterious tiny crickets that swiftly darted around when we exposed the host colony under the rock, then called *Myrmecophila* but now called *Myrmecophilus.* This endemic population of *M. acervorum* in the old limestone quarry was a rare find, because the cricket was not common in the surrounding limestone habitats.

My father had cultured *M. acervorum* together with a variety of host ant species in his laboratory, and he studied their interactions with the host ants and tried to follow their life cycle. Although my father was not a professional zoologist, he had studied biology in addition to his training as a physician. As a relatively young surgeon in 1940, he was drafted to serve as war surgeon in World War II on the East Front in Finland-Karelia. In his notes he mentioned that he took with him a small ant colony and several ant crickets in a boxed formicarium, stowed away with his surgical instruments. How much he was able to watch his beloved ant crickets at the war front I do not know, but two years after this horrible war ended, he published a paper on *Myrmecophilus acervorum* in which he summarized all his observations made over several years.

It was due to my father that I first had interest in myrmecophiles, and since these early experiences I have become even more fascinated by these creatures, studying how each evolved their ways to exploit particular niches in ant colonies. But most of these enigmatic myrmecophiles are hard to find in large enough numbers to make experimental studies possible. This brought me back to the old limestone quarry, and indeed, for some of my study objects I could collect enough specimens for my work; in other cases I had to explore other venues. After I had published several papers on the behavioral mechanisms that underlie the interactions between ants and their guests, I emigrated to the United States, where I focused on a variety of other behavioral-physiological and ecological projects.

In 1990 my friend and colleague at Harvard University, Edward O. Wilson, and I published a major monograph on ants; one extensive chapter deals exclusively with myrmecophiles. Ever since, Ed kept "pushing" me to write a book on this subject, but so many other duties and pressing research and other book projects prevented me from doing so.

Then, in 2014, Christina Kwapich joined our Social Insect Research Group at the School of Life Sciences of Arizona State University as postdoctoral fellow.

Besides several other projects, she was becoming increasingly captivated by myrmecophiles, after having uncovered new spider, worm, and beetle associates during her excavations of *Pogonomyrmex* seed-harvesting ant nests as a graduate student with Walter Tschinkel. And so, she embarked on studies of myrmecophilous predatory spiders and *Myrmecophilus* crickets in Arizona. Together with our colleague and friend Robert Johnson, we commenced on a (still ongoing) research project on these myrmecophiles. So, when I finally decided to follow Ed Wilson's repeated advice to write a book on myrmecophiles, focusing mainly on the physiological and behavioral mechanisms underlying the interactions between ants and their guests, I invited Christina, who is now an assistant professor in the Biology Department of the University of Massachusetts Lowell, to join me.

At the beginning of our collaboration on this book, we agreed that we would not write a monograph on myrmecophiles. The focus of this work is instead the behavioral natural history of a selected group of myrmecophiles and critical analysis of the mechanisms of interactions between myrmecophiles and their host ants. Unavoidably, therefore, a large body of literature on myrmecophiles will not be cited in this book. We both reviewed this large body of literature and made the choice of studies most appropriate for our synthesis. Each one of us took main responsibility for certain chapters, but we both worked on multiple drafts of all chapters.

We agreed to dedicate this book in memoriam to Karl Hölldobler, who was my first teacher in myrmecology and who made substantial contributions to the study of myrmecophiles, and to my friend, the late Ulrich Maschwitz, a superb naturalist, myrmecologist, and tropical biologist with whom I had many stimulating discussions about symbiotic interactions in ants. We hope this book encourages and stimulates an interest in the guests of ants. Although the old quarry has been destroyed, it exists in my mind and in the pages of this volume.

BERT HÖLLDOBLER

The GUESTS *of* ANTS

1 Superorganisms: A Primer

ALL ANTS LIVE in societies. They are social insects, and they play a significant role in almost all terrestrial ecosystems. In fact, the ants are the little creatures that run the world. This is how our fellow myrmecologist Edward O. Wilson once characterized these formidable insects. Indeed, ants are among the most ecologically dominant organisms of this planet. Although the approximately 14,000 ant species known to science make up only about 2% of all insect species, their biomass (dry weight) constitutes 30%–45% of the entire insect biomass.

This tremendous ecological success of ants is due to their social organization, which is based on division of labor and communication. Whereas solitary animals can at any moment be in only one place and can be doing only one or two things, an ant colony can be in many places by deploying its workers and can be doing many different things because of the size of the worker cohorts and their division of labor. In evolutionarily advanced ant societies, this division of labor system is so densely knit and interdependent that it is justified to call such a social organization a superorganism. Not surprisingly, such social systems are prone to parasitism and exploitation by foreign organisms, and, indeed, ants are hosts of many freeloaders and parasites. In this book we focus on one such group, the so-called myrmecophiles. However, before we begin to explore the multifold interactions of myrmecophiles with their host ants, we present a brief introduction to the sociobiology of ants.

Ants live almost everywhere in the terrestrial environment, but they are especially abundant in tropical rainforests. A German research team, Ernst J. Fittkau and Hans Klinge (1973), working in the forests near Manaus in Northern Brazil, estimated that ants make up about 40% of the insect biomass in a Neotropical

1-1   Once in a year, a mature ant colony usually produces winged reproductive males and females that depart from the mother colony for the mating flight. The picture shows the large chestnut-brown, winged females and the much smaller black, winged males of the honey ant *Myrmecocystus mendax* leaving the nest for their nuptial flight. (Bert Hölldobler).

rainforest, and, considering all social insects (all ant species, all termite species, and some bee and wasp species), the social insects together constitute about 80% of the entire insect biomass. In some temperate hardwood forests, ants make up an average of 4.87 grams of dry biomass per m^{-2}, with an average abundance that is nearly ten times greater than all other leaf-litter-, soil-, and wood-dwelling macro-invertebrates combined (King et al. 2013). This enormously skewed relationship between the number of social insect species and their overwhelming biomass might not be true in all terrestrial habitats, but in most cases the ants dominate. The reason for their tremendous evolutionary success is cooperation and communication. Without communication there would be no cooperation. This is true for any society, even for organelles that cooperate within a cell or for interacting cells and organs that cooperate within an organism. One might say, without communication there would be no life.

Mating and Colony Founding

Social insects in general, and particularly the ants, are masters of cooperation (Hölldobler and Wilson 1990, 2009). The more advanced ant societies are characterized by a division of labor system, which is divided into two distinct functional parts. One concerns the division of labor among the sterile worker castes and the other is key to the colony reproduction. Usually only one or a few individuals in an ant society reproduce. These individuals are "queens." Their main function is to mate once in their life with one or several males. This usually happens during the mating flight, when the young, winged females and the winged males leave the colony (Figure 1-1) and, with other alates from different colonies of the same species, gather in the air in particular areas, such as above treetops, or on particular spots on the ground or low bushes.

At these communal mating sites an amazing mating frenzy takes place, which rarely lasts longer than an hour and might reoccur with different groups of individuals the next day or a few days later. After this mating event, the males' life usually ends, but before death, many of them succeed in delivering their sperm

to the females, who store the sperm in a special internal sperm pocket, the so-called spermatheca; though a male lives only a relatively short period, his sperm lives on in a kind of "sperm bank," which the reproductive females carry inside their bodies. The females, after mating, break off their wings, and from that moment on, their life will be almost entirely inside the incipient nest, which they dig in the ground, or in a tree trunk, or construct under a leaf in the tree canopy. The choice of habitat for nest construction differs, depending on the particular ant species. In many cases, the young queen raises her first offspring on her own. She lays eggs from which the first-instar larvae hatch (Figure 1-2).

The young queen feeds the young brood using up body reserves stored in a richly endowed fat body in which she carries valuable nutrients she received in her mother's colony. She also digests her own wing muscles, not needed anymore, and

SUPERORGANISMS

1-2   Young queen of the Australian green tree ant, *Oecophylla smaragdina,* founding a new colony. The queen feeds the developing larvae with her body reserves. (Bert Hölldobler).

1-3   An incipient colony of the honey ant species *Myrmecocystus mexicanus.* The first workers were raised by the queen, but now they take over in nursing their sister larvae and help them during the pupation process. (Bert Hölldobler).

in some species, young queens collect food in the surroundings of the incipient nest. This is, however, risky, and in the most "advanced" societies, the queen raises the first offspring in total isolation. Once the larvae have fully grown and completed the final larval instar, they pupate. During this pupal period the metamorphosis to an adult takes place. At its completion, the fully developed ant emerges from the pupal case. Although this freshly eclosed ant is also female, she looks very different from her mother queen. She is usually much smaller, with slightly built alitrunk (the body part adjacent to the head) and a relatively small gaster (the last body part, which includes most segments of the abdomen). In fact, this ant belongs to a different caste; she is a worker. More and more workers will eclose from their pupal stage in the next days and weeks, increasingly taking over all the maintenance activities of the growing colony. They will continuously extend nest structures, leave the nest to collect food, and care for the increasing number of freshly laid eggs and developing larvae. The mother queen, on the other hand, will from now on almost exclusively dedicate her existence to laying eggs (Figure 1-3).

As the colony grows, newly eclosed workers will be distinctly larger in size than the first workers, but in most species, they will still be smaller than the queen. In some ant species, the collective of workers consists of several subcastes (Figures 1-4, 1-5). This is impressively exhibited in the leafcutter ant genus, *Atta,* wherein the worker population consists of minors (very small workers), media (middle-sized workers), majors (larger workers), and super-majors (very large workers) (Figure 1-6). It is important to note that the smaller workers do not go on to become larger workers. In ants, as is the case in all holometabolous insects, the individual that emerges from the pupa has reached its terminal body size (Figure 1-7). The only developmental stage during which ants grow is the larval stage.

Division of Labor among Worker Ants

Let us now consider the division of labor among the worker ants. There are a multitude of tasks an ant colony must perform day in and day out. The queen has to be groomed, fed, and protected, and eggs gently removed from the queen.

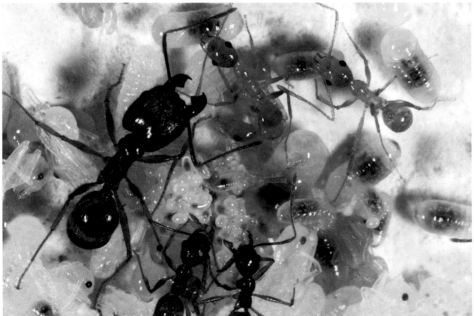

1-4 Once the founding queen has raised the first workers, the colony grows swiftly. A queen of *Pheidole desertorum* is shown in the upper picture surrounded by the first workers, eggs, larvae, and pupae. The lower picture shows the first square-headed soldiers and a few newly eclosed, still lightly colored workers. (Bert Hölldobler).

1-5 Minor and major workers of *Carebara urichi* from Costa Rica. The gigantic majors (soldiers) defend the nest and cut up larger prey items. They can also serve as living "storage containers," because they store nutritious reserve in their gaster. (Bert Hölldobler).

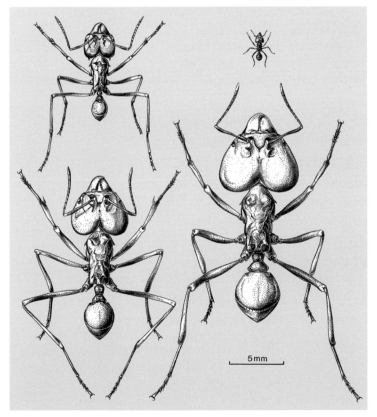

1-6 Worker subcastes of the leafcutter ant *Atta laevigata.* (Turid Hölldobler-Forsyth, ©Bert Hölldobler).

5mm

1-7 Super-major and minim (the smallest worker subcaste) of the leafcutter ant *Atta cephalotes*. Although both individuals are sisters, they are very different in size and equally different is the spectrum of tasks they perform inside and outside the nest. (Courtesy of AlexWild/alexanderwild.com).

This is not a small task, as in some species, such as the leaf cutter, *Atta laevigata,* a fully fertilized queen may at times lay 100–850 eggs in one hour. The hundreds of thousands of larvae that hatch out of the eggs inside the ant colonies must be fed. The often-large nest structures have to be continuously extended or renovated, cave-ins have to be repaired, and garbage removed. Food must be collected, such as honeydew from extrafloral nectaries, or excretions from hemipteran insects, like aphids or coccids. Leafcutter ants harvest leaf fragments, from which they produce a kind of humus inside the nest, on which they culture a particular fungus, which they eat and feed to their larvae (Figure 1-8).

Other species, such as mound-building wood ants (*Formica polyctena*), collect honeydew, and hunt and retrieve insects and other small arthropods that must be processed to be fed to larvae inside the nest. This process occurs at such a high volume that surroundings of wood ant mounds resemble green islands during pest insect outbreaks. In fact, forest entomologists report that the inhabitants

1-8   Brood chambers and fungus are intimately intertwined, as shown in this view of the fungus garden of a mature *Atta sexdens* colony. (Bert Hölldobler).

of a single large *Formica polyctena* nest can retrieve 100,000 insects in one day. This adds up to ten million prey insects in one summer (see references in Gösswald 1990). These numbers are astonishing, and the effect is striking: trees visited by wood ant foragers remain protected from insect herbivores (Figure 1-9).

1-9 An almost two-meter-high nest mound of the red wood ant *Formica polyctena* in Finland. Hundreds of thousands of ants and many myrmecophiles are housed in these nests. (Bert Hölldobler).

The physical aptitude of ants and the consequences of their division of labor are truly amazing. Although the foragers of one harvester ant species in the Southwest region of North America, *Veromessor pergandei,* are less than 8 mm long, they travel up to 40 meters each day to harvest seeds. At a modest egg-laying rate of 650 eggs per day, colonies of this species collect enough seeds to produce 3.4 kg of dry worker biomass per hectare of suitable Sonoran Desert habitat each year, equivalent to the biomass of a live human infant, or twenty-seven giant kangaroo rats (Kwapich et al. 2017). Ant colonies achieve such feats of production through species-specific patterns of labor allocation (such as brood care, cleaning, and patrolling the colony) and ratios of workers belonging to specialized morphological castes.

In general, ant species that live in large colonies, with thousands or hundreds of thousands of worker ants, also control large areas around their nest. They

1-10   A nearly constructed leaf-tent nest of the African weaver ant *Oecophylla longinoda*. One weaver ant colony lives in many such tent nests spread over the canopy of several major trees (upper). The weaver ants are very territorial. They fiercely defend their territory against conspecific intruders. The lower image shows three resident ants attacking a conspecific foreigner. (Bert Hölldobler).

are highly territorial. That is, they defend their territories against foreign intruders of their own species, or they attempt to aggressively extend their own colony's home range and wipe out smaller colonies of their species in their neighborhood. Although the territorial tactics and strategies can be quite different, depending on species and resource distribution, in all species we have investigated it was always the older workers that engaged in territorial battles. In fact, in the African weaver ants (*Oecophylla longinoda*) that live in the tree canopies of forests in Africa, the colonies designate special so-called barracks nests, where particular old workers are housed, which readily and aggressively rush out when their colony is threatened by a neighboring foreign *Oecophylla* colony (Figure 1-10).

SUPERORGANISMS

Territories of ant societies are part of the colony phenotype (extended phenotype). They are defended cooperatively by the workers of the owner colony. Because of the division of labor between the reproductive individuals and the usually sterile workers, fatalities caused by territorial defense have a different qualitative significance for social insects compared to those for solitary animals. The death of a sterile, and mostly older, worker represents an energy or labor debit, rather than the destruction of a reproductive unit. A worker death might more than offset its costs by protecting resources and the colony itself (Hölldobler and Lumsden 1980).

This brings us back to the basic organization of the division of labor among worker ants. As a rule (with some exceptions) one can say the youngest workers are nurses and take care of the queen and ant brood. The somewhat older workers are engaged in nest maintenance and food processing; the still older workers are patrollers and foragers; and the oldest workers, although they may also be engaged in foraging, when needed, are first in the risky line of defense. The system of division of labor is called age polyethism. Somewhat more complex is the division of labor system in those species that have morphological worker subcastes, for example, the already mentioned leafcutter ants. Although a certain age polyethism also exists, the special features of the morphological subcastes constrain the breadth of the tasks each subcaste is able to perform efficiently. The tiny minim subcastes in leafcutter ants usually handle eggs and small larvae and take care of the symbiotic fungus. Older minims and the somewhat larger minors often ride as hitchhikers on the leaf fragments carried by the medium-sized workers to the nest. The main task of hitchhikers is to clean the leaf fragments of microorganisms and spores that could be harmful to the symbiotic fungus and, most importantly, to fend off parasitic phorid flies that seek to attack the defenseless leaf-carriers. Obviously, the minors are very good at their tasks, but would perform miserably at leaf-cutting, and indeed this is not a task they perform. In a very insightful laboratory study with leafcutter ant colonies, Edward O. Wilson discovered that gardening is achieved by means of an intricate assembly line, in which the leaf fragments are processed and the fungus reared in steps, with each of the steps accomplished by a different worker sub-caste (Figure 1-11). The function of the members of the largest cast, called super-majors, is primarily defense (see Figures 1-6, 1-7). They serve as bodyguards for the precious queen and fend off other threats to the colony. Occasionally, they can be observed cutting hard vegetable material.

Division of Labor in Reproduction

The key feature of ant societies is the reproduction division of labor between the reproducing queen or queens and the multitude of usually sterile workers. Such social systems are called eusocial. The late Charles Michener, one of the greatest comparative bee naturalists, defined the fundamental tenets of eusociality: "Cooperative brood care, differentiation of colony members into fertile reproductive castes and sterile non-reproductive castes, an overlap of generations such that offspring assist their parents in brood care and other tasks involved in colony maintenance" (Michener 1969). This is a widely accepted definition of the eusocial insects, such as all ants, certain bee and wasp species, all termites, several other eusocial arthropods, and even a few vertebrates. However, as entomologist and sociobiologist Raghavendra Gadagkar and other sociobiologists and evolutionary biologists have pointed out, all three fundamental traits are not equally well developed in all cases. This should not surprise us, because nature cannot be neatly sorted into separate drawers. Nevertheless, the term *eusociality* is useful, if we use it exclusively for such social aggregations wherein pronounced division into one or a few fertile reproductive individuals and many sterile nonreproductive members is exhibited.

In the insect family Formicidae (ants) we find species where the nonreproductive members still have the full reproductive potential but, in the presence of a fertile queen, remain sterile. Should the queen wane in her fertility or die, individuals among the workers aggressively compete for reproductive ranks until one or a group of reproducing workers have established themselves. These egg-laying workers (which have mated with some of their brothers) are now called

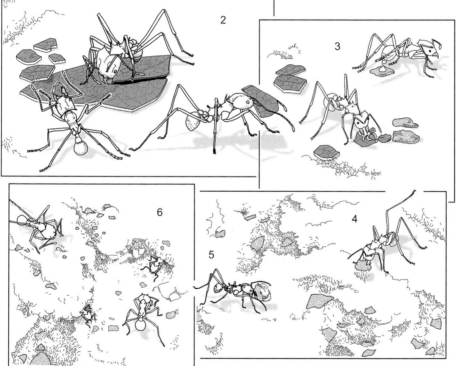

1-11 The "assembly line" by which colonies of *Atta cephalotes* create a fungus garden with fresh-cut leaves and other vegetation. (Courtesy of Margaret Nelson).

gamergates. They are not called queens, because the queen caste is also morphologically defined. A gamergate, however, looks like any other worker but is a reproductive individual in the colony. This social organization is found in, among others, the ant genus *Harpegnathos* (Figure 1-12). Colonies of this genus experience phases of social quiescence and phases of social revolutions that break out when a gamergate's fertility wanes or a gamergate dies.

1-12 The upper picture shows a worker of the Asian jumping ant *Harpegnathos saltator*. It is almost the same size as the queen (lower), except that it does not feature the heavier-built alitrunk that accommodates the wing muscles in the virgin queen. (Bert Hölldobler).

The situation is very different in evolutionarily advanced (or derived) eusocial organizations, such as colonies of the weaver ants, the leafcutter ants, the huge colonies of the army ants, the *Solenopsis* fire ants, or the mound-building red wood ants. In leafcutter ants, African weaver ants, and army ants, normally only one queen resides in a colony. In the mound-building wood ants, some species produce colonies that also have only one queen (monogynous), while others can have a number of queens (polygynous), but in all colonies the number of workers can be hundreds of thousands or even millions. The workers in these species may have rudimentary ovaries, and in the absence of the queen they may be able to lay viable eggs, but these eggs will not be fertilized, because workers

SUPERORGANISMS

in these species have no functional spermatheca. If larvae hatch out of these eggs, they develop into males, because in all hymenopteran species (and all ants belong to the insect order Hymenoptera), males develop from unfertilized eggs. They are haploid (have only one set of chromosomes from their mother). Having said this, it is important to point out that in the presence of a highly fertile queen a worker laying viable eggs is extremely unlikely, and should it happen, the eggs will usually be killed by fellow workers, no matter whether the egg layer is a full sister, a half sister, or a quarter sister. It has been demonstrated in a number of ant species that the workers recognize the fertility status of their queen, based on a particular hydrocarbon blend on the queen's body, and in certain species that have huge nests or contain many subnests (polydomous), the queen's fertility signal is also on queen-laid eggs, and, in this way, workers spread the message of a fertile queen when they carry eggs to distant brood chambers. It is important to note that non-queen members of species with this kind of advanced eusociality will never be able to take over a queen's role as a fully functional reproductive. In the evolutionary biology of these species, "the point of no return" has been reached. Justifiably, we call such ant colonies superorganisms (Hölldobler and Wilson 2009).

If we were to accept this concept of a superorganism, with its codicil of little or no reproductive competition among nestmates, many of the poneromorph societies, to which the formerly mentioned *Harpegnathos* belong, might not be considered superorganisms because intracolony reproductive competition is indeed conspicuously common. Nevertheless, among themselves, the thousands of social insect species display almost every conceivable grade in division of labor, from little more than competition among nestmates for reproduction status to highly complex systems of specialized subcastes. The level of this gradation at which the colony can be called a superorganism is subjective; it may be at the origin of eusociality or at a high level "beyond the point of no return," in which within-colony competition for reproductive status is greatly reduced or totally absent. One can argue that, at this level, the colony as a whole becomes a major target of natural selection. That is, between-colony selection is stronger than within-colony selection. To adopt the jargon of Richard Dawkins, the colony serves as a "vehicle for genes." The mature colony produces thousands of virgin queens and males every year. Colony fitness is theoretically measured based on the number of young queens per colony that succeed to produce new incipient colonies that survive and eventually reach the stage of maturity, so that they themselves will be able to produce reproductive females and males on their own. Of course, included in this calculation of colony fitness is the number of males of this colony that succeed to mate with females that will be successful in colony foundation.

The often-complex nest structure of such colonies results from the collective actions of many individuals, and thus represents an extended phenotype of the cooperating groups—the superorganisms—that build them. Only a few studies have provided quantitative descriptions—in particular, those of Walter Tschinkel (2021), who has also produced three-dimensional casts of nest structures by pouring a thin slurry of dental plaster or molten aluminum into nests. Similar comparative studies of the nests of leafcutter ant species were conducted by Luiz Forti and his team in Brazil. Interestingly, although each cast produced by Walter Tschinkel or Luiz Forti differs, the scientists were able to document species-specific structural features (Figure 1-13).

Obviously, these have to be considered collective, species-specific traits, and many more examples of such species-specific colony traits can be listed, such as the extensive horizontal foraging trunk routes aboveground that can extend more than 250 meters from the nest in certain leafcutter ant species (e.g., *Atta colombica*) (Figure 1-14). These features can be considered part of the extended phenotype of the ant superorganism.

Let us briefly summarize: a superorganism is a colony of individuals in which one or few reproduce and the majority remain sterile, and is self-organized by division of labor and united by a closed system of communication. One can say that the eusocial insect society possesses features of organization analogous to the properties of a single organism. The colony is divided into reproductive castes (analogous to gonads) and sterile worker castes (analogous to somatic tissue),

1-13   A mature nest of the fungus-growing ant *Atta laevigata* in Brazil was excavated after six tons of cement and 8,000 liters of water were poured into the nest to preserve the structure in petrified form. (Courtesy of Wolfgang Thaler).

which are further divided into specialized labor groups (organs). Nevertheless, among the thousands of eusocial species, we can find almost every conceivable grade in the division of labor, from hierarchical organization with competition among nestmates for reproductive status and poorly developed division of labor to the highly complex cooperative networks with specialized worker subcastes. The level of this gradient at which the colony can be called a superorganism is perhaps subjective. It may be at the origin of eusociality, or at a higher level in which within-colony competition for reproductive status is greatly reduced or absent. In our view, insect societies with considerable reproductive competition

1-14   The long trunk route of a leafcutter ant colony, along which *Atta* workers carry yellow leaf fragments harvested from a flowering tree or bush. (Courtesy of Hubert Herz).

among nestmates and, as a consequence, poorly developed division of labor among workers, may have some incipient superorganism traits but do not deserve to be called fully functional superorganisms. In true superorganisms, the size dimorphism (morphological skew) between reproductive individuals (queens) and sterile individuals (workers) is large, and reproductive division of labor is deep and not plastic. Although workers receive all their genes from the queen and her mates, they exhibit very different phenotypes because, during their larval development, social and environmental influences cause different genes to be turned on and expressed in workers than in queens and males. The phenotypic plasticity continues during adult ontogeny. From the behavioral interactions of hundreds of thousands, or even millions, of workers, colony-specific traits emerge that are part of the collective colony phenotype of the superorganism.

Obviously, such an ant superorganism provides many physical and social niches for parasites. Although no ant species appears to be totally free of parasites, parasitic species that exploit the social "acquisition" of a society are usually fewer in poneromorph ant societies. This contrasts with the numerous socially parasitic arthropods associated with such ant species whose colonies can be considered true superorganisms.

Parasites inside the Superorganism

Behaviorally parasitic symbioses in ants are usually subdivided into two distinct categories: the behavioral parasitism among different ant species, and the behavioral parasitism of other organisms (such as worms, beetles, butterflies and moths, wasps, flies, mites, snakes, and other vertebrates), living close to or inside the ant colony. Socially parasitic ants are usually called "social parasites," and other parasitic or, in some cases, mutualistic organisms in ant societies are called "myrmecophiles" (from Greek *myrmekes* = ants, *philia* = friendship), or they are often designated as guests of the ants. Indeed, in the most evolutionarily derived grade of such myrmecophilous behavior, one gets the impression that ants really "love" their guests. However, as we will see, many myrmecophiles "furtively" hoodwink their hosts. The advantage of these relationships is in most cases entirely on the myrmecophile's side.

In recent years some scientists have chosen to call all organisms that exploit the host ants' social attainments "social parasites" (Thomas et al. 2005). At first glance, this appears to be logical, but it can be confusing, and we think the old myrmecologists had good reasons for making a distinction between the types of social exploitation. The evolutionary origins of socially parasitic ants and myrmecophiles are quite different. One can say that the socially parasitic ants "derived from within," that is, almost all social parasitic ants are closely related to their host ant species (Hölldobler and Wilson 1990; Buschinger 2009; Rabeling 2020; Degueldre et al. 2021), although there are a few exceptions (see Maschwitz et al. 2004; Witte et al. 2009; Fischer et al. 2020). In contrast, the evolutionary pathways of myrmecophilous parasites originated "from outside," that is, they are derived from ancestor organisms that lived outside of ant nests and do not share a direct evolutionary ancestry with their host ants. Fossil evidence suggests that the symbioses between ants and their myrmecophiles are almost as ancient as ants themselves. Close inspection of ninety-nine-million-year-old Burmese amber revealed a fossil clown beetle, *Promyrmister kistneri,* with well-developed myrmecophilous adaptations that would have allowed it to exploit the ants of its day, now presumed to be long extinct (Zhou et al. 2019).

While we agree there should be a distinction between ant social parasites and other parasites that exploit the social organization of ant colonies, we have decided not to follow the traditional subdivision of myrmecophilous organisms into "synechtrans," myrmecophiles that are mostly predators and are treated as hostile by the ants; "synoeketes," myrmecophiles that live as scavengers in or near the ant nests and are mostly ignored by the ants; and "symphiles," myrmecophiles that are tended and even fed by the ants as if they were members of the ant colony (Wasmann 1894; Wheeler 1910; Donisthorpe 1927; Wilson 1971; Hölldobler and Wilson 1990). As we will see, many times these stereotyped definitions do not reflect reality, and we therefore decided not to continue to adhere to these categorizations (see also Mynhardt 2013). (We also need to advise the reader that we do not cover the huge and extremely interesting topics of trophobiosis between ants and hemipteran species, or interactions between ants and plants.)

The Jesuit monk Erich Wasmann pioneered the scientific study of myrmecophiles. In 1894, Wasmann counted 1,246 species of myrmecophilous arthropods, and the doyen of British myrmecology, Horace Donisthorpe, enumerated 1,392 species of myrmecophiles in his book *The Guests of British Ants,* published in 1927. Donisthorpe estimated that perhaps 5,000 myrmecophilous arthropod species might exist worldwide. Without even attempting to provide an exact count, we can confidently state that nowadays about 10,000 myrmecophile species are known, most of them beetles (Coleoptera), including species of thirty-five beetle families (see Parker 2016 for a more recent review), at least five families of butterflies (Lepidoptera), sixteen families of Hymenoptera, fourteen families of flies (Diptera), twenty-five families of Hemiptera, one family each of Orthoptera and Thysanura, and three families in Blattodea. In addition, six spider families, four mite families, and one tailless whip scorpion family have been identified as myrmecophiles (Glasier et al. 2018). The book *The Guests of Japanese Ants* by M. Maruyama, T. Komatsu, S. Kudo, T. Shimada, and K. Kinomura (2013) provides a gorgeously picturesque account of the amazing diversity of myrmecophilous species in Japan, and Paul Schmid-Hempel published a major monograph, *Parasites in Social Insects* (Schmid-Hempel 1998). In 1990, Bert Hölldobler and Edward O. Wilson, in their book *The Ants,* published a list of all arthropod orders, families, and many genera of myrmecophiles known up to 1990. The table listing many of the known genera and families in this coffee-table book comprises fourteen pages and shows that the greatest abundance of myrmecophilous species is found in Coleoptera (beetles). The excellent work of a new generation of investigators, including Joseph Parker (Parker 2016; Maruyama and Parker 2017); Munetoshi Maruyama, Rosli Hashim, Christoph von Beeren, and Volker Witte (Maruyama et al. 2010a,b); Daniel Kronauer (2020); and Thomas Parmentier (Parmentier et al. 2015a,b; 2016a,b; 2017a,b), to name a few, provide new insights concerning the diversity and phylogeny of myrmecophilous beetles. Jean Paul Lachaud and colleagues edited two volumes of the journal *Psyche* that are entirely devoted to the parasites of ants (Lachaud et al. 2012; Lachaud et al. 2013).

In this book we do not aim to describe and analyze the enormous biodiversity of myrmecophilous species. Instead, we focus on the description of behavioral mechanisms that enable the myrmecophiles to coexist with and, in a considerable number of cases, exploit their host ant species. In the chapters that follow, we review the experimental analyses of the interactions of the myrmecophiles and ants. We investigate how the so-called guests of ants, over the course of evolution, have broken the communication code of their ant hosts and have thereby been able to coexist as socially parasitic invaders with their host ant species by exploiting host resources—for example, by inserting themselves into the social food flow normally shared among nestmates.

The ant superorganism provides many niches for parasites, but at the same time, we may also view the ant colony as an ecosystem, as an ecological island, in order to better understand certain aspects of the biology of symbionts. The ant colony and its surroundings are richly structured into many diverse micro-habitats, such as foraging trunk routes, refuse areas, peripheral nest chambers, food storage chambers (not limited to seeds, fungus, and desiccated insect corpses), brood chambers (with separate areas for pupae, larvae, and eggs), queen chambers, and, of course, the bodies of adult and immature inhabitants of the nest. Ants are ecosystem engineers, and their colonies are hidden wellsprings of biodiversity that can be considered in an ecological framework.

2 Inside and on the Bodies of the Ants

ONE OF THE major niches for myrmecophiles in the ant superorganism are the bodies of the individual ants themselves. Some of these body dwellers are just hitchhikers, like the many species of mites that ride on army ants in the forests of the American tropics, or those that cling beneath the heads of ants and steal food directly out of their mouths. Other parasites penetrate and invade the bodies of the hosts and manipulate from inside the ants' behavior to their own advantage.

Indeed, a multitude of organisms reside on and within the bodies of individual ants. Some are beneficial to their hosts, as part of a mutualistic relationship, while others are just commensals. However, a great variety of these organisms can be considered true parasites. The diversity and number of such co-inhabitants is enormous: they include cestodes (Buschinger 1973; Trabalon et al. 2000; Beros et al. 2015), nematodes (Poinar 2012; Csösz 2012), trematodes, fungi, single-cell parasites, and others (see de Bekker et al. 2018).

Mutualistic Symbionts: The Case of *Blochmannia*

In 1887 Friedrich Blochmann described "bacteria-like structures" in the tissues of the midgut and ovaries of the carpenter ant *Camponotus ligniperdus*. Subsequent investigations have revealed these bacteria are gram-negative rods of variable length, and large numbers of them are packed into bacteriocytes (or mycetocytes) intercalated between the normal epithelial cells of the ants' midgut (Dasch et al. 1984; Schröder et al. 1996). The same bacteria are also found in the cytoplasm of oocytes of queens and workers, and transmission of the bacteria is obviously vertical, that is, female and male offspring inherit the bacteria from their mothers (Kolb 1959; Schröder et al. 1996).

A comparative genetic analysis of thirteen *Camponotus* species and their endosymbiotic bacteria revealed that the phylogenies of the bacteria and hosts deduced from the sequence data show a high degree of congruence, which strongly suggests co-speciation or parallel evolution of the bacteria and their host ants and provides further evidence for a maternal transmission route of the symbionts (Sauer et al. 2000). Based on the molecular characterization, it was proposed that these bacterial endosymbionts of the ant genus *Camponotus* be assigned to a new genus, *Blochmannia,* with the species named according to their hosts—for example, *Blochmannia floridanus, B. ligniperdus, B. herculeanus,* and so on (Schröder et al. 1996; Sauer et al. 2000) (Figures 2-1, 2-2). Subsequent work strongly supported the proposal of parallel evolutionary trends among the bacterial symbionts and their host ant species (Degnan et al. 2004, 2005; Wernegreen et al. 2003, 2009; Ramalho et al. 2017a,b; see also Sameshima et al. 1999). In fact, Sameshima et al. (1999) showed that *Blochmannia* occurs in other genera of the tribe Camponotini, *Polyrhachis* and *Colobopsis,* and Jennifer Wernegreen and her colleagues (2009) identified *Blochmannia* in the Camponotini genera *Calomyrmex, Echinopla,* and *Opisthopsis.* Wernegreen and colleagues (2009) state that *"Blochmannia* is nestled within a diverse clade of endosymbionts of sap-feeding hemipteran insects, such as mealybugs, aphids, and psyllids." According to their analyses, "a group of secondary symbionts of mealybugs are the closest relatives of *Blochmannia."* These findings strongly suggest that ancestral Camponotini ants acquired the symbionts from their hemipteran trophobionts, from where they obtained their food.

Indeed, a previously conducted phylogenetic analysis of a set of conserved protein-coding genes shows that *Blochmannia floridanus* (and most likely also the other *Blochmannia* species) are phylogenetically related to *Buchnera aphidicola,* which lives in aphids (Baumann 2005). In fact, phylogenetic evidence suggests

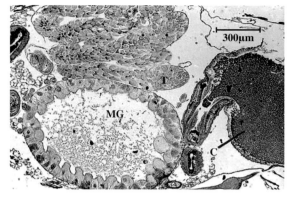

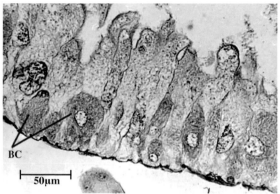

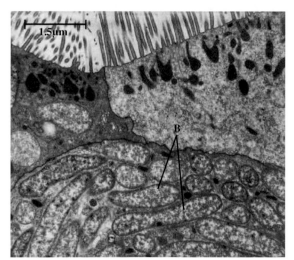

2-1 Sagittal section through the gaster of a male of the carpenter ant *Camponotus herculeanus,* showing (upper) the midgut (MG), the crop (C), and testes (T). The middle image depicts a close-up of the midgut epithelium, with the bacteriocytes (BC) (also called mycetocytes), intercalated between the midgut cells (enterocytes) (Bert Hölldobler). The lower picture is an electron-microscopic image of part of a bacteriocyte containing the rod-shaped bacteria *Blochmannia* (B) and the enterocytes with the microvilli on the apical surface. (Courtesy of Roy Gross).

2-2 All investigated *Camponotus* species are hosts of *Blochmannia* species. The endo-symbiont of *Camponotus floridanus* queen and workers, shown in the upper picture, is *Blochmannia floridanus*, and in *Camponotus socius* (lower), Blochmannia socius. (Bert Hölldobler).

that the ancestor of *Blochmannia* was horizontally transferred from hemipterans (with which many formicine ant species foster trophobiotic relationships) to the most recent common ancestor of the tribe Camponotini approximately 51 million years ago (Gil et al. 2003; Wernegreen et al. 2009; Ward et al. 2016). These endosymbiotic bacteria have extremely reduced genomes and include some strains of *Buchnera aphidicola* with a small genome of only about 450 kb (Gil et al. 2002). Likewise, *Blochmannia floridanus* also has a very reduced genome of only about 700 kb in size (Gil et al. 2003). Remarkably, *Blochmannia* lacks the known genes that code for the biosynthetic pathways of nonessential amino acids, while retaining genes that code for biosynthetic pathways of essential amino acids.

It was not known whether the *Camponotus* ants gain any benefits from housing *Blochmannia* bacteria in their midgut epithelium until Heike Feldhaar, Roy Gross, Evelyn Zientz, and their collaborators were able to experimentally demonstrate the significance of these endosymbionts in the developing host ant larvae and pupae, and for the nurse ants' brood care potential. They followed the expression of several genes related to nitrogen metabolism, as well as genes involved in the synthesis of amino acids in *Blochmannia* symbionts of *Camponotus floridanus,* and found that, in the early developmental stages of the host ants, symbiont gene expression was quite low, and increased only in the last larval instars. A peak of gene expression "related to nitrogen recycling" was detected throughout the pupation phase, and the expression "of biosynthesis pathways for aromatic amino acids was elevated only during a short phase in pupation" (Zientz et al. 2006). These results indicate that the endosymbiotic bacteria may play a significant role during the developmental processes of the host ants. However, perhaps even more exciting is the discovery that *Blochmannia* bacteria appear to affect the brood-tending ants' ability to nurse the ant larvae. Zientz et al. (2006) provided the first evidence with experimental worker groups that were treated with antibiotics, but the decisive proof of the endosymbiont's significance for the success of brood-tending ants was delivered by Feldhaar et al. (2007).

Here is a brief description of their experimental protocol: Normally, *Camponotus floridanus* colonies cultured in the laboratory were fed with cockroaches, honey water, and so-called Bhatkar agar (a mix of agar solution with chicken egg, honey, and vitamins). This standard diet was also provided to the control worker groups. Meanwhile, the first series (A) of experimental worker groups received a special chemically designed diet, which contained, among other substances, all essential amino acids for ants (Feldhaar et al. 2007). A second series (AR) of experimental worker groups received the same special diet as those in A, except that, in this case, the antibiotic 2% rifampicin was mixed into the food every other week. In a third series (B), the workers also received an artificial diet, like in A, except the essential amino acids were omitted and instead nonessential amino acids were added. Finally, a fourth series (BR) received the same diet as B; however, 2% rifampicin was added every other week. Each worker group received equal numbers of eggs and first-instar larvae, in three installments over a time period of eight weeks, and the groups' success at raising pupae was recorded over a period of twelve weeks.

The results were striking: Workers fed with artificial diet (A) or without essential amino acids (B) raised larvae to pupation as successfully as the workers of control groups did (fed with cockroaches, honey water, and Bhatkar agar). However, the workers treated with antibiotics and fed with a diet that was lacking the essential amino acids (BR) failed to raise the larvae to pupation, whereas antibiotic-treated worker groups that were fed with a diet containing essential amino acids (AR) were able to raise larvae to pupation almost as well as the control groups did. These wonderful experiments convincingly demonstrate that the *Blochmannia* endosymbionts provide their host ants essential amino acids lacking in their diet, and that they also play a role in nitrogen metabolism.

Although these results represent only the beginning of our understanding of the complex symbiotic relationship between *Blochmannia* and *Camponotus,* it appears justified to say that the endosymbionts play at least a "role in nutritional upgrading, i.e. enhancing the nutritional value of food resources," and thereby probably enable *Camponotus* to exploit food sources that would be useless to the ants without the endosymbionts' support (Feldhaar et al. 2007). Similar conclusions were reached in studies with the herbivorous turtle ants (*Cephalotes*), where Hu et al. (2018) discovered that gut-associated bacteria transform urea (possibly also from bird droppings) into essential amino acids that are used by their host ants (see also Moreau 2020).

As we have seen, the "cooperation" of *Blochmannia* spp. with its host *Camponotus* spp. is especially significant in brood-tending ant workers, and indeed, Florian Wolschin and his collaborators (2004) have found that the bacteria proliferate during the pupal phase and immediately after eclosion of the adult ants. In older workers the number of bacteria in the midgut bacteriocytes decreased significantly. This pattern was found in *Camponotus floridanus, C. herculeanus,* and *C. sericeiventris*. In fact, in the last species, workers more than three years old could be investigated. Whereas workers a few months old had bacteriocytes densely packed with bacteria, hardly any *Blochmannia* could be detected in the old workers. Incidentally, the same pattern could be found in *Camponotus herculeanus* males, which live inside the colony for more than eight months. Young males had many filled bacteriocytes intercalated in their midgut epithelium. In this phase the males develop a massive, fat body and exhibit some social behavior such as exchanging food by trophallaxis with fellow males, young workers, and virgin queens. After the winter pause, the males use up their fat body and eject the *Blochmannia* endosymbionts into the gut lumen (Hölldobler 1966). Once the *Camponotus* males are ready for the mating flight, no *Blochmannia* can be found inside them (Wolschin et al. 2004). An age-dependent degeneration of the midgut bacteriocytes was also found in *C. floridanus* queens, who, at several years of age, no longer harbored bacteria in their midgut bacteriocytes but still carried them in their ovaries (Sauer et al. 2000; Wolschin et al. 2004).

In conclusion, the presence of bacteria in the bacteriocytes of the midgut decreases with age, whereas the bacterial population in the ovaries is dependent on the reproductive state of the ant hosts. Considering that in *Camponotus,* as in most ant species, division of labor is based on age polyethism, where the young workers usually feed the larvae and queen, their especially rich endowment with the endosymbiotic bacteria is perfectly plausible. The endosymbionts are not significant for the older, nonreproductive workers but may play a significant role in ant workers that nurse the immatures.

Recently, new research has been published on the role of the bacteriocyte dynamics on the development of *Camponotus* ants (Stoll et al. 2010), and the ways that the endosymbiotic *Blochmannia* affect the embryonic development of the host ants (Rafiqi et al. 2020). In the context of our current discussion, the work by Sinotte et al. (2018) is of special interest. To elucidate the symbionts' effects on development and disease defense, the authors, by feeding the ants with antibiotics, diminished or completely depleted *Blochmannia floridanus* in colonies of *Camponotus floridanus*. These experiments revealed several effects, although it is not always clear whether these are truly causal effects or artifacts. Colonies treated with antibiotics had fewer workers, and the workers were of a smaller body size, and a lower major-to-minor worker caste ratio. This suggests that the endosymbionts play a crucial role in development, which was recently demonstrated by the research group led by Ehab Abouheif (Rafiqi et al. 2020). Although workers of treated colonies exhibited extenuated cuticular melanization, they showed higher resistance to the entomopathogen *Metarhizium brunneum*. It appears that the symbiont reduces the ants' ability to fight infection, despite the availability of melanin, which is known to positively affect the immune system. The authors conclude:

> The primary endosymbiont *Blochmannia* provides essential nutritional supplementation to its host, *Camponotus,* which in the case of *C. floridanus,* facilitates growth and cuticle maturation in individuals and may contribute to colony polymorphism, a trait that is key to the success of the genus. The symbiont also imposes a critical trade-off in that it increases the ants' susceptibility to fungal infection and thus secondarily raises the cost of pathogen transmission within the colony. The extensive coevolutionary relationship between the mutualistic partners and ubiquity of the symbiont

throughout the genus implies that the bacteria's benefits to the ants' life history, ecology, and evolution, outweigh the costs to their individual and social immunity. (Sinotte et al. 2018, 1)

We have chosen to focus on studies of *Blochmannia* these past few pages because research in recent years has allowed us to begin to understand the function of these mutualistic endosymbionts. However, there are many more microorganisms closely associated with ants.

For instance, consider the gut symbionts of herbivorous ants that appear to be closely linked with their host ants' evolution (Russell et al. 2009; Russell et al. 2017; Martins and Moreau 2020). The gut of the turtle ant genus *Cephalotes* is fitted with a special proventricular filter that is populated with symbiotic bacteria during oral and anal trophallaxis. The porosity of the filter partitions microbial communities between the crop and midgut, much like the filter used in some hemipterans, while allowing the passage of dissolved nutrients for digestion (Lanan et al. 2016). In fact, the symbiosis between bacteria and ants obviously had major effects on the evolutionary ecology of many ant species. The work by Corrie Moreau and her colleagues has provided many comparative case studies that suggest that gut-associated bacteria play a significant role in the evolution of ant herbivory (Russell et al. 2009; Pringle and Moreau 2017). They have also suggested that environmental factors may play a role in structuring the microbiota in ants (Ramalho et al. 2019) and have demonstrated that symbioses across the ant species are not evenly distributed and that some lineages seem to be without any bacterial symbionts (Russell et al. 2017).

Furthermore, there exists a rich literature on symbiotic external microorganisms in leafcutter ants, colony hygiene in leafcutter ants, and their role in "agricultural pathology," mainly based on the discoveries and work by Cameron Currie and his collaborators (Hölldobler and Wilson 2011). Finally, Ronque et al. (2020), in a very recent study conducted in the laboratory of Paulo Oliveira, demonstrated that the bacterial communities on the bodies of four fungus-growing ant species (*Mycocepurus smithii, Mycetarotes parallelus, Mycetophylax morschi,* and *Sericomyrmex saussurei*), collected from three different environments in the Brazilian Atlantic rainforest, differed both by species and among colonies of the same species, whereas bacterial communities from nest workers and foragers did not differ or differed only slightly within each ant species. We can only speculate about the biological function of these patterns. (For other studies of the bacterial communities on the cuticula of ant species, see Birer et al. 2020.)

Internal Parasites and Parasitoids That Affect the Ants' Behavior

Parasitoids develop inside a living host and, ultimately, kill it. In contrast, parasites may also live on or inside the host, but generally benefit from the host's continued survival. In social insects, and particularly in ants, internal parasites and parasitoids can have multiple effects on the infected ant's appearance and physiology and its behavior and interactions with nestmates. We will first consider how tapeworms can affect the host ant's appearance and behavior.

Tapeworms: Cestodes

Temnothorax nylanderi is a myrmicine ant species, the colonies of which live in cavities of decaying wood and consist of one queen and about 10–300 relatively small workers (2–3 mm). The workers and queen usually have a reddish-yellow color. In 1972, Luc Plateaux, who studied many aspects of these ants, discovered that several colonies contained some workers that had a bright whitish-yellow color and looked almost like "albino-ants." It was soon revealed that those workers were parasitized by the cestode worm (tapeworm) *Anomotaenia brevis*. In a follow-up study, Marie Trabalon, Luc Plateaux, and their collaborators (2000) discovered that this parasite not only induces the unusual pigmentation of the adult ants but also affects morphological features. It precipitates smaller body size and reduction of the size of head, eyes, and legs, and increases the size of petiole and postpetiole. Most importantly, it affects a change of the colony's cuticular hydrocarbon profile. Among ants, colony-specific blends of hydrocarbons are used as nestmate recognition cues (see Chapter 3). Though the nonparasit-ized and parasitized workers carry the same cuticular hydrocarbons, infection leads to a quantitative change of thirteen compounds in the hydrocarbon blend. The authors report that the presence of a single cysticercoid (the larval stage of the tapeworm) is enough to cause the described changes in the host ants; how-ever, the higher the number of these parasites inside the host ants' bodies, the larger the difference between afflicted ants in comparison to nestmates free of parasites. This difference might explain the occasional antagonistic behavior of healthy ants toward their parasitized nestmates.

How do the changed features in the parasitized ants affect the fitness of the parasite? This question was investigated in the laboratory of Susanne Foitzik (Beros et al. 2015). It was found that tapeworm-infected workers of *Temnothorax nylanderi* were less likely to escape when attacks against the nest were simulated. Nevertheless, their longevity was significantly higher than that of their unin-fected nestmates. In other words, "inactivity in battles helps individual sur-vival." Interestingly, the tapeworm-infected workers from foreign colonies elicit stronger aggression in *T. nylanderi* workers than do healthy non-nestmates (provided there is no infected nestmate in their own colony). If, however, ants in their own colony are also afflicted by this tapeworm parasite, the aggression of noninfected workers toward foreign conspecific ants is markedly lower. Al-though there is no direct proof, it is suggestive that the alteration of the hydro-carbon profile of infected ants might play a key role in these changes of social interactions. Whether or not these behavioral modifications evolved to the ben-efit of the parasite or are just incidental symptoms of a colony affliction remains an open question. We think, however, that the hypothesis proposed by Beros et al. (2015) that the host ants' behavioral and structural modification benefit the parasite's fitness is very plausible. They state,

Our findings suggest that the observed parasite-induced alterations in in-fected host individuals could benefit the parasite's survival and transmission success. The reduced escape response of infected *T. nylanderi* workers in response to nest attacks would presumably increase the tape worms' transmission into the definitive avian host, which preys upon ant brood or beetle larvae. Causes of the lower escape rate might be, in addition to the lower activity of infected workers, their smaller eye and body size and shorter legs. The higher survival of infected ants would extend the parasite's time period for transmission because predation events by woodpeckers might be rare. (Beros et al. 2015, 5–6)

As Sara Beros and colleagues point out, there are other possible reasons for the higher survival of the infected ants: for example, it has previously been shown (Scharf et al. 2012) that healthy nestmates provide more than average care to the infected workers, which in turn might cause a considerable cost for the colony as a whole and might be the reason that uninfected workers of a colony with parasitized nestmates have a shorter life span than workers of colonies free of the tapeworm parasites. Nevertheless, such colony-level reactions can hypothetically be explained as the parasites affecting their "extended host" (the ant colony) in such a way as to optimize their own survival, thereby securing their transmission into their definitive host.

Eelworms: Nematodes

Nematode worms, in the family Mermithidae, are another common parasite of ants. Ants infected as larvae often develop morphological anomalies as adults. These so-called mermithergates (Wheeler 1907) are easily recognized by their distended gasters, and may also suffer a reduction in head size, a loss of defensive-gland function, enlarged trachea, shortened wings, and aberrant sexual organs (reviewed by Schmid-Hempel 1998). In one such case, larvae of the yellow meadow ant *Lasius flavus* are infected when they consume earthworm flesh containing the encysted form of the mermithid *Pheromermis villosa* (Kaiser 1986, 1991). As adults, infected alates depart on foot in search of water rather than participating in mating flights. The mermithid parasite then erupts from its host

2-3   The trap-jaw ant *Odontomachus haematodus* reveals a mermithid nematode within its gaster. (Courtesy of Alex Wild / alexanderwild.com).

alate's gaster, in the presence of moisture, and molts into its adult form. The chance of a worm finding a mate and completing its life cycle on a damp patch of ground is surprisingly high, as one in twelve gynes are infected by *P. villosa*. Parasitism in ants by mermithids is quite common. Figure 2-3 depicts the trap-jaw ant *Odontomachus haematodus,* the gaster of which is crammed full of such a parasite.

A similar, but undescribed, mermithid species is known to castrate gynes of the fire ant *Solenopsis geminata* and drive them to pond margins up to 40 meters away. McInnes and Tschinkel (1996) demonstrated that the worm's yearlong development is synchronized with that of its host, coinciding with a pulse of macrogyne (large queen) production that occurs only in early summer. As many as five worms can be found coiled inside the gaster of a single macrogyne. Here, worms may reach their maximum size, an astonishing 15.5 cm. Although the worm can infect all castes, it does not appear during phases of strict worker production or in colonies that adopt an alternative reproductive strategy, by rearing microgynes (small queens) late in the year. Six to 32% of macrogynes are infected each season, suggesting that the alternative reproductive strategy of *S. geminata* may even be driven by pressure from mermithid parasites.

In addition to drawing their hosts to water, some mermithids influence phototactic and general locomotory behavior in their hosts, and even induce behavioral changes in uninfected nestmates. Kwapich found that infected callows of the Florida harvester ant, *Pogonomyrmex badius,* travel to the upper strata of their nests, where older workers gather them and carry them up to 20 cm away from

2-4 A callow *Pogonomyrmex badius* worker. The opened gaster is devoid of fat body and distended by a three-centimeter-long, curled-up worm. (Christina Kwapich).

the nest entrance. The callows wander haplessly, their gasters devoid of fat body and distended by three-centimeter-long, curled-up worms (Figure 2-4). If they find their way back to the nest, they are repeatedly ejected by their sisters (Kwapich in prep. a). These findings suggest that nestmates detect parasitized individuals, either because of their aberrant behavior or through changes in surface chemistry, as is the case for cestode-infected *Temnothorax nylanderi,* described earlier (Trabalon et al. 2000).

Many nematodes are parasites of ants (Poinar 2012), but the newly discovered tetradonematid nematode genus and species *Myrmeconema neotropicum* is a very special case (Poinar and Yanoviak 2008). The only known ant host species is the Neotropical arboreal myrmicine ant *Cephalotes atratus.* Stephen Yanoviak and his coworkers made a remarkable discovery. Although the entire body of the workers of *C. atratus* is black, occasionally specimens with a reddish-brown gaster (this is

2-5   Two workers of the ant *Cephalotes atratus;* the one with the red gaster is infected by the nematode *Myrmeconema neotropicum,* and the other ant is free of the parasite. The lower picture depicts an infected *C. atratus* worker typically displaying its red gaster. (Courtesy of Steve Yanoviak).

the posterior region of the abdomen) can be found. Like previous collectors, Yanoviak and his colleagues first thought they had found a new *Cephalotes* species or some sort of developmental aberration, but dissections of the gaster revealed "hundreds of transparent eggs, each housing a small, coiled worm," which was subsequently recognized to represent a new genus of the nematode family Tetradonematidae (Figure 2-5) (Poinar and Yanoviak 2008). Only one other tetradonematid species (*Tetradonema* sp.) has been found in ants. It is a parasite of the fire ant *Solenopsis invicta* and was collected from a site in Mato Grosso (Jouvenaz et al. 1988). Except for a somewhat enlarged gaster, no other morphological changes or behavioral aberrations were observed in the infected fire ants.

The situation is markedly different in *Cephalotes atratus,* a species of arboreal ant. Here the parasite induces several alterations of the ants' appearance and behavior, whereby the color change of the gaster is the most conspicuous. Surprisingly this color change is not caused by a change of pigmentation. Just the contrary, the gaster cuticle of the infected ant becomes increasingly "translucent amber, which, in combination with the yellowish nematode eggs held inside, causes a bright red appearance" (Yanoviak et al. 2008). The more nematode eggs are inside the gaster, the more intense is the red color. In general, infected ants are on average 10% smaller but 40% heavier than healthy ants of similar size. This increased weight is obviously due to the parasite load. In addition, the postpetiole-gaster junction is significantly weakened in parasitized ants, which has the effect that the gaster can easily be separated from the ant. Infected ants also exhibit striking behavioral alterations. They almost continuously hold the gaster in an upright position when they walk, and they hardly show a defensive or escape response when disturbed by the observer (Figure 2-6).

It is truly striking how closely the gasters of parasitized ants resemble some of the berries fruit-eating birds forage on. Yet, Stephen Yanoviak and his colleagues state, with some frustration, "Despite hundreds of hours of observing and recording bird and ant behaviors in the tree crowns, we lack direct observation of bird predation on the infected or healthy *C. atratus.*" However, in our view, the authors assembled a large amount of data concerning the foraging of ants,

the degree of parasite loads in individuals and colonies, and conducted predation experiments with birds, all of which enabled them to present the most parsimonious reconstruction of the parasite transmission process.

The *Cephalotes atratus* foragers retrieve dead insects, and they may also feed on extrafloral nectaries. However, the major part of their diet they carry into their nests is bird droppings (Corn 1980). The feces collected by *C. atratus* can contain nematodes that will not pass the filter of the proventriculus between crop and midgut of the ant, but instead are directly fed to the larvae with the ant's other crop contents. The nematodes develop in the ant larvae and, after ants' pupation, the worms migrate into the gaster. The worms then mate in the freshly eclosed young (callow) ants. "Hundreds of developing nematode embryos within the gravid female worms cause increased reddening and modified behavior, probably by sequestering food and exoskeletal compounds from the adult ant" (Poinar and Yanoviak 2008). All this time, the infected worker stays inside the nest, but once the gaster has reached its peak color, the ant becomes an outside worker, that is, the ant changes, like her contemporary uninfected nestmates, from brood care and nest maintenance to foraging. The infected ant positions her red gaster upright and presumably will eventually be consumed by a frugivorous bird. All the direct and circumstantial evidence assembled by Yanoviak et al. (2008) strongly supports this proposed life cycle of *Myrmeconema neotropicum*.

Flukes: Trematodes

The trematode *Dicrocoelium dendriticum* (the lesser liver fluke) is a parasite of all grazing ruminant animals, such as cattle and sheep, but also wildlife. Its complex life cycle was first described by Krull and Mapes (1952, cited in Carney 1969). The trematode spends its adult life inside the bile ducts and liver of its hosts. After mating, females produce numerous eggs, which are discharged with the hosts' feces. From the eggs hatch the miracidia (singular: miracidium), the first larval stage, either when they are still embedded in the feces or after the eggs and miracidia have been taken up by the land snail *Cochlicopa lubrica,* which feeds on the ruminant feces. Here, the miracidium develops into a sporocyst that asexually produces cercaria (another larval stage) inside the snail, which are subsequently ejected with the snail's slime. Eventually, ants of the genus *Formica* or *Camponotus* feed on the excreted snail slime, and thereby the cercariae of the lesser liver fluke enter the body of the second intermediate host, the foraging ant. Most of these cercariae penetrate the pharynx and crop walls and become motionless encysted larvae inside the body of the ant host. They are now called metacercariae. In this stage, they wait until they get into their definitive host, which is, as we already learned, one of the grazing ruminants.

But how can an encysted, motionless metacercaria inside an ant's body invade the body of a cow, sheep, or deer? This puzzle was resolved by W. Hohorst and his collaborators (Hohorst and Graefe 1961; Schneider and Hohorst 1971). They discovered that a few cercariae do not become encysted metacercariae, but instead they migrate to the head of the host ant, and one of them invades the ant's brain, where it settles in the suboesophageal ganglion. Daniel Martín-Vega and

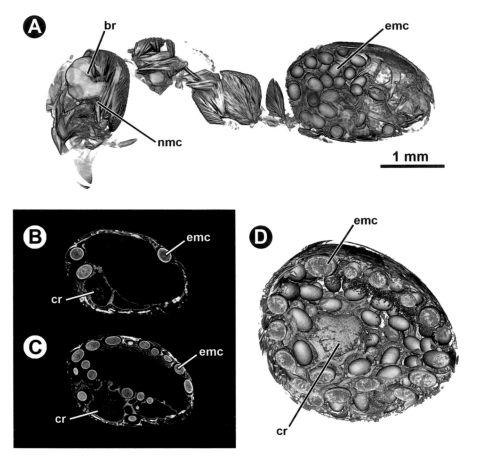

2-7　Figure (A) shows a false-colored 3D volume rendering of an infected *Formica aserva* ant worker in sagittal section displaying encysted (emc) and non-encysted (nmc) metacercariae. The non-encysted metacercaria in the ant brain is also called "brain worm." (B) depicts micro-CT-based virtual sagittal section of an ant gaster harboring six encysted metacercaria, and in (C) the gaster is shown that contained ninety-eight encysted metacercariae. Figure (D) pictures false-colored 3D volume rendering of an infected ant gaster in sagittal section, showing scar-like marks (arrow) on the crop (cr) surface caused by the metacercaria that penetrated the crop wall. (Courtesy of Daniel Martín-Vega).

colleagues (2018) were able to beautifully visualize this with the application of noninvasive micro-CT scanning (Figure 2-7).

Apparently, this "brain worm" induces its *Formica* hosts (or *Camponotus* hosts, see Carney 1969) to leave the nest when the daily temperature decreases. They attach themselves with a firm mandible grip on the top of grass leaves or other herbaceous vegetation, where they stay throughout the night and cool morning.

Of course, this is the period when the final host grazes, and the infected ant's behavior increases the chances that a grazing animal will ingest the ant loaded with metacercariae. Inside the definitive ruminant host, the metacercariae develop into adult liver flukes. There, they mate and produce fertilized eggs, and the cycle begins again.

Fungi

Many fungi are associated with insects. Besides some mutualistic symbionts, the best known of which are found in the fungus gardens of fungus-growing ants (Hölldobler and Wilson 2011), most fungi are parasitic (Araújo and Hughes 2016). Many of the most astounding such fungal parasites are found in the genus *Ophiocordyceps* (formerly called *Cordyceps*). These are ascomycete fungi of the family Ophiocordycipitaceae of the order Hypocreales. They parasitize larvae and adults of many insect species of Coleoptera, Diptera, Hemiptera, Hymenoptera, Lepidoptera, and Orthoptera (Steinhaus 1946, cited in Van Pelt 1958). Though they are ferocious parasites of insects, at least one species became a most valuable commodity for humans. The species *Ophiocordyceps sinensis* occurs on high-elevation meadows of the eastern Himalayas, Qinghai-Tibetan Plateau, and the Hengduan Mountains (Wang and Yao 2011; Wang et al. 2019), where it is called Yarsagumba (literally meaning "caterpillar fungus"). It is a parasite of caterpillars of the moth genus *Thitarodes* (Hepialidae) and has become a much-demanded, expensive natural product used in traditional medicine. Collecting of this fungus by humans is so intense that, combined with the negative effects caused by climate change, the survival of this species is seriously endangered (Cannon et al. 2009; Yan et al. 2017; Hopping et al. 2018).

For us myrmecologists, the more fascinating *Ophiocordyceps* species are those that attack adult ants of several species, because they manipulate the ants' behavior in ways to ensure efficient spread and propagation of the fungus's spores. This fungal parasite of ants has long been recognized by naturalists and was first described as a parasite of the leafcutter ant *Atta cephalotes* in 1865 in Brazil.

However, about 150 years later, David Hughes convincingly demonstrated on the basis of the accurate drawing accompanying the first report that the host ant is not *Atta,* but most likely the carpenter ant *Camponotus sericeiventris.* Indeed, according to Evans, Elliot, and Hughes (2011), two infected specimens of *C. sericeiventris* have subsequently been found in Brazil.

The best-investigated fungal ant parasite is *Ophiocordyceps unilateralis.* Originally it was believed that *Ophiocordyceps unilateralis* is a highly host-specific fungus parasitizing one or a few *Camponotus* species. However, more recent molecular phylogenetic work has revealed that *O. unilateralis* rather represents a so-called core clade, which contains a monophyletic group of twenty-three species (Araújo et al. 2018). The *Ophiocordyceps* species infecting ants are widespread within tropical forests worldwide, with few reports from temperate ecosystems (Araújo et al. 2018). The following few cases illustrate the remarkable parasitic interactions of the fungus with its ant hosts.

The individual species in this *Ophiocordyceps unilateralis* sensu lato (s.l.) group exhibit relatively high host ant specificity, but host switching can occur. In any case, the fungus needs ants to reproduce; it is an obligate, directly transmitted parasite. The behavior, morphology, ecology, and physiology of these parasitic interactions have been studied in detail by David Hughes and his collaborators (Andersen et al. 2009, 2012; Pontoppidan et al. 2009; Hughes et al. 2011; Hughes 2013; de Bekker et al. 2014; Hughes et al. 2016; Araújo et al. 2018).

The foraging ants incidentally get contaminated with spores outside their nest. The fungal spores that attach to the cuticle of a prospective host ant species grow a specialized cell, the so-called appressorium, that is used to penetrate the cuticle of the ant. Once inside, the parasite colonizes the ant's body, and the head of the ant fills up with hyphal bodies. In this way the initial fungal colonization of the host ant occurs inside the ants' nest. But once the time has come for the fungus to prepare for its own propagation, it takes complete control of the ant's behavior. The fungus "hijacks" the ant's central nervous system, as David Hughes expresses it, making the ant leave the nest. Although the ant's locomotion is not very directed and somewhat sluggish, it eventually climbs up a grass leaf or foliage of the herbaceous undergrowth and firmly clenches a leaf with its mandibles, so that the jaws deeply penetrate plant tissue. In other cases, the infected ant might grasp a leaf stalk or another piece of the vegetation, but they will always firmly anchor their body there. This so-called death grip is the "beginning of the end of the ant's life. The mandibular muscles deteriorate, and the jaws remain locked in position and the ant stays firmly latched onto the vegetation even after it has perished." Hughes et al. (2011) demonstrated the presence of "a high density of single celled stages of the parasite within the head capsule of dying ants" and propose that these are likely responsible for the muscular atrophy. Even after the host is deceased, the fungus still needs at least another fourteen days or so to develop further inside the ant's corpse. Finally, it grows a "spore-dispersal structure," called the stroma, originating from the ant's frontal thorax (pronotum) or between alitrunk and petiole (Evans 1982; Evans and Samson 1984; Evans et al. 2018) (Figure 2-8).

At this stage, most of the head, alitrunk, and gaster are filled with fungal hyphal tissue. On the stroma develop the fruiting bodies, or ascomatal cushions, with several ascomata, each containing asci, in which ascospores are formed. Once mature, the asci burst, shooting out the ascospores, some of which eventually hit another host ant worker, thus commencing a whole new cycle (Figure 2-9).

David Hughes and his collaborators (2011) suggest, based on some fossil evidence, that this kind of host manipulation by a parasitic fungus can be traced back forty-eight million years, to the Eocene. During this long evolutionary history, a high host specificity apparently developed, yet there is some host switching possible (Hughes et al. 2009; de Bekker et al. 2014; de Bekker et al. 2017; de Bekker et al. 2018). The host specificity is obviously affected by many intrinsic and extrinsic factors, but, as the work by Charissa de Bekker and her coworkers (2014) suggests, one crucial factor might be the parasite's chemical matching with the host's central nervous system. The authors conducted a series of ex vivo experiments with the North American *Ophiocordyceps unilateralis* s.l., which

2-8   *Ophiocordyceps* fungus grows two "spore-dispersal structures" (stroma) from the dorsal pronotum and petiole of a *Camponotus* worker, which was killed by the fungus. (Courtesy of Alex Wild / alexanderwild.com).

infects carpenter ants. Using cultures of this fungus, they tested the ability of the fungal tissue to affect three *Camponotus* and one *Formica* species; two of them were found to be infected by the fungus in nature (*C. castaneus* and *C. americanus*), and the other two ant species (*C. pennsylvanicus, Formica dolosa*) were never seen parasitized by *Ophiocordyceps*. The authors focused on secondary metabolites that the fungus produces in the presence of the ants' brain tissue. The results are astounding: the authors found that the compounds guanidinobutyric acid (GBA) and sphingosine were enriched when the fungus was growing in the presence of its natural target ants, but not when exposed to brain tissue of the other two ant species. These results are indeed astounding, but also puzzling, because GBA is not known from insects. It has been isolated from calf brains and other mammal tissue, and it is a metabolite of yeast (*Saccharomyces cerevisiae*), but it is hard to imagine how it could affect the ant's brain. Perhaps it interferes with the insect neurotransmitter γ-aminobutyric acid (GABA) receptor in the brain. In rats, rabbits, and cats, guanidino compounds induce seizures and convulsions (Hiramatsu 2003). Of course, the authors

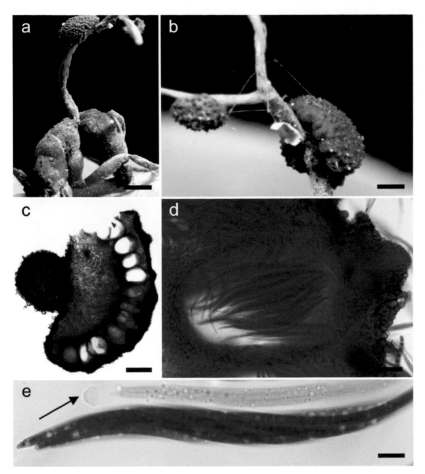

2-9 *Ophiocordyceps camponoti-rufipedis:* (a) single stroma, characteristic of *Ophiocordyceps unilateralis* sensu lato, with two lateral ascomatal cushions or plates arising from the dorsal pronotum of *Camponotus rufipes* (the red-legged ant), firmly attached to a leaf vein (bar = 0.8 mm); (b) detail of fertile region showing the immersed to partially erumpent ascomata within the cushions, with the short necks or ostioles visible (bar = 0.4 mm); (c) section through an ascomatal cushion showing the mainly immersed arrangement of ascomata (bar = 150 μm), and (d) detail of asci within chamber (bar = 25 μm); (e) asci, clavate in shape and with prominent refractive cap (arrow, bar = 7.5 μm). (Courtesy of David Hughes).

do emphasize that up to now these findings are just correlations and, based on the current data, any causal conclusion can only be speculative. This also concerns the sphingosine, which is part of the sphingolipids, which are important structural components of all membranes and, according to Acharya and Acharya (2005), are significant as second messengers during development and differentiation in the fruit fly *Drosophila melanogaster.* In fact, a sphingosine from another fungus that preys on insects has been isolated, and apparently this compound facilitates the invasion of the fungus into the host insect (Noda et al. 2011).

Species of the ant genus *Camponotus* seem to be most frequently reported as hosts of *Ophiocordyceps,* but there are also recorded cases from *Formica, Oecophylla,* and *Ectatomma;* and among myrmicine ants *Daceton armigerum;* and the leafcutter ants *Acromyrmex* and *Atta* (Hughes et al. 2009; Evans et al. 2018; personal observations by B. H.). However, as David Hughes has pointed out

INSIDE AND ON THE BODIES OF ANTS

repeatedly, *Ophiocordyceps* can kill many ant species, but each *Ophiocordyceps* species grows the stroma only on specific host species.

Scuttle Flies: Phoridae

The fly family Phoridae comprise about 4,000 species worldwide. They are called hump-backed flies, because the dorsal part of their thorax resembles a humped back, or scuttle flies, because when disturbed they exhibit rapid escape runs instead of taking to the wing. Other phorid species are called sewer flies, as they feed on decaying organic material such as animal cadavers and rotting vegetables and deposit their eggs on such substrates. There they often develop huge populations and can become a public health problem. A large number of the phorid species are parasitoids of spiders, millipedes, and insects and, in particular, of ants. Ant species from most ant subfamilies have some kind of association with phorid flies, and in many cases, a high species-specificity exists between the myrmecophilous flies and their ant hosts.

Besides some early observations by Lubbock, the myrmecologist Erich Wasmann was among the first who recognized and described the myrmecophilous parasitic behavior of the phorid *Pseudacteon formicarum* (Wasmann 1918). About that time, the young Thomas Borgmeier, a German-born Franciscan monk in Petropolis (Brazil), while watching ants, was puzzled by the strange behavior of phorid flies hovering above the ants. Through the Jesuit entomologist Hermann Schmitz, Borgmeier was introduced to the work of Erich Wasmann, and in 1922 he investigated the natural history of the new species, the phorid *Pseudacteon borgmeieri,* which he discovered and which was described by Hermann Schmitz. Borgmeier not only became an authority on the army ants of the Neotropical region, he also accomplished the major monographs on the taxonomy and revision of phorid flies in the Neotropics, North America, and Indo-Australia and, finally, a catalogue of the Phoridae of the world (Wirth et al. 1978).

These were the foundations on which future generations built their enormously extended taxonomic, ecological, and natural history work on phorid myrmecophiles, first of all the dipterologist Ronald Henry L. Disney from

Cambridge University, who says of himself he became "obsessed" with scuttle flies. He is the author of the monograph *Scuttle Flies: The Phoridae* (1994) and the world authority for this insect group. The next generation of phorid systematists is already in full swing, exemplified by the impressive work of Brian V. Brown from the Natural History Museum of Los Angeles County.

The published work on interactions of phorid flies with their ant hosts is overwhelming, and various aspects have been reported and synthesized in numerous articles (Feener and Brown 1997; Hsieh and Perfecto 2012; Brown et al. 2017), especially those studies that deal with phorid parasitism of agriculturally important ant species such as the fire ants of the genus *Solenopsis* (Orr et al. 1995; Gilbert and Morrison 1997; Morrison et al. 1997; Porter 1998a,b; Chen and Fadamiro 2018) and the leafcutter ants of the genus *Atta* (Feener and Moss 1990; Elizalde et al. 2012; Folgarait 2013; Braganca et al. 2016).

In addition to sharing a few studies that impressively illustrate the complexity and diversity of myrmecophily in phorid flies, we must begin with an almost anecdotal observation by Irenaeus Eibl-Eibesfeldt and his wife Eleonore, which occurred during their visit to the William Beebe Tropical Station in Trinidad. The great behavioral physiologist Donald Griffin and his wife, the ethologist Jocelyn Griffin-Crane, pointed out to the Eibl-Eibesfeldts that often tiny ant workers of the leafcutter ants *Atta* ride as hitchhikers on the leaf fragments the ants carry along their foraging routes into their nest. They mentioned that no one knows why they do this. So, the Eibl-Eibesfeldts commenced to find the answer, and indeed, based on their observations and simple but revealing experiments, they published a paper in 1967 in which they propose that these minis serve as guardians, which protect the defenseless leaf-carriers from attacks by the parasitoid phorid flies swarming along the ants' foraging trails. The flies attempt to land on the ant and, after localizing the right spot, within a fraction of a second inject an egg into the body of the ant. The Eibl-Eibesfeldts observed the hitchhiking mini caste patrolling the edges of the leaf fragment and fending off approaching flies with gaping mandibles (Figure 2-10). This hypothesis was

2-10 *Atta cephalotes* worker carries a leaf fragment with minim nestmates riding as hitch-hikers on the transported leaf. The hitchhikers patrol with gaping mandibles along the edges of the leaf fragment, defending the leaf-carrier from attacks by phorid flies. (Bert Hölldobler).

later challenged by Stradling (1978), who suggested that hitchhikers collect plant sap seeping out of the cut leaves, and it may be energetically more efficient if the minis ride home as hitchhikers instead of moving on their own. However, this hypothesis was put to rest by the subsequent quantitative studies by Donald Feener Jr. and Karen Moss (1990) who studied phorid-ant interactions in *Atta colombica*.

They found that females of the phorid species *Apocephalus attophilus* attack leaf-carriers of *A. colombica* and deposit eggs in the head capsules of the ant. It is not entirely clear whether the flies always aim at soft areas of the head, such as the posterior opening (foramen magnum) covered only by the membranous cervix (Feener and Brown 1993). Some other reports suggest that the parasitoid flies use the membranous connection of the leg coxa with the thorax or the posterior gaster (Folgarait 2013). Most likely there are also species-specific differences in the modes by which eggs are injected into the ant's body, and many of these parasitoids have specially modified, sclerotized ovipositors that make them into sharp egg-injection devices. In any case, the phorid larva appears to develop inside the head, and the adult fly emerges through the mouth of the afflicted ant. According to Feener and Moss (1990), in the leafcutters of the genus *Atta,* the fly requires leaf fragments to stand on during oviposition, and apparently only leaf-carriers are susceptible to parasitic attacks—at least this seems to be the case along the trail. A wealth of quantitative data collected by these authors convincingly demonstrates that the presence of hitchhikers significantly reduces the probability that the fly lands on the leaf fragment, and should it succeed in touching down, the presence of hitchhikers significantly reduces the time the fly stays on the leaf. The "defensive effect" depends on the size of the carried piece of vegetable and probably also on the number of the hitchhikers on it.

It appears that the frequency of hitchhiking in the tropical leafcutter ant species is often correlated with the occurrence of phorid fly attacks. There is, however, one caveat that seems to contradict this assessment. In central Texas, where *Atta texana* is attacked by the phorids *Apocephalus wallerae* and *Myrmosicarius texanus,* hitchhiking appears to be uncommon, and according to Feener and Moss (1990), Deborah Waller never saw phorid parasites attacking leaf-carriers of *Atta texana.* The authors also point out that hitchhiking has not been observed in the leafcutting genus *Acromyrmex,* and they suggest that in this genus the phorid parasitoids employ a different attack strategy (see also Folgarait 2013) (Figure 2-11).

Although hitchhiking by mini workers clearly serves in the protection of the leaf-carrier against phorid attacks, Timothy Linksvayer and his collaborators (2002) provide solid observations suggesting the hitchhikers may also serve in cleaning the leaf fragments of germs and spores of other fungi that can be harmful to the cultivated fungus inside the nest (see also Griffiths and Hughes 2010). Other studies suggest that the mini workers remove trichomes on the leaf surface. For more details concerning the multifunctionality of hitchhiking behavior in *Atta,* see Vieira-Neto et al. (2006).

In general, the various phorid parasitoids can markedly differ in their attack strategies. *Apocephalus attophilus* are usually found flying along the foraging trail of *Atta colombica* during the dry season. Only worker ants returning to the nest with a leaf fragment will be attacked. The parasitoid phorid *Neodohrniphora curvinervis* attacking *Atta cephalotes* exhibits a quite different behavior. They pursue a "sit-and-wait" strategy, and they attack only outbound foragers whose head width is at least 1.6 mm or greater.

Luciana Elizalde and colleagues (2012) analyzed the host-searching and oviposition behaviors of thirteen phorid parasitoids of leafcutting ants, and they found considerable variation and specificity in the phorids' attack behaviors. Some show ambush behavior like the already mentioned *Neodohrniphora curvinervis;* others actively search for hosts, specializing on different task groups, such as foragers or refuse workers, where some show preference for specific host body parts for injecting their eggs. It appears that the phorid flies attacking leafcutter ants are quite specific in choosing their host species; though several species may parasitize the same ant species, each fly species specializes in attacking different worker sizes (subcastes) (Braganca et al. 2016). There have been various suggestions about how phorids localize and identify their host ant species using species-

2-11   A worker of an *Acromyrmex* leafcutter ant species is attacked by a phorid fly of the genus *Apocephalus* "snaking" its abdomen to the right spot for injecting an egg into the ant's body. (Courtesy of Alex Wild / alexanderwild.com).

specific chemical cues, but rigorous experimental evidence is still lacking. We address this question again in a subsequent section.

Because the phorids often negatively affect the ant colonies' foraging activity, they have been considered a valuable biological control agent against leafcutter ants. Such interference was, in fact, demonstrated in laboratory experiments with the parasitic phorid *Neodohrniphora* sp. attacking *Atta sexdens* (Braganca et al. 1998).

Previously, Donald Feener (1981) demonstrated in natural populations that the phorids can have a very negative effect on afflicted ants in their competitive interspecific interactions. The myrmicine ant *Pheidole dentata* has two worker subcastes, the so-called majors (large workers with squarish heads and massive mandibles) and minors (of smaller size, with slender oval-shaped heads). As Edward O. Wilson (1976) has shown in laboratory experiments with *Pheidole dentata* colonies, when workers of fire ants of the genus *Solenopsis* venture too close to the home range of the *Pheidole* colony, the minor *Pheidole* workers recruit majors in defense of the nest territory. They will fiercely attack and dismember the *Solenopsis* intruders, thereby preventing them from moving back to their nest and recruiting a massive fire ant invasion into the *Pheidole* colony's home range. This specific alarm-recruitment, performed by *Pheidole* minors in response to the encounter with *Solenopsis* intruders, Wilson termed enemy-specific alarm-recruitment. This behavior may have evolved because of intense nest site and food competition between *P. dentata* and the aggressive, mass recruiting *Solenopsis* colonies. Don Feener observed that *P. dentata* (and six other ant species in the study area in central Texas) are attacked by phorid flies, but each ant species appears to be afflicted by one specific phorid parasite. In *P. dentata,* apparently only the majors are attacked by the *Apocephalus* species, and only the majors react with frantic escape behavior when approached by flies, seeking refuge in the leaf litter (Figure 2-12). As Feener describes it, "Once under cover, major workers periodically extend their heads from under the leaf litter. If this action does not attract phorids, major workers slowly venture out to resume their previous

2-12 *Apocephalus* flies hovering over a soldier ant of *Pheidole dentata*. The lower picture shows the phorid flies about to attack the soldiers that attempt to retrieve pupae. (Courtesy of Alex Wild / alexanderwild.com).

tasks. However, a single phorid hovering near several hiding places often holds major workers at bay for more than an hour" (Feener 1981, 861). Thus, the mere presence of phorid flies can seriously derail the alarm-recruitment in *P. dentata* against *Solenopsis* invasion. Indeed, Feener was able to test this assumption in a series of field experiments. He provoked alarm-recruitment of majors by exposing *P. dentata* minors, foraging at an artificial prey bait, to *Solenopsis* workers. In the absence of the phorid parasites, the number of recruited soldiers at the bait remained high, but the recruitment effect was significantly attenuated when *Apocephalus* flies were hovering over the bait or along the trail established by *Pheidole* workers between nest and bait. Though the presence of phorids did not affect the minors' alarm-recruitment activity, majors responding to the recruitment signal and leaving the nest were instantaneously spotted by the phorids, which chased them into hiding. These observations clearly demonstrate that phorid parasites can seriously affect interspecific competition in ant ecosystems.

Feener reports similar interactions between the carpenter ant *Camponotus pennsylvanicus* and the phorid *Apocephalus pergandei*. *Camponotus* workers on a bait, sensing the presence of the phorid flies, will hurriedly abandon the bait, which will then be exploited by other coexisting ant species not affected by phorid parasitoids.

Perhaps inspired by these results, Donald Feener and Brian Brown (1992) set out to study the interaction of the phorid parasite genus *Pseudacteon* attacking the fire ant *Solenopsis geminata* in Costa Rica. They found that three *Pseudacteon* species were attracted to foraging trails of *S. geminata,* that the presence of only one *Pseudacteon* fly was "sufficient to elicit a defense response from 100 or more foraging ants," and that the entire foraging activity along a recruitment trail is seriously encumbered by a relatively small number of phorids hovering along the trail.

These results raised again the previously disputed argument about whether phorids could be used as a biological control agent against the imported fire ant

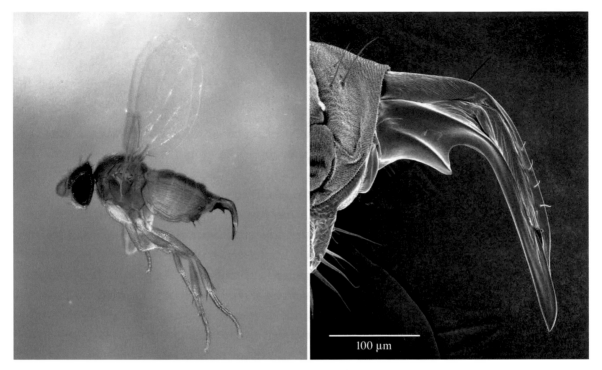

2-13   The figure on the left shows the phorid *Pseudacteon curvatus* female with the curved ovipositor ready for an attack. The picture on the right is a scanning microscope image of the ovipositor of *P. curvatus*. (Courtesy of Sanford Porter).

Formica, the myrmicine *Myrmica*, and the dolichoderine *Tapinoma*. However, in a thorough investigation by Weissflog et al. (2008), Donisthorpe's extended list of possible hosts of *P. formicarum* could not be confirmed. The authors validate the high host specificity of this parasitoid in accordance with other *Pseudacteon* species (see Disney 1994; Porter 1998a,b). Weissflog et al. (2008) demonstrated that in Central Europe *P. formicarum* parasitizes *Lasius niger* and *L. emarginatus*, whereas in Northern Italy (Liguria) it is exclusively found with *L. emarginatus* and does not attack *L. niger* that were brought from Central Europe for experimental tests. The authors suggest that Donisthorpe may have partly confused *P. formicarum* with *P. brevicauda*, a parasitoid of *Myrmica* (Disney 2000).

Donisthorpe (1927) made coincidental observations that suggest the flies employ odor cues to identify their host and suggested that they may perceive the odor of formic acid in the venom gland of formicine ants. The latter was confirmed by Andreas Weissflog and Ulrich Maschwitz and their collaborators

2-14 A worker of the fire ant *Solenopsis invicta* trying to defend itself against attacks by a *Pseudacteon* fly. The image below depicts *Pseudacteon* emerging from the head of a decapitated fire ant worker. (Courtesy of Sanford Porter).

2-15   The picture on the left shows the phorid *Pseudacteon formicarium* ovipositing into the gaster of the ant *Lasius emarginatus*. The two *Lasius* workers are engaged in antagonistic interactions, which appears to attract the *Pseudacteon flies*. The picture on the right is a scanning electron microscopic image of the ovipositor of *Pseudacteon formicarium*: ov = ovipositor tip; eg = egg-laying groove. (Courtesy of Andreas Weissflog).

(Maschwitz et al. 2008). These authors present the most detailed analysis of the behavioral mechanisms the phorids employ for identification and localization of their host ant species. They found that *Pseudacteon formicarum,* which attack *Lasius niger,* are rarely observed at undisturbed *L. niger* nests, but were clearly attracted to disturbed nest mounds. Obviously, agitated workers rushing out of the nest discharge secretions from their venom gland that contain formic acid. However, this defense secretion did not repel the parasitoid scuttle flies but appeared to attract them, as Donisthorpe (1927) had already observed. Maschwitz et al. (2008) used synthetic formic acid diluted to the naturally occurring concentration and presented it on filter paper. Indeed, it proved to be a powerful attractant. In an area where many flies were hovering around, they could collect 200 flies on the test paper within twenty minutes. Flies approached the test paper closely and flew in search loops, apparently searching for worker ants. Here is the authors' concise description of the phorids' attack behavior after they localized the ants:

> They approached the ants and pursued them from behind. Hovering about them, at a distance of less than 1 cm, they dived down on the gasters of the ants. At this point two possible reactions could be observed: Either the flies contacted the workers only very briefly and left them immediately (failed oviposition) or the flies remained 1–3 seconds on the gaster surface and deposited an egg. Without moving their wings and their legs drawn to the body they "stood" above the ant's gasters, inserted their ovipositor in the intersegmental fold between gaster segment 2 and 3, their body axis parallel to that of the ant and their head pointing forward. After successful insertion of the ovipositor the flies left and continued the search behavior. (Maschwitz et al. 2008, 132) (Figure 2-15)

But how do the phorids recognize their host ant species? Maschwitz et al. (2008) tested several of the exocrine glandular secretions, such as the contents of the Dufour's gland, rectal sac, mandibular glands, metapleural glands, and

hemolymph. None of these elicited any attraction in *P. formicarum* flies, except venom gland secretion and formic acid. They also tested acetic acid with negative results. When formic acid was tested on filter paper dummies about the size of an ant, the flies that were attracted did not land or attack these dummies. The flies apparently only react to moving objects. Interestingly, flies close by tried to attack *Lasius* workers moving beneath a glass plate. Obviously, the flies' visual perception of the host ant species plays an important role, yet this alone is not enough for host identification, because workers of a *Crematogaster* species, of about the size of *L. niger,* were briefly pursued by the flies but not attacked. Thus, although visual cues such as size, color, and movement are important parameters in host identification, most likely chemical cues, such as species-specific

INSIDE AND ON THE BODIES OF ANTS

cuticular hydrocarbon profiles, also play a key role (though this has not been studied). The long-distance attractant emitted by the host species and exploited by the parasitoid *P. formicarum* is formic acid.

However, other *Pseudacteon* species do parasitize ant species that do not produce formic acid—for example, the myrmicine genus *Solenopsis*. Porter (1998b) reports that in some cases *Pseudacteon* parasitoids can be attracted to *Solenopsis* mounds over distances of 10–20 m or even longer ranges, but we do not know anything about the nature of this attraction cue or putative kairomone.

In a subsequent study, Lloyd Morrison and Joshua King (2004) demonstrated that *Pseudacteon tricuspis* were frequently attracted to disturbed *Solenopsis invicta* colonies but were almost never seen at baits where ants of the host species forage. However, when some disturbance was experimentally initiated by releasing conspecific worker ants from a foreign colony—invariably, skirmishes and some fighting among the ants broke out—considerably more of the parasitoid scuttle flies were attracted. The authors speculated that the phorid flies might be attracted by the alarm pheromone discharged by fighting ants, but to our knowledge, no experimental evidence has yet been obtained. Chen and Fadamiro (2007) conducted behavioral tests in a Y-tube olfactometer and reported the *Pseudacteon tricuspis* flies were attracted to odors extracted from the ant's (*S. invicta*) whole body, or head and thorax, but significantly less so to odors extracted from the gaster. The ants' trail pheromones were not effective. Data from electroantennograms taken from the flies' antennae were in accordance with the result obtained from the behavioral test. These results support the previous assumption that the phorid parasitoids find their hosts by odor.

The other well-studied example of host identification is the parasitoid scuttle fly *Apocephalus paraponerae,* which parasitizes workers of the giant bullet ant *Paraponera clavata* (Brown and Feener 1991). This phorid has been repeatedly reported to hover near the nests of *Paraponera* nests (Borgmeier 1958; Janzen and Carroll 1983), but attack behavior has never been observed. Brian Brown and Donald Feener found out that the phorids, contrary to previous reports, are "rarely or never attracted to disturbed or undisturbed *P. clavata* nests or to aggregations of foraging workers." However, they have observed *A. paraponerae* in large numbers around injured workers of *P. clavata*. They found both male and female phorids were attracted to forty injured *Paraponera* workers, where males fed on tissue or oozing hemolymph and the females oviposited. The females, as in most parasitoid phorids, have specially adapted ovipositors; the authors observed the oviposition process and noted that one ant worker is often attacked by several ovipositing *A. paraponerae* females. Because the phorids are almost exclusively attracted to injured *Paraponera* workers, Brown and Feener conducted a series of field experiments. They found that freshly crushed ants attracted significantly more flies than either freshly killed ants (ants exposed to hydrogen cyanide, HCN) or crushed HCN-killed ants. In addition, whole-body extracts with diethyl ether of *Paraponera* workers have attracted significantly more phorids than did the control substance (pure diethyl ether).

One might wonder whether phorids encounter enough injured *Paraponera* workers for oviposition. Brown and Feener refer to the work of Jorgenson et al. (1984) on territoriality in *P. clavata* resulting in many injured ants, and they cite Michael Breed's observations of encountering aggressive interactions between neighboring *Paraponera* colonies approximately every second day during his studies in La Selva (Costa Rica). Breed also noticed that such skirmishes attracted many phorid flies. Whereas crushed bodies of leafcutter ants *Atta* or army ants *Eciton* attracted none of the many parasitoid phorids, *Apocephalus paraponerae* appears to be attracted exclusively to injured *Paraponera clavata* workers (Brown and Feener 1991).

In certain ant species, apparently those with especially large workers, injured specimens seem to be especially attractive. This is indicated by an anecdotal observation reported by Disney and Schroth (1989), who noted that females of *Megaselia persecutrix* were observed hovering over an injured worker of *Dinomyrmex (Camponotus) gigas* (cited by Brown and Feener 1991). *Paraponera clavata* and *Dinomyrmex gigas* workers belong to the largest ants known, and although their phorid parasitoids are much smaller than are their hosts, they are among the larger phorid parasitoids known. Brian Brown (2012) described the smallest phorid fly *Euryplatea nanaknihali* from Thailand. At only 0.40 mm in length, it is most likely the smallest fly known. According to Brown, the other *Euryplatea* species (*E. eidemanni*) measures 1.10 mm. It was found attacking the myrmicine *Crematogaster impressa*, which is about three times as long as the fly. Although the host species of *E. nanaknihali* is not known, Brown suggests it might attack the smallest workers of *Crematogaster rogenhoferi*, which measure about 2 mm in length. Indeed, there is a striking correlation between parasitoids' body length and host body length, as was particularly well documented for *Pseudacteon* species that attack *Crematogaster* species (Brown 2012). We can generally conclude that host size is an important parameter for the phorid parasitoids during host selection (see also Feener 1987; Fowler 1997).

Although scientists devoted most of their attention to parasitoid phorids, there are many other myrmecophilous scuttle fly species that are not parasitoids. Anyone who is lucky enough to encounter an army ant column in the Neotropics or the deserts of the Southwest of North America will notice many phorid flies moving among the ants' trails. Carl Rettenmeyer and Roger Akre (1968) sampled 300 colonies in the Neotropics and collected 3,900 females and 490 males of scuttle flies. Most were found in refuse deposits of the bivouac nests or at the end of emigration or raid columns. The number of scuttle flies per colony varies greatly, and sometimes can exceed 4,000. Most of the flies are scavengers, feeding on refuse, booty, and dead workers, but also on ant brood within the bivouacs. The females of these phorids are wingless, whereas the males have large wings. They seem to fly to other colonies for mating opportunities, whereas the females move with the ants from bivouac to bivouac. However, it has also been observed that wingless females are transported in copula by their flying male partners to a new host colony. Most of these commensal phorids were found with *Eciton* and *Labidus* army ant species, whereas *Neivamyrmex* colonies have fewer phorid commensals.

As Thomas Borgmeier first noted, commensal phorids have membranous ovipositors, tubular organs through which eggs are deposited. They do not need egg injection devices, because they deposit their eggs in the refuse piles of their host ants. It has been suggested, and some of the field observations support the hypothesis, that the oviposition and larval development of the commensal scuttle flies is somehow synchronized with the army ants' stationary phase. This would allow the phorid larvae to complete their development and pupation in the refuse area of the bivouac, a temporary nest constructed of and by migratory ants, before the colony emigrates. During the nomadic phase, which lasts about two weeks, the colony usually emigrates every night. Phorids developing in the refuse deposit in such colonies would emerge from pupae long after the colony has left the site.

Although some of the myrmecophilous scuttle flies seem to feed on the ant brood inside the bivouac, most of them live in refuse deposits and cause no harm, in fact, might even be beneficial. Nevertheless, army ants are also afflicted by parasitoid phorid species, as documented by Brown and Feener (1998).

The lifestyle of these phorid myrmecophiles is remarkably diverse. Most are parasitoids; others, as we have learned, feed on the ant brood of army ants; again others are garbage consumers in the ants' refuse deposits; and as Brown et al. (2017) recently reported, some phorid parasitoids are "ant baby killers." They observed "females of the phorid *Ceratoconus setipennis* attacking workers of the ant *Linepithema humile* carrying brood and ovipositing directly onto the brood in the nest." They observed similar behavior in an unidentified phorid species of the *Apocephalus grandipalpus* group ovipositing on the brood of a *Pheidole* ant species.

Probably the most bizarre case of myrmecophilous adaptation in Phoridae was discovered by Ulrich Maschwitz and his student Andreas Weissflog in the rainforest of Malaysia in 1994. They collected an entire colony of the *Aenictus gracilis* army ants (also called driver ants). A total of 57,100 ants interlocked with their tarsi were clustered in a densely packed bivouac in the middle of which resided a non-physogastric queen (a queen in the nonreproductive stage, therefore her gaster was not swollen as it is when the queen is in the egg-producing stage). During sorting of the entire catch, K. Rościszewski, also a collaborator of Maschwitz, found eighty phorid larvae, and 104 specimens of an unidentified wingless adult insect. "The unidentified insects looked like ant larvae, yet they had vestigial legs and the head resembled that of a fly." After much consultation, it was concluded that this might be a highly aberrant species of the Phoridae. This was confirmed by the doyen of the systematics of Phoridae, Ronald Henry Disney. At least for myrmecologists, this was a sensational discovery, which was published by Weissflog, Maschwitz, Disney, and Rościszewski (1995) in the journal *Nature*. Subsequently Disney (1996) described this creature as a new genus and species and named it *Vestigipoda myrmolarvoidea*. It became the type species of that genus. In the meantime, five additional *Vestigipoda* species were discovered in Malaysia, one of which is named *Vestigipoda maschwitzi* (Disney et al. 1998; Maruyama et al. 2008). They all were found in colonies of *Aenictus* army ant species (Figure 2-16).

Not much is yet known about the biology of these *Vestigipoda* myrmecophiles. The flies' eggs closely resemble those of the ants. Adults obviously mimic ant larvae, and all found amid the ant brood were females, the abdomen filled with eggs. Maruyama et al. (2009) report that the surface cuticular hydrocarbon blends detected on the *Vestigipoda* adults and *Aenictus* larvae shared two peaks, and the authors suggest "that these two peaks were the substances that made the ants recognize them as their larvae." At this stage this is, of course, just a suggestion, because no experimental evidence exists in support of this speculation. We think similarities in the hydrocarbon profiles might help the myrmecophiles to be ignored or not recognized as foreign, but most likely do not elicit brood adoption and tending behavior in the ant workers (see our discussion in Chapter 4).

Presumably, the *Vestigipoda* larvae also live with the host ants' brood, and both the adult ant-larva mimics and phorid larvae may be tended by the ants and may prey on ant brood. *Vestigipoda* males are unknown, but they clearly exist, because

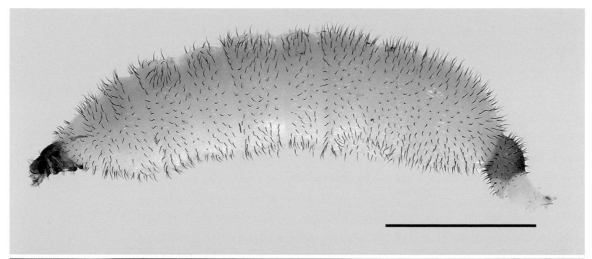

2-16 The aberrant scuttle fly *Vestigipoda longiseta*. The lower picture demonstrates that the adult fly matches perfectly with the shape of the larvae of its *Aenictus* host ants (bar = 1 mm). (Courtesy of Munetoshi Maruyama).

2-17   The phorid fly *Metopina formicomendicula* (microscopic photograph, upper) is illustrated riding on the host ant *Solenopsis fugax*. The fly rapidly strokes the ant's mouthparts with its forelegs to elicit regurgitation of food. (Photo: Karl Hölldobler; illustration E. Kaiser based on observation sketches by K. Hölldobler; ©Bert Hölldobler).

Weissflog et al. (1995) have found spermathecae full of sperm in *Vestigipoda* females. Like in other phorid species with wingless females, the winged males will search for *Aenictus* colonies and somehow find access to young unmated females. However, all this is unknown and awaits new discoveries.

A completely different myrmecophilous behavior has been described by Karl Hölldobler (1928) in the scuttle fly *Metopina formicomendicula,* which lives inside the nests of the thief ant *Solenopsis (Diplorhoptrum) fugax.* The tiny fly spends most of its time on the bodies of its host ants. It mounts a thief ant worker and rapidly strokes the head and mouthparts of the ant with its forelegs (Figure 2-17). The worker usually responds by slightly raising its head, opening the mandibles, and regurgitating a droplet of food, which is then quickly imbibed by the fly. The ants occasionally attack the phorids when they move through the nest, but rarely harm them, because the flies are too swift and elusive. When resting, the *M. formicomendicula* frequently ride on the queen, where they are usually ignored by the workers. According to Disney (1994), Borgmeier observed a similar behavior in the phorid fly *Allochaeta longiciliata,* which solicits food from its host ants *Acromyrmex muticinodus.* This is a surprising find, because the host species is a fungus-growing ant and, to our knowledge, regurgitation of crop contents is rare (confirmed by Flavio Roces, personal communication). However, it is known that leafcutter ants do imbibe plant saps, and the fly's persistent and forceful stimulation may "force" regurgitations in the host ants.

Finally, Wheeler (1910) describes the behavior of the myrmecophilous scuttle fly *Metopina pachycondylae* which live with the ponerine ant, *Pachycondyla harpax.* We cannot improve on Wheeler's vivid description: "Its small larva clings to the neck of the ant larva by means of a sucker-like posterior and encircles its host like a collar. Whenever the ant larva is fed by the worker with pieces of an insect placed on its trough-like ventral surface within reach of its mouth parts, the larval *Metopina* uncoils its body and partakes of the feast; and when the ant larva spins a cocoon it also encloses the *Metopina* larva within the silken web" (Wheeler 1910, 412).

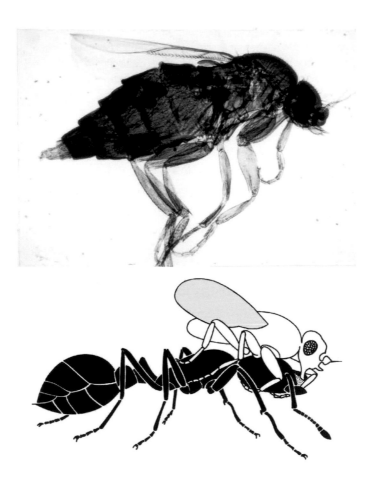

Metopina pupates within the ant pupal cocoon, and after the ant's eclosion, which occurs before the *Metopina* adult is ready to emerge, the ant's empty cocoon with the *Metopina* puparium is transported by the ants to the refuse site, where the fly eventually emerges and escapes. Apparently, nothing is yet known about how the adult mated female *Metopina* deposits her eggs inside the *Pachycondyla* nest.

Tachinid Flies

There are many more endoparasites known in ants, in particular endoparasitic wasps, some of which we visit in the next section. However, it is not our intention to produce a full account of parasites on and in the bodies of ants, a wide field that has previously been reviewed in Paul Schmid-Hempel's splendid book *Parasites in Social Insects* (1998). Yet, one example of the fly family Tachinidae

2-18 The myrmecophilous tachinid fly *Strongygaster globula* develops as an endoparasite inside the gaster of a colony-founding queen of *Lasius niger*. When the last-instar larva of the parasite leaves the body of the queen, the queen tends the larva and pupa shown in the upper picture. After the adult fly has emerged from the pupal case (lower picture), it must leave the nest of the queen quickly. (Turid Hölldobler-Forsyth; ©Bert Hölldobler).

deserves special mention. Larvae of most members of this family are parasitoids of insects, organisms that develop inside a living host and, ultimately, kill it. The larvae of the tachinid fly *Strongygaster globula* (formerly called *Tamiclea globula*) develop as endoparasites inside the gaster of colony-founding queens of *Lasius niger* and *L. alienus* (Gösswald 1950). The behavior of the infected queen is not noticeably affected, except that she is unable to lay eggs. When the last-instar larva of the parasite leaves the host's abdomen through the cloaca, it quickly pupates and is groomed and tended by the queen. In contrast, the fly imago, after eclosion from her pupal case, is not treated amicably by the queen and has to leave the incipient ant nest quickly (Figure 2-18). The ant queens infected with *Strongygaster* die after the parasitoids leave the nest. Histological investigations by Gösswald and his coworkers revealed that all life-essential organs inside the infested queen remained intact, yet the queen could not produce her own eggs, because the ovaries were totally degenerated, and the flight muscles and fat body entirely metabolized. All these reserves were apparently used up by the developing parasitic fly inside the queen's body.

Parasitic Wasps

There are nine families of wasps, of which some or all species exploit ants as predators, social parasites, or parasitoids (Kistner 1982; Godfray 1994, 2007; Schmid-Hempel 1998; Lachaud and Pérez-Lachaud 2012). Although many species have been recognized as parasites or parasitoids, relatively little is known about their life cycle or behavioral interactions with their hosts. Here we discuss a few exemplary case studies.

Braconidae

The braconid tribe Neoneurini (Euphorinae) are parasitoids of adult ants. They feature a curved, hook-shaped ovipositor that, when exerted, is forward pointing (Huddleston 1976; Durán and van Achterberg 2011). Supposedly, female Neoneurini deposit their eggs through the anal opening of their host ants. With few exceptions, they parasitize formicine species (Shenefelt 1969; Marsh 1979; Yu et al.

2007; cited in Durán and van Achterberg 2011). It has been proposed that formic acid produced in the ants' poison glands may serve as an attractant (kairomone) for the parasitoids (van Achterberg and Argaman 1993; Durán and van Achterberg 2011). However, no behavioral analysis is yet available.

José-Maria Gómez Durán and Cornelius van Achterberg (Durán and van Achterberg 2011) provide the following information for neoneurine genera and their recorded host ant species: *Elasmosoma* species are mainly found with *Formica* species, and occasionally with *Lasius niger* and species of *Camponotus; Kollasmosoma* species parasitize ants of the genus *Cataglyphis;* and species of *Neoneurus* have been found only with *Formica* species. Oviposition in the braconid genus *Elasmosoma* has been observed by several scholars, dating back to Forel's first report in 1874 (see Durán and van Achterberg 2011), and in some cases adult *Elasmosoma* wasps have been reared in *Formica* nests. However, the precise oviposition behavior remained unknown. Some assumed that the parasitoid females deposit their eggs by puncturing the intersegmental membrane in the posterior region of the gaster, whereas others suggested the eggs are deposited through the anus. Durán and van Achterberg (2011) were able to make precise observations of the oviposition behavior of *Elasmosoma luxemburgense* on *Formica rufibarbis*. José-Maria Gómez Durán succeeded catching on video the entire behavioral sequence of approaching, alighting, grasping the host ant, and inserting the ovipositor. Apparently, the *F. rufibarbis* workers, which were approached by the wasps near the nest entrance, were agitated, and circumstances indicated that fighting with other ant species had recently occurred. Two to three *E. luxemburgense* females hovered 1–3 cm above the ants, repeatedly attacking the *Formica* workers by approaching them from behind and diving down on the rear end of the gaster. "The ants were aware of these attacks, turning around and chasing the wasps with open mandibles." A total of fifty attempts of oviposition were recorded, forty of which were successful. The entire process of one attack lasted an average of just 0.73 seconds.

These video recordings made it possible to analyze all the details of the behavioral pattern that precedes the actual oviposition. The parasitoids approached the ants' gasters (metasoma) "with the hindlegs extended in curved shape. The forelegs are darted forward and when alighting the curved hindlegs brace the rear end of the ant's gaster." The authors describe the details of the alighting behavior in the following way, slightly altered by us for better understanding: The frame-by-frame analysis of the film clip shows that the wasp's head hits with open mandibles the posterior margin of the ant's first gaster tergite. This causes a slight structural deformation between first and second tergites. Presumably, this structural modification between both tergites is used by the wasp to secure a firm grasp with the mandibles. Specific tarsal modification of *Elasmosoma,* such as vestigial tarsal claws and enlarged pulvilli (singular pulvillus, cushion-like pads on the feet), may be adaptations to effect this grasping behavior. The precise moment of the ovipositor insertion could be detected because the wasp exhibits a conspicuous downward movement with its metasoma apex (tip). Usually, one such movement occurs during oviposition, but occasionally two to three such movements could be noted.

Similar observations were made with *Kollasmosoma sentum,* which parasitizes the formicine ant *Cataglyphis ibericus.* This wasp also attacks its host ant from behind. However, oviposition occurs only rarely through the ant's gaster tip, and is more frequently initiated through the intersegmental membrane of the ant's dorsal or ventral gaster segments. The wasps' alighting tactic is somewhat more versatile than that of *Elasmosoma.* They approach the ant in either the horizontal or the vertical flight position, probably depending on the gaster posture exhibited by the ant. The authors describe many more details concerning the wasps' behavior to outmaneuver the ant's defensive reactions. For that we refer the reader to the original publication and film clips (Durán and van Achterberg 2011).

We would like to add one more example of a braconid parasitoid genus, *Neoneurus,* the species that appears to have perfected a raptor technique when attacking host ants (Shaw 1993). The female's forelegs are strikingly modified with a compressed femur, a robust shortened tibia with sharp tubercles and spines and enlarged tibial spur, and shortened tarsus with large tarsal pads (pulvilli). Scott Shaw, who studied *N. mantis,* reports that the wasps oviposit on their host ant's (*Formica podzolica*) metasoma (or alitrunk, which is also not quite correctly called the thorax). However, Durán and van Achterberg (2011) observed that *N. vesculus* oviposit in the mesosoma (gaster) of *Formica cunicularia*. Whereas the braconid genera mentioned before hover above the ants before oviposition, *Neoneurus* employ two different tactics. Either the wasps perch on grass stems or tree trunks a few centimeters above or on the ground, waiting for approaching ants to attack, or they conduct hovering flights following host ants that emerge from the nest entrance. The wasps seem to preferably attack the ants when they climb up a tree trunk (Shaw 1993; Durán and van Achterberg 2011).

We focused on the above-discussed studies because they primarily investigate the behavior of the parasitoids attacking their host ants. The final study of braconid parasitoids we cover emphasizes not so much the behavioral mechanisms but rather the ecological impact exerted by the parasitoids on their hosts. Douglas S. Yu and Donald Quicke (1997) discovered that a braconid species, *Compsobraconoides* (Braconinae), is an ectoparasitoid of queens of three dolichoderine *Azteca* species. The *Azteca* queens colonize the ant plant *Cordia nodosa* (Boraginaceae) in southeast Peru. Like hundreds of tropical plant species that have evolved structures to house and feed ants, *Cordia nodosa* have special domatia (singular domatium, a swollen hollow stem at the base of leaves that creates small chambers) that house *Azteca* ants that protect the plant from herbivory. Yu and Quicke discovered *Compsobraconoides* larvae feeding on paralyzed *Azteca* queens inside the domatia. Pupating wasp larvae and cocoons protected behind a silken tent at the distal end of the domatium were also found. The authors note that "the adult wasp emerges by boring a small hole in the domatium wall . . . that is located directly below the cocoon / silk tent." They hypothesize that parasitoid wasps play an important role in the ecology of the *Azteca–Cordia nodosa* symbiosis.

Besides *Azteca,* two other ant species are associated with this plant, the myrmicine *Allomerus octoarticulatus (demerarae)* and the formicine *Myrmelachista* sp. Whereas *Azteca* and perhaps also *Myrmelachista* have a straight mutualistic relationship with the host plants, which provide shelters for the ants and in turn receive protection against herbivore attacks, the *Allomerus* workers may protect the leaves that contain the domatia, but they attack and destroy the floral buds of the host plants, "acting as a castration parasite" (Yu and Pierce 1998; Yu et al. 2001). In *Azteca,* several queens may start colonies in different domatia in the same plant, but as these colonies grow, they fight to death, and one colony will eventually take over the entire plant (Perlman 1993; Choe and Perlman 1997; Yu and Davidson 1997). However, in situations where the parasitoid *Compsobraconoides* wasps are abundantly present, the wasps will attack many colonizing *Azteca* queens, thereby increasing the probability that the harmful *Allomerus* queens will succeed in establishing colonies in *Cordia* plants (Yu and Quicke 1997; Yu and Pierce 1998).

Ichneumonidae

The Hybrizontinae were previously called Paxylommatinae and considered to be a subfamily of the Braconidae. However, they are nowadays assigned to the Ichneumonidae (Yu and Horstmann 1997; Yu et al. 2007). It has long been known that they develop inside ant nests (Donisthorpe 1915; Donisthorpe and Wilkinson 1930) and that they are associated with a number of ant species. For *Hybrizon buccatus* alone, several *Formica* and *Lasius* species, *Myrmica* species, and the dolichoderine *Tapinoma erraticum* have been listed as potential host species. The records are based mostly on observations of wasps hovering over the nest entrance or workers of various ant species. Similar observations exist for another ichneumonid hybrizontine species, *Ghilaromma fuliginosi* hovering over *Lasius fuliginosus* workers. The report by Cornelis van Achterberg (1999) of females of *Hybrizon buccatus* "diving at *Formica rufa* workers ant during spring ant wars in the dunes near The Hague," suggests that the parasitoid wasps are attracted to accumulations of agitated ants, and formic acid might, indeed, serve as an attractant for the parasitoids. Yet, Myrmicinae and Dolichoderinae, which had also been listed as possible hosts, do not produce formic acid. Whereas the braconid wasps discussed above are undoubtedly parasitoids of adult ants, and the wasp larvae develop within the body of their host ants, there are contradicting reports for the ichneumonid *Hybrizon buccatus*. Donisthorpe and Wilkinson (1930) assumed that *H. buccatus* was a parasitoid of adult ants, because they found naked pupae of the wasp among the pupae of the host species *Lasius alienus*. Other authors (Watanabe 1984; Marsh 1989; cited in Durán and van Achterberg 2011) suggested that *Hybrizon* spp. most likely are parasitoids of ant larvae. To our knowledge, the video recordings by José-Maria Gómez Durán are the first direct demonstrations of *H. buccatus* parasitoids attacking larvae of their host ants *Lasius grandis*. Females of the wasp were observed hovering about 1 cm above a trail along which the ants commute back and forth between two nest entrances. "Even in the absence of ants on the trail, specimens of *H. buccatus* found the precise location of the trail and stayed hovering over it" (Durán and van Achterberg 2011). Despite many video recordings, even catching individual wasps quickly approaching and even touching the ant, no oviposition could be recorded. The careful analysis of all video recordings revealed that the wasps "pounce" down at the ants, apparently checking whether they are carrying larval brood. In two cases, where ants carried last-instar larvae, wasp ovipositing into the transported ant larvae could be recorded. "The wasp grasped the larva with its forelegs and placed its body in vertical position over the adult ant. When the metasoma began to bend toward the larva, the middle legs seized the adult ant's head, and wings were folded until the oviposition finished." The entire oviposition process lasted about a half second. Apparently, *H. buccatus* females ignore smaller ant larvae being carried by their adult nest mates.

Such a curious oviposition behavior has, to our knowledge, been observed in only one other parasitoid genus, *Smicromorpha,* which belongs to the wasp family Chalcididae. In his revision of the Indo-Australian Smicromorphinae,

Ian D. Naumann (1986) reports early observations by the Australian entomologist F. P. Dodd, which have been communicated by Girault (1913; cited by Naumann 1986), that *Smicromorpha doddi* is a parasitoid of the Australian green tree ant *Oecophylla smaragdina*. This arboreal weaver ant employs its silk-producing larvae for binding leaves together during the construction of the leaf-tent nests. Reportedly, females of *S. doddi* oviposit on the silk-spinning larvae held between the mandibles of adult ants outside the nest (Figure 2-19). Although these observations are mostly anecdotal, Naumann found a close concordance of the distribution of the wasps with that of *O. smaragdina* in Northern Australia. Further direct evidence for the association *Smicromorpha* spp. with *Oecophylla* was provided by Christopher Darling (2009), who discovered a new smicromorphine species in Vietnam, *Smicromorpha masneri,* which he obtained from leaf-tent nests of *O. smaragdina* reared in a greenhouse. In addition, *Smicromorpha* wasps have been observed hovering around leaf-tent nests of *Oecophylla* colonies in nature.

Eucharitidae

The relatively small wasp family Eucharitidae contains three subfamilies: Oraseminae, Gollumiellinae, and Eucharitinae. They are obligate parasitoids of ants that specialize on attacking ant brood (Clausen 1941; Heraty 2000; Baker et al. 2020). However, eucharitid females do not attack the ant larvae directly; instead, they deposit their eggs in or on tissues of flower buds or leaves, often on young shoots near extrafloral nectaries. The orasemines (Figure 2-20) feature specialized ovipositors with which they puncture and form cavities inside the plant tissue and deposit one or several eggs (Heraty 2000).

The minute first-instar larvae that hatch there are called planidia (singular planidium), a term used in all parasitoids with highly motile first-instar larvae, in contrast to the "stationary" instars that feed and grow on or within the host. They are only slightly longer than one-tenth of a millimeter, with an unusually hard, sclerotized cuticle. Shortly after hatching from their eggs, the planidia become active, finding their hosts. Although it has never been possible to follow

2-19 A worker of the weaver ant *Oecophylla smaragdina* holds a last-instar larva in her mandibles and moves it back and forth while the larva releases a continuous thread of silk from the labial gland openings on the head. (Bert Hölldobler).

2-20 A wasp female of the genus *Orasema* deposits an egg into the tissue of a plant leaf. (Courtesy of Alex Wild / alexanderwild.com).

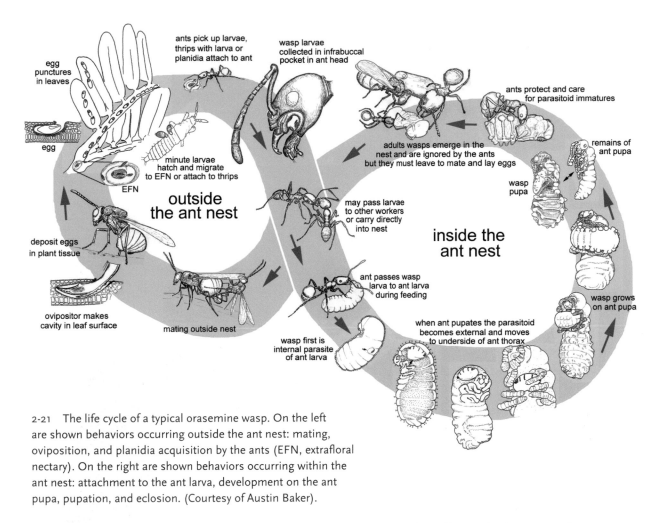

2-21 The life cycle of a typical orasemine wasp. On the left are shown behaviors occurring outside the ant nest: mating, oviposition, and planidia acquisition by the ants (EFN, extrafloral nectary). On the right are shown behaviors occurring within the ant nest: attachment to the ant larva, development on the ant pupa, pupation, and eclosion. (Courtesy of Austin Baker).

The labels in the figure read:

- egg punctures in leaves
- egg
- deposit eggs in plant tissue
- ovipositor makes cavity in leaf surface
- minute larvae hatch and migrate to EFN or attach to thrips
- EFN
- **outside the ant nest**
- mating outside nest
- ants pick up larvae, thrips with larva or planidia attach to ant
- wasp larvae collected in infrabuccal pocket in ant head
- may pass larvae to other workers or carry directly into nest
- wasp first is internal parasite of ant larva
- ant passes wasp larva to ant larva during feeding
- when ant pupates the parasitoid becomes external and moves to underside of ant thorax
- adults wasps emerge in the nest and are ignored by the ants but they must leave to mate and lay eggs
- ants protect and care for parasitoid immatures
- remains of ant pupa
- wasp pupa
- **inside the ant nest**
- wasp grows on ant pupa

exactly the way and means of how the planidia infiltrate the host ant colony, it has been suggested that they are passively carried by foraging ants into the host ants' nests, either by attaching themselves to prey retrieved by the ants or staying close to extrafloral nectaries and thereby being picked up by the ants and transported in the infrabuccal pocket inside the ant's mouth cavity (Figure 2-21) (Herreid and Heraty 2017).

Baker et al. (2020) state that some species of the Oraseminae "tend to specialize on certain plant structures for oviposition," and some species "have very specific plant hosts (e.g., *Orasema simulatrix* only oviposits near leaf extrafloral nectaries" of desert willow, *Chilopsis linearis*), whereas others, such as *Orasema simplex* "oviposit onto leaves of at least eight plant families" (Varone and Briano 2009; cited in Baker et al. 2020). It has also been suggested that planidia of some

2-22 *Orasema wheeleri* in the brood nest of *Pheidole bicarinate*. The upper picture shows the wasp pupa in a pile of ant pupae. The freshly eclosed *Orasema* adult is tolerated by the host ants, although some workers exhibit mandible gaping indicating some degree of antagonism (middle picture). Once the wasp's wings have fully hardened (lower picture), it will leave the host ant's nest for mating. (Courtesy of Alex Wild / alexanderwild.com).

species carry substances attractive to host ants, but to our knowledge, no experimental evidence exists in support of this suggestion (Heraty and Barber 1990; Heraty et al. 2004; Baker et al. 2020). The most detailed observations of the Oraseminae are reported and reviewed by Austin Baker, John Heraty, and their colleagues (Baker et al. 2020).

Inside the host ants' nest, the returning forager worker passes the planidia to the ant larvae, where the parasites attach and feed "in a state of semi-arrested development." The planidium cuts itself through the larval cuticle and settles "on the dorsal thoracic region" of the ant larva, where it feeds and sometimes expands more than a hundred times its original size without changing instars (Clausen 1941; Heraty 2000; Baker et al. 2020). In this phase, the wasp is an internal parasite until the host larva reaches the prepupa instar and the developing wasp becomes an external parasite. Sticking to the ventral side of the pupa's thorax, it nourishes itself by extracting resources from the developing ant pupa. As the parasitoid larva grows and pupates, it is groomed and tended by the host ants. Meanwhile, the ant pupa shrivels and, although it is still alive, will not be able to complete its development. The emerging adult wasp is ignored by the host ants and eventually leaves the host ants' nest (Figure 2-22).

There is very little known about the behavioral mechanism that enables the parasitoid to be tolerated by its host ants. One study by Vander Meer et al. (1989) suggests that a species of *Orasema* that is a parasitoid of *Solenopsis invicta* passively acquires the colony-specific hydrocarbon profiles of their host ants, but the parasitoid larvae do not produce their own hydrocarbons. However, the adult wasps feature their own specific hydrocarbons and, in addition, still retain residual amounts of the host hydrocarbons. Whether the acquisition of the hosts' cuticular hydrocarbons is coincidental or serves the function of chemical disguise, or "identity theft," is not clear.

Based on collection data, it is proposed that the Oraseminae (which consists of thirteen genera and eighty-nine described species; Heraty 2017, cited in Baker et al. 2020) are specialized parasitoids of the ant subfamily Myrmicinae (Clausen

1941; Heraty 2000, 2002; Lachaud and Pérez-Lachaud 2012), which includes some of the world's most invasive ant genera, such as *Pheidole, Solenopsis, Wasmannia,* and *Monomorium.* One species, *Orasema minuta,* has been found in nests of two myrmicine ant genera, *Pheidole* and *Temnothorax* (Baker et al. 2020). Lachaud and Pérez-Lachaud (2012) also list *Tetramorium,* as well as other ant subfamilies, as possible hosts of orasemine wasps. Most collecting records and documented prevalence in all biogeographic regions report *Pheidole* as the most common host genus associated with orasemine species. These facts and further biogeographic and phylogenetic studies strongly suggest that *Pheidole* is the ancestral host for the Oraseminae (Baker et al. 2020). John Heraty (2002) proposed that, generally, the phylogeny of Eucharitidae is correlated with that of the host ant subfamilies and relates to differences in oviposition of the parasitoids, the phoretic behavior of the planidia, and the ways the host larvae and pupae are parasitized.

Whereas wasps in the subfamily Oraseminae attack primarily myrmicine ants (with a few possible exceptions, see Lachaud and Pérez-Lachaud 2012), the Eucharitinae parasitize many other ant species from the subfamilies Formicinae, Ectatomminae, Myrmeciinae, and Ponerinae. Despite variation in host preference, all eucharitid parasitoids share a similar developmental sequence, wherein the first larval instar is a special planidium, the sole function of which is to use foraging ants as vectors for invading the host ants' nest, where they target ant larvae. Though planidia of Oraseminae burrow into the ant larva and first live as endoparasitoids, eucharitine planidia attach externally to the host larva and remain ectoparasitoids throughout their development. They too move to the posterior ventral thoracic region of the host once the ant larva has proceeded with pupation. In contrast, the larvae of the eucharitid subfamily Gollumiellinae are endoparasitic in larvae and pupae of the formicine genus *Nylanderia (Paratrechina)*, that is, they make their entire development inside the host ant larvae and pupae (Darling and Miller 1991; Darling 1992, 1999; cited in Heraty and Murray 2013; Lachaud and Pérez-Lachaud 2012).

In some cases, several conspecific parasitoids have been found in a single host pupa (Heraty and Barber 1990; Lachaud and Pérez-Lachaud 2012; Peeters et al. 2015), and "concurrent parasitism" has been reported for *Ectatomma tuberculatum*, which is simultaneously parasitized by the eucharitine species *Dilocantha lachaudii*, *Isomerala coronata*, and *Kapala* sp. (Pérez-Lachaud et al. 2006), and for *Ectatomma ruidum*, parasitized by two *Kapala* species, *K. iridicolor* and *K. izapa* (Lachaud and Pérez-Lachaud 2009). A typical life cycle of a eucharitine wasp has been beautifully documented by Jean-Paul Lachaud and Gabriela Pérez-Lachaud (2012) using as an example the species *Dilocantha lachaudii* (Figure 2-23).

However, as usual, there are exemptions to the rule: John Heraty and Elizabeth Murray (2013) studied the life history of the eucharitine *Pseudometagea schwarzii*, which is a parasitoid of the formicine *Lasius neoniger*. They confirmed a previous finding by Ayre (1962), that the planidium of *P. schwarzii* undergoes internal growth-feeding inside the larva of the host ant, thus it must be considered an endoparasitoid. Heraty and Murray argue, based on phylogenetic analyses, that the ectoparasitoid mode in Eucharitidae is the ancestral trait and that the derived endoparasitoid mode has evolved independently three times within the Eucharitidae.

Thanks to the profound investigations primarily by John Heraty and his collaborators, and additional studies by Jean-Paul Lachaud and Gabriela Pérez-Lachaud, Christopher Darling, and other entomologists, much information has been gained concerning the systematics, phylogeny, and life cycle of eucharitid parasitoids. Much less is known about the role these parasitoids play in ecosystems. However, one such study stands out, which we summarize below.

The plant *Leea manillensis* (Leeaceae, formerly Vitaceae) is a large, up to six meters high, shrub, very common in the Philippines and other parts of Southeast Asia. All *Leea* species, of which representatives also occur in Australia and New Guinea, are endowed with extrafloral nectaries and also produce so-called food bodies (also called pearl bodies) that are richly loaded with lipids, proteins, and

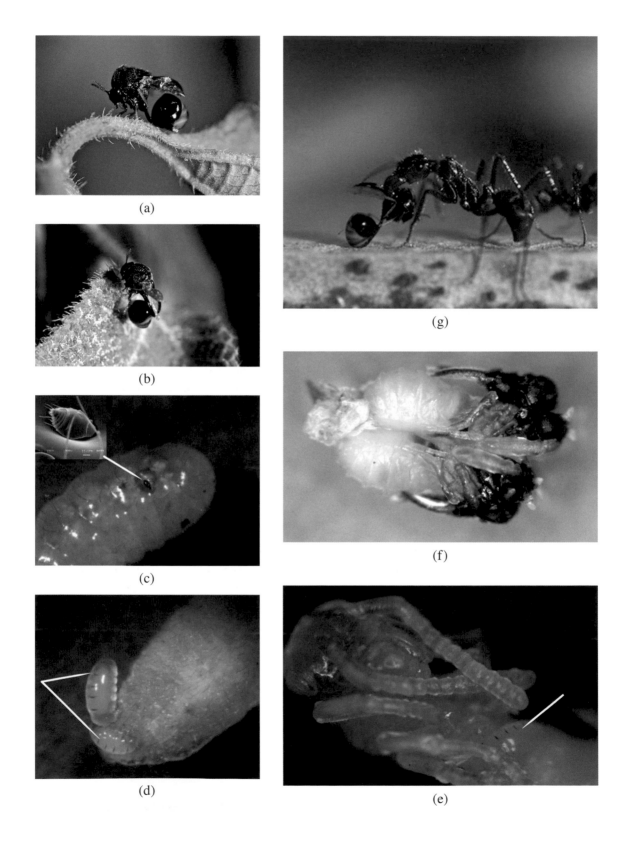

(a)

(b)

(c)

(d)

(e)

(f)

(g)

2-23   Life cycle of a typical eucharitid wasp.

a. Female of *Dilocantha lachaudii* (Eucharitinae) ovipositing on *Lantana camara*
   (Verbenaceae).
b. *D. lachaudii* female with eggs scattered on the leaf surface.
c. Planidium (white pointer) attached on a larva of the ant *Ectatomma tuberculatum*.
     Insert: scanning electron microscopic picture of a planidium.
d. Two *D. lachaudii* swollen planidia (white pointer) feeding on an *E. tuberculatum* larva.
e. Second-instar larva (white pointer) relocated after host pupation.
f. Two *D. lachaudii* pupae from a single host pupa. The host cocoon has been removed.
g. *E. tuberculatum* worker transporting a recently emerged *D. lachaudii* female.
     (Courtesy of Jean-Paul Lachaud and Gabriela Pérez-Lachaud).

carbohydrates. These plant products are sought after by ants that imbibe the nectar and harvest the food bodies. It has been hypothesized that the foraging ants provide protection for the *Leea* shrubs against herbivore predation. Yet, although suggestive, the direct proof of such mutualistic relationship between ants and *Leea* remained inconsistent. A quantitative study by Christoph Schwitzke, Brigitte Fiala, Eduard Linsenmair, and Eberhard Curio (2015) revealed the reasons for these inconsistencies. These authors were able to demonstrate that *L. manillensis* that were not attended by ants indeed suffered a significantly greater loss of leaf area than shrubs on which ants visited extrafloral nectaries. But, they wondered, why are so many *Leea* plants vacant of visiting ants and badly affected by herbivory? Particularly the fresh shoots, which are the sites of active extrafloral nectaries located at the abaxial side (lower surface) of the young leaves, were attacked by herbivores.

In this context, the work by Carey et al. (2012), who studied eucharitid wasps in southwestern Arizona and reported females of *Orasema simulatrix* oviposit almost exclusively near the extrafloral nectaries of the Bignoniaceae *Chilopsis linearis,* is of special interest. Nectary conditions varied, and filled nectaries were three times as likely to be chosen for oviposition by *O. simulatrix* than partly filled or empty nectaries. Most likely, the ants are more prone to visit well-filled nectaries than empty ones. The myrmicine *Pheidole desertorum* was the dominant ant species that visited the extrafloral nectaries in the tree canopy and is most likely also the primary host species of *O. simulatrix*. Circumstantial evidence suggests that the eggs or freshly hatched wasp larvae (planidia) are taken up by the foraging ant workers with the nectar and thus carried into the *Pheidole* colony.

It has frequently been reported that conditions of extrafloral nectaries can vary in many plants, and many factors can negatively affect the extrafloral nectary's productivity and thereby indirectly influence the ants' foraging and protective efficiency in plant-ant mutualism (see Lange et al. 2017). However, the interfering effect of parasitoid wasps in such mutualistic ant-plant relationships was not recognized until the study by Schwitzke et al. (2015). These authors investigated the ecological relationship between the plant *Leea manillensis* and the eucharitine wasp *Chalcura* sp. This parasitoid wasp places her eggs into the stipules' extrafloral nectaries. In contrast to several other eucharitid parasitoids, which exhibit a relatively high host ant specificity, at least at the subfamily level, planidia of *Chalcura* sp. seem to attack a broad range of ant species, although the authors emphasize that they do not know how the parasitoids fare inside the ant colonies. Schwitzke et al. (2015) demonstrated that the hatched planidia had a dramatic impact on the ant-plant-herbivore system. They write, "53% of the plants' shoots were infested with planidia. All of the common ant visitors at the extrafloral nectaries strictly avoided the shoots after eclosion of the wasp larvae." Soon the planidia left the shoots and dispersed. The ants attempted to dodge any contact with the planidia, and they attacked adult wasps, but eventually abandoned the shoots entirely. The authors conclude "the avoidance of planidia directly resulted in a lasting lack of ants on the shoots, which then caused an almost tenfold increase in complete shoot losses compared to unaffected shoots." Thus, the primary "enemies" of *Leea* plants are not so much the herbivores but the wasp parasitoids that interfere with the ant plant protectors.

As we already discussed in the section about parasitoid phorid flies, these wasps present another example demonstrating how parasitoids affect the foraging behavior and possible competitive interactions in ant communities (see Feener 2000). On the other hand, because of the frequently substantial parasitism rates observed in some eucharitid parasitoids of the genus *Orasema* infesting the myrmicine ant species *Wasmannia auropunctata* and *Solenopsis invicta*, John Heraty (1994) hypothesized that the wasps may modify colony demographics with potentially strong fitness consequences and may therefore serve as an effective means for biological control of these invasive species.

Diapriidae

The Diapriidae are a family of parasitoid wasps that attack larvae and pupae of a variety of insects, in particular flies. The species are distributed worldwide and subdivided into three subfamilies: Ambrositrinae, Belytinae, and Diapriinae (Sharkey et al. 2012). Several species of the Belytinae, but mainly of the Diapriinae, are associated with ants (Masner 1959, 1993; Huggert and Masner 1983; Masner and Garcia 2002; Lachaud and Pérez-Lachaud 2012; Loiácono et al. 2013), yet not much is known about their interactions with ants. Only a few diapriid wasps are demonstrated parasitoids of ants. In fact, the first direct evidence for a diapriid ant parasitoid was provided by Jean-Paul Lachaud and Luc Passera (1982), who reared the diapriid wasp *Plagiopria passerai* from cocoons of the formicine *Plagiolepis pygmaea*. Particularly, the subfamily Diapriinae includes several species that have been revealed to be true ant parasitoids (Loiácono et al. 2002, 2013; Masner and Garcia 2002; Fernández-Marín et al. 2006; Pérez-Ortega et al. 2010). Marita S. Loiácono (1985) reared *Szelenyiopria lucens* (Diapriini), collected in Uruguay, from larvae of the fungus-growing ant *Acromyrmex ambiguus*. She reports that up to three wasps developed in one mature ant larva. *Szelenyiopria pampeana* were also found in larvae of the ant *Acromyrmex lobicornis* (which is also the host of the parasitoid *Trichopria formicans*), whereas other *Szelenyiopria* species—for example, *S. reichenspergeri*—are associated with army ant species (*Eciton quadriglume* and *Neivamyrmex legionis*); however, parasitoidy has not been confirmed (Loiácono et al. 2013).

Two other fungus-growing ant genera were reported to be parasitized by diapriinine wasps. Fernández-Marín et al. (2006) provided the first description of the biology of *Acanthopria* spp. and *Mimopriella* sp., which attack the larvae of the fungus-growing ants *Cyphomyrmex minutus* and *C. rimosus,* respectively. The studies were conducted in Puerto Rico and Panama. Between 27% and 53% of the *C. minutus* colonies were parasitized by one morphospecies of *Acanthopria* with considerable variation among populations. The within-colony prevalence of *Acanthopria* in Puerto Rican *Cyphomyrmex* populations was about 16%, and in

the Panamanian population approximately 34%. Remarkably, 70% of *Cypho-myrmex rimosus* colonies were infested with one *Mimopriella* species and four morphospecies of *Acanthopria* sp. Although the authors did not find a correlation between colony size of the host ants and the proportion of parasitized larvae of *C. minutus,* in *C. rimosus* a negative correlation could be determined. To our knowledge, this is the first population study of parasitoid infection by diapriid wasps. The *Acanthopria* larvae, which occupy almost the entire body of their host larvae, do not alter the external appearance of the host larvae, except that the parasitized larvae have a grayish color instead of the creamy complexion of healthy ant larvae. The host ants do not seem to discriminate between parasitized and healthy larvae. When forced to relocate to new nest sites, all the larvae, including the infected ones, are carried by the host ant workers to the new nest. Adult wasps, in contrast, are treated with hostility and may even be killed by the ants. A second, similar, study by Pérez-Ortega et al. (2010) describes the parasitism of the fungus-growing ant *Trachymyrmex* cf. *zeteki,* which are attacked by several diapriinine wasp species. The study was conducted on a single population in Panama. From a large number of excavated *Trachymyrmex* nests, six wasp morphotypes were reared: two of the genus *Mimopriella,* one *Oxypria,* two *Szelenyiopria,* and one *Acanthopria.* The mean parasitism of ant larvae per colony was almost 34% in one season and approximately 27% in another season. The data from these two studies strongly suggest that the diapriid parasitoids play a significant role in the population biology of these attine ants.

2-24   Specimens of the diapriid wasp *Apopria coveri* are shown from lateral and dorsal view. (Courtesy of the late Michael C. Thomas). The pictures on the right side depict the host ant *Neivamyrmex opacithorax.* (Jen Fogarty/AntWeb.org).

In 1983, Lubomír Masner and his collaborator, the late Lars Huggert, two of the leading experts concerning the systematics and phylogeny of diapriid wasps, proposed that diapriid ant parasitoids evolved from ancestors that parasitized Diptera living in the kitchen middens or refuse sites of ants. They noted that diapriid females, "in search for potential hosts have progressively integrated with ants." Although ant parasitism has been demonstrated in only a minority of all known myrmecophilous diapriid wasps, many more species have been found in ant nests, which obviously are not parasitoids. These myrmecophiles often exhibit morphological and behavioral adaptations to their host ants. For example, the diapriinine *Apopria coveri* has been collected from bivouac nests of the army ant species *Neivamyrmex opacithorax* and *N. nigrescens* (in fact, the latter *Neivamyrmex* species was later identified by Mark Deyrup as *N. texanus;* Stefan Cover, personal communication) "in the very center of the colony, deep underground, in total darkness" (Masner and Garcia 2002). The authors state that the wasps were wingless and had no ocelli, and therefore they assumed that the life cycle and dispersal strategies are closely synchronized with their host ants. Presumably *A. coveri* disperse with the budding colonies of their host ants, and they strikingly look like the *Neivamyrmex* hosts (Figure 2-24). In fact, one of us (B. H.) collected this *N. opacithorax* colony with queen with Stefan Cover in Florida, but we did not notice the diapriid myrmecophile at that time. Only later did Stefan recognize that there was a somewhat strange-looking *Neivamyrmex* among the worker ants, and of course, he realized this is a diapriid wasp that strikingly resembles the host ants. Stefan sent the specimens to Lubomír Masner, who determined that it was a new species of the genus *Apopria* and named it after its discoverer, Stefan Cover.

There are many more examples of diapriinine myrmecophiles, especially in army ants, that feature special morphological adaptations to their host ants. Loiácono et al. (2013) lists, among other species, the diapriinine wasp *Asolenopsia rufa,* which was found with the army ant species *Neivamyrmex carettei* and *N. sulatus,* and the wasp *Notoxoides pronotalis* that lives with the army ant *Eciton dulcium.* Similarly, the wasp *Bruchopria pentatoma* features traits that appear to be adaptations to its life as a myrmecophile of the myrmicine *Solenopsis richteri.*

2-25   The diapriid myrmecophilous wasp *Bruesopria americana* with its host species *Solenopsis molesta.* (Courtesy of Alex Wild / alexanderwild.com). The lower picture shows the diapriid wasp *Solenopsia imitatrix* with its host species *Solenopsis fugax.* (Courtesy of Claude Lebas).

Specimens of these wasps, which had the wings removed (presumably bitten off by the ants), have been found in *Solenopsis* nests (Masner and Garcia 2002). Loiácono et al. (2002) studied *Bruchopria* individuals with and without wings, and they reasoned that apterism (winglessness) is either "caused by autotomy or by bites of the host ants." The authors argue that wasps without wings more easily move into nest mound galleries and chambers. Females of *Bruchopria pentatoma* and *B. hexatoma* have been found with *Solenopsis* and in nests of the fungus-growing ant *Acromyrmex lundii.*

Wing (1951) describes the new diapriid genus and species *Bruesopria americana* and *B. severi,* which appear to be myrmecophiles of the thief ant *Solenopsis* (subgenus *Diplorhoptrum*) *molesta* in North America (Figure 2-25). For the Nearctic region, Huggert and Masner (1983) list, among others, the genera *Auxopaedentes* and *Bruesopria,* both associated with the thief ants of *Solenopsis* (subgenus *Diplorhoptrum*). These diapriinine genera are related to the Palearctic genus *Solenopsia imitatrix,* first discovered and described by Erich Wasmann (1899), and *Lepidopria pedestris* and *Trichopria inquilina,* both described by Jean-Jacques Kieffer (1911 and 1904). These species are associated with the thief ant *Solenopsis fugax* (Wasmann 1891; K. Hölldobler 1928; Lachaud and Passera 1982; Borowiec 2013). Their behavior inside the thief ant nests has been better studied than that of most other diapriid myrmecophiles. In particular, *Solenopsia imitatrix* received more attention than the other species (Figure 2-25). The first behavioral observations date back to Wasmann (1891). He concluded that the wasps are passively tolerated by the host ants. Janet (1897a) observed more active interactions between *Solenopsia* wasps and ants. The wasps were seen tapping the head of the host ants, and the ants were observed licking the wasps, though no trophallaxis could be observed. Karl Hölldobler (1928) cultured the three diapriid species in *Solenopsis fugax* colonies in formicaria but paid more attention to *Solenopsia imitatrix.* He describes the host ants frequently licking the wasp, mainly the areas between mesametasoma, between the head and mesosoma, and the antennae. He also observed trophallaxis and noted that the wasps were frequently carried around inside the nest by host ant workers.

Jean-Paul Lachaud (1980, 1982) conducted a behavioral study of *Solenopsia imitatrix* and *Lepidopria pedestris* and described the tactile communication between the myrmecophile wasps with their host ants and observed trophallactic behavior between *L. pedestris* and *S. fugax* workers. He also investigated the structure of the integumental exocrine glands of the wasps, the secretions of which might play a role in the interspecific relationship between the ants and myrmecophiles,

INSIDE AND ON THE BODIES OF ANTS

species of the new genus *Larvamima* (*L. marianae, L. schneirlai, L. carli, L. cristata*), which are mimics of army ant larvae. The mites have an elongated shape and are apparently undetected as larval mimics within the colony. Carl Rettenmeyer reports about this ant-larva-like mite in his PhD thesis (1961). He observed them being carried by the *Eciton hamatum* workers in the typical army ant fashion, held underneath the carrier's body. Mites were seen "walking upon ant larvae, eliciting no abnormal reactions from the larvae or adults in the vicinity. Based on cheliceral structure, it is hypothesized that the mites feed on their larval hosts" (Elzinga 1993). Most likely these *Larvamima* species not only have the physical appearance of a host ant larva but also are protected by a brood pheromone like allomone.

Like the phorid fly *Metopina formicomendicula,* certain mites also exploit the ants' social response to tactile stimulation of the ants' mouth parts. The best-known examples are species of the antennophorid mite genus *Antennophorus*

2-27 The myrmecophilous mite *Antennophorus*, the species that lives in nests of the formicine genus *Lasius*. (Courtesy of Taku Shimada).

(Figure 2-27), of which ten species are described (Trach and Bobylev 2018). All of them live with the formicine ants of the genus *Lasius* (Haller 1877; Janet 1897b; Wasmann 1902; Karawajew 1906; Wisniewski and Hirschmann 1992). The mites ride on the bodies of the ants. Charles Janet observed that when more mites occupy the body of one ant, they tend to adjust their position in such a way as to produce a balanced load. Nevertheless, often the workers try to remove the mites from their bodies, usually to no avail.

The *Antennophorus* live on food regurgitated by the host ants in response to the mites' intensive stroking of the workers' mouthparts with their long, antenna-like forelegs. In addition, the mites often insert themselves in a trophallactic food exchange between two ant nestmates, imbibing the regurgitated liquid as it passes between the ants (Figure 2-28).

It has also been observed that mites sitting on the gaster of one worker solicit food from another ant. The mites preferably mount newly eclosed workers and are placed mostly on the underside of the head (Figure 2-29). This is advantageous because young workers usually stay inside the nest, and they are attractive to older workers, who frequently feed them by regurgitation, offering the mites many opportunities to participate in the food exchange.

Nigel Franks and his coworkers (1991) presented a quantitative ethogram of the interaction between *Antennophorus grandis* and *Lasius flavus;* they summarize their findings as follows:

> The mites tend to occur on the smaller nurse workers and receive food from them at an extremely high frequency. *Antennophorus grandis* also frequently gain food when one ant is donating food to the one, they are riding upon. The mites seem to inhibit the ability of their host worker to show most social behaviors such as tending ant larvae. The mites frequently move from one host worker to another. For these reasons the mites may have a larger impact upon their host colony than their relative rarity first suggests. The ants do not seem to have any specific defense against these parasites. (Franks et al. 1991, 59)

INSIDE AND ON THE BODIES OF ANTS

2-28   The myrmecophilous mite *Antennophorus* sp. stimulating the mouthparts of the host ant *Lasius morisitai* to elicit regurgitation of food by the ant. The lower picture shows how *Antennophorus* mites partake in the trophallaxis between two host ant workers. (Courtesy of Taku Shimada).

2-29　The preferred place for *Antennophorus* is the underside of the ant's head. The picture shows *Antennophorus* sp. and host ant *Lasius talpa*. (Courtesy of Kinomura Kyoichi).

When the authors removed mites from host workers, they were never attacked after reintroducing them to workers, even from foreign colonies. This was also the case with mites that were prevented from begging by amputating their front legs. When such injured mites were returned to the colonies, "on two out of five occasions, the mites were carried to the ants' brood pile and placed among the larvae. Even though this was in the minority of cases, these preliminary findings may suggest that the mites may have the odor of the ants' brood. This hypothesis would help explain why the mites are not attacked, even by alien workers, as ant larval brood can be accepted even by non-congeneric ants" (Franks et al. 1991, 69).

Beetles: Coleoptera

Alexey Tishechkin, Daniel Kronauer, and Christoph von Beeren (2017) recently discovered an especially interesting adaption in the hitherto unknown histerid beetle of the subfamily Haeteriinae, *Nymphister kronaueri,* which is associated with army ants and is endowed with an exceptional mechanism of phoresy. This beetle is relatively small, not larger than the gaster of a medium-sized worker of the host ant species *Eciton mexicanum* s. str. Like all true army ants, *E. mexicanum* frequently emigrate to new bivouac sites. This small beetle would probably have great trouble moving on its own along the ants' emigration column, as many other myrmecophiles do—for example, staphylinid beetles, which are closely associated with the army ants (Akre and Rettenmeyer 1968; Rettenmeyer et al. 2011). Instead, *N. kronaueri* attaches itself between the petiole and postpetiole (the waist between alitrunk and gaster) of the host ant, usually a medium-sized worker. The attached beetle closely resembles the gaster of the ant worker; it appears as if this ant has two gasters (Figure 2-30). One might be tempted to speculate that this resemblance serves as a mimetic protection mechanism (Batesian mimicry); however, Alexey Tishechkin and his colleagues point out, emigrations usually take place at night and "hence, it seems unlikely that nocturnal predators, which barely rely on visual cues, have caused selection pressures of *N. kronaueri* to resemble a host gaster." Could it serve tactile mimicry in interactions with host ants?

The authors suggest that similar morphological characters as exhibited by *N. kronaueri* have been found in another histerid genus, *Ecclisister* beetles that show host specificity with *Eciton burchellii* (Tishechkin et al. 2017). It attaches itself to

2-30   The Haeteriinae beetle *Nymphister kronaueri* attaches itself between petiole and postpetiole of army ant workers. It resembles the ant's gaster. (Courtesy of Daniel Kronauer).

the underside of major workers' heads. They speculate that the dense cover with mechano-receptive setae in both genera is perhaps a convergently evolved adaptation to their intimate life with army ants, because this association requires a high sensitivity to tactile stimuli. The authors argue, "The frequent host contact might also be related to the second characteristic shared among these beetles, the integumental micro-sculpture," because the integumental micro-sculpture of *N. kronaueri* and *Ecclisister* "is fairly similar to that of their host ants," as August Reichensperger (1924), the leading coleopterologist of his time, already noted for several myrmecophilous histerid beetles. Indeed, the so-called tactile mimicry has been discussed repeatedly by myrmecologists investigating myrmecophilous adaptations (Wasmann 1925; K. Hölldobler 1947, 1953; Kistner 1979, 1982; Hölldobler and Wilson 1990), yet the experimental proof remains elusive.

Another interesting case of phoretic behavior, which might, however, be ectoparasitism, was discovered by Auguste Forel (1894). He noted that some species of the dermestid, thoricitine beetle genus *Thorictus* attach themselves on the host ant's antennal scapes. According to Escherich (1898b), the beetle has been seen attached to the legs as well as antennae. The phoretic *Thorictus* species seem to be quite host-specific and were found mainly with species of the ant genus *Cataglyphis*. For example, *T. foreli* has been reported with *C. bicolor,* *T. panciseta* with *C. savignyi,* and *T. castaneus* with *C. bombycina,* but in the last case the beetles were mostly seen walking freely in the host ants' nest chambers. *Thorictus castaneus* beetles have a more pronounced trichome structure than the other species. In fact, most myrmecophilous *Thorictus* species do not live as attached ectoparasites but move freely in the nest area and refuse piles of their hosts, where they feed on dead ants (Figure 2-31). Such *Thorictus* species have also been found in nests of other ant genera—for example, *Messor barbarus* (Forel 1894).

Wasmann (1898a,b) argues, based on circumstantial evidence, that species attached to the antennal scape of their host ants are true ectoparasites; they puncture with the pointed lacinia of their lower jaw the cuticle of the antennal scape and imbibe hemolymph oozing out of the wound. It has also been claimed that *T. foreli* and *T. panciseta* partake in the regurgitated liquids that two nestmates

INSIDE AND ON THE BODIES OF ANTS

2-31   Different species of the thorictine beetle *Thorictus* sp. are specialized to spend much of their life attached to the legs as well as antennae of their host ants, but other species move freely inside the ants' nest and feed on the discarded garbage. (Courtesy of Pavel Krásenský).

exchange during trophallaxis (Gösswald 1985), in a fashion similar to what we just described for the mite *Antennophorus,* but no unequivocal documentation is yet available.

Numerous organisms rely on the bodies of ants, and many more remain to be discovered. This varied roster of associates can affect both the morphology and behavior of their unsuspecting hosts, shrinking their heads or driving them to sites where they will be consumed for the benefit of their internal guests. Many organisms are so intimately affiliated with ants that they occupy the very cells of their hosts, which cannot live without them. In turn, some parasites have come to resemble parts of ant anatomy, even conceding to function as ant feet (tarsi). Although it is the individual ant that is afflicted by endo- and ectoparasites, we can see evidence of an "immune response" at the superorganismal level. In some cases, the most harmful invaders, like phorid flies, are met by a subcaste of workers specialized in defense against these parasites. In chapter 3, we explore how the ant colony discriminates between self and foreign, and how myrmecophilous intruders are able to surmount the social immune barriers of their host ant colonies.

3 Recognition, Identity Theft, and Camouflage

Nestmate Recognition

Like any regular organism that protects itself with an immune system that discriminates between self and foreign, ant superorganisms are endowed with "social immune barriers," allowing nestmates to enter while repelling non-nestmate intruders. It is one of the major challenges of myrmecophiles to overcome these hurdles when seeking acceptance in an ant society. In fact, recognition is a profoundly important form of communication in all social insects, including the recognition of alien species, of members of other colonies of the same species, and of nestmates belonging to various castes and immature stages.

Consider nestmate recognition. Just as a human being identifies another person by scanning face and body form, and perhaps the sound of the voice, an ant classifies another ant by the bouquet of odors on and around its body. The transaction occurs in a split second. When two ants meet in or outside the nest, each sweeps its antennae over part of the body of the other. They are testing body odors. If they belong to the same colony and hence possess familiar odors, they move on with no further response. On the other hand, if the encountered ant is from another colony of the same species, and the species is one of the large majority in which colony boundaries are recognized, the intruder is treated very differently. The foreigners will be treated with hostility and, in the extreme, be attacked or killed. At intermediate levels, often depending on specific contexts, foreigners are variously avoided, threatened with open mandibles, nipped, and dragged out of the colony's home range and dumped (Hölldobler and Wilson 1990, 2009).

What are the colony recognition labels (also called "discriminators")? In general, they are hydrocarbons contained in the waxy coating that covers the epicuticle of the body (for reviews see Singer 1998; Lenoir et al. 1999; Howard and Blomquist 2005; Hefetz 2007; Drijfhout et al. 2009; Martin and Drijfhout 2009a,b; van Zweden and d'Ettore 2010; Nehring et al. 2011; Sturgis and Gordon 2012; Sprenger and Menzel 2020).

All insects are endowed with cuticular hydrocarbons. The original function of these waxy compounds is protection against infection and desiccation. Insects exhibit species-specific differences in the kinds and relative amounts of cuticular hydrocarbons (see, e.g., Lucas et al. 2005; Greene and Gordon 2007; Martin and Drijfhout 2009a; Kather and Martin 2015; Guillem et al. 2016). For example, the compounds can differ in the carbon-chain length, or in whether they are saturated (no double bonds: alkanes) or unsaturated (double bonds: alkenes), or in the number and position of double bonds or possible methyl branches.

It was a key discovery in the late 1970s and early 1980s, by Ralph Howard, Gar Blomquist, Jean-Luc Clément, Robert Vander Meer and their coworkers, that in termites and social Hymenoptera (ants, social bees, and social wasps), these differences in the cuticular hydrocarbon blends or mixtures not only exhibit species-specific patterns, but also encode colony specificity that possibly underlies nestmate recognition. In addition, the hydrocarbon blends encode information concerning the fecundity status of reproductive individuals in an insect society (Peeters and Liebig 2009; Liebig 2010; Smith and Liebig 2017; Funaro et al. 2018, 2019) and the age class and labor role of the worker castes (Morel et al. 1988; Bonavita-Cougourdan et al. 1993; Wagner et al. 1998; Kaib et al. 2000; Liebig et al. 2000; Greene and Gordon 2003). From studies with the fruit fly (*Drosophila melanogaster*) and cockroaches (Schal et al. 1998) we learned that the cuticular hydrocarbons are produced in subcuticular glandular cells, such as glandular hypodermal cells, and especially in glandular cells (without duct cells) embedded in the fat body, the so-called oenocytes. The hydrocarbons are transported from these cells by high-density hemolymph proteins to various target tissues such as the cuticle, the ovaries (the eggs are coated with hydrocarbons), and various storage organs, such as the postpharyngeal gland and Dufour's gland, although these structures also have glandular epithelia and may produce hydrocarbons on their own.

The postpharyngeal gland, which is unique to ants, is a large organ located in the head. It appears to play a key role in creating the specific colony odor. Bagnères and Morgan (1991) found that the postpharyngeal glands contain the same characteristic hydrocarbons as identified on the epicuticle, and Soroker et al. (1994, 1995a,b, 1998) and Soroker and Hefetz (2000) present evidence that the main function of the postpharyngeal gland, in the context of the colony odor, is to store hydrocarbons. The hydrocarbons are spread over the body of each ant as she grooms herself and her nestmates by licking motions of her tongue. At the same time, during grooming, the active ant appears to load cuticular hydrocarbons from the groomed nestmate into her own postpharyngeal gland, where they are admixed with other hydrocarbons collected during previous grooming actions. Thus, during each grooming and trophallaxis event, contents of the postpharyngeal gland are transferred to nestmates and gathered from the groomed ant. Victoria Soroker, Abraham Hefetz, and their collaborators were the first to recognize this unique function of the postpharyngeal gland. They named it the "Gestalt" organ for nestmate recognition (Soroker et al. 1994). Indeed, each individual ant produces her own genetically encoded hydrocarbon mixtures, and the individual variation among nestmates most likely depends on how closely workers in a colony are genetically related. However, the mixing of all individual hydrocarbon mixtures with the aid of the postpharyngeal gland invariably creates a collective blend of hydrocarbons, produced and spread in the colony in a more or less homogeneous state. Genetic differences among nest-mates are "erased," no within-colony kin recognition or nepotism is possible, and, in fact, has never been explicitly demonstrated (Hölldobler and Wilson 2009). In this way an odoriferous "Gestalt" is created to which colony members predictably respond (Crozier and Dix 1979).

From an ethological point of view, we can hypothesize that, in the course of social evolution, the protective barrier on the insect cuticle has been co-opted to also function as a recognition or identification cue in insect societies and, in some cases, to serve as "honest signals" (Liebig et al. 2000; Liebig 2010; Smith et al. 2008, 2009). Incidentally, this is a striking example of ritualization in social insects. In evolutionary behavioral biology, ritualization is the transformation of a morphological or physiological trait, or a "no-display" behavior, into a signal function. The original function is maintained, but over the course of evolution it receives a second function in social communication (Huxley 1966). In fact, signals in animal communication often do not originate de novo; instead they evolve by ritualization, that is, "co-option" of already existing traits.

But can we really claim that the cuticular hydrocarbon profiles function as a discrimination signal or cue? Many papers have been published describing the specificity of cuticular hydrocarbon blends among colonies of numerous ant species, but in most cases the inference that these colony-specific hydrocarbon profiles serve as colony discriminators is based on correlations. Laurence Morel, Robert Vander Meer, and Barry Lavine (1988) reported the first experimental evidence that cuticular hydrocarbon blends affect colony recognition, and Sigal Lahav, Victoria Soroker, Abraham Hefetz, and Robert Vander Meer (1999) were the first to present experimental behavioral evidence demonstrating hydrocarbons from the postpharyngeal gland function in colony recognition (or discrimination) in ants. For their analysis they used the formicine ant *Cataglyphis niger* and, since the cuticular hydrocarbon blends are identical with those of the postpharyngeal gland, Lahav et al. (1999) used the contents of these glands for their experiments.

By separating the hydrocarbons from the other lipid constituents, the authors unequivocally demonstrated that blends of hydrocarbons exclusively function as colony odors or discriminators. The other lipid fractions of the postpharyngeal gland elicited no behavioral response. However, application of alien hydrocarbon mixtures onto ants elicited aggression in nestmates comparable to that exhibited in encounters with foreign ants treated with solvent only. Treating an alien ant with a nestmate's hydrocarbon mixture reduced the aggression exhibited by fellow nestmates against the foreigner. Interestingly, nestmates treated with alien hydrocarbons elicited more aggression in nestmates than occurred when alien workers were treated with nestmate discriminators. According to Lahav et al. (1999), "this implies that ants are more sensitive to label / template differences than to similarities." Indeed, it is generally assumed that recognition or discrimination in ants is based on the comparison of a received signal or cue with a neural reference or learned (imprinted) template. In this respect, it is also noteworthy that ants treated with foreign hydrocarbon mixtures do not exhibit aggression toward nestmates, but do so against foreigners, even though they are covered in the foreigners' hydrocarbons. This demonstrates that treatment with foreign hydrocarbons does not affect the template of the treated ant. The neurobiological and sensory physiological aspects of nestmate recognition and non-nestmate discrimination were long not studied and were therefore based mainly on hypothetical models.

In a subsequent study, Akino et al. (2004) analyzed the cuticular hydrocarbons of colonies of *Formica japonica* and essentially confirmed the results obtained by Lahav et al. (1999) with colonies of *Cataglyphis niger*. Akino et al. (2004) applied crude extracts of cuticular hydrocarbons to glass dummies and reported that workers showed aggressive behavior "against the dummies treated with one worker equivalent of isolated non-nestmate hydrocarbons, while they paid less attention to control dummies [glass dummies untreated or contaminated with solvent] and dummies treated with nestmate hydrocarbons." Like Lahav et al. (1999), they found that only the hydrocarbon fraction of the cuticle lipids elicited discrimination behavior. Interestingly, Akino and his coworkers also elic-ited asymmetric responses to dummies coated with artificial "cocktails" of synthesized hydrocarbons that resembled the natural colony-specific hydro-carbon mixtures of a foreign colony. Dummies coated with artificial "cocktails" resembling the hydrocarbon profiles of their own colony were treated indiffer-ently by nestmate workers.

As behavioral biologists, we have to address a few critical aspects in this study. Apparently, the bioassays with glass dummies were not conducted double-blind, nor were they video-recorded. From our own experience, we know how easily involuntary biases can potentially distort the observational records. In addition, the authors do not describe how the hydrocarbon mixtures were applied onto the dummies. They report that one ant equivalent of hydrocarbons in 20 μl sol-vent were used for each dummy. This is a relatively large amount of solvent, and as glass dummies do not absorb the liquid, the application of the total amount must have taken some time, before the dummies could be tested. For compar-ison, Lahav et al. (1999) used 0.7–0.8 ant equivalents in 1 μl solvent for each ap-plication onto live ants. This is easily comprehensible, especially because the au-thors describe how they applied the hydrocarbon extracts, and all bioassays were videotaped. In any case, the study by Akino et al. (2004) confirms the find-ings by Lahav et al. (1999). Although they used cuticular hydrocarbon blends, they also demonstrated that the hydrocarbon contents in the postpharyngeal gland are very similar to those found on the ants' cuticle.

An excellent study of graded behavioral actions toward dummies with hydro-carbon blends of different volatility was published by Andreas Brandstaetter, Annett Endler, and Christoph Kleineidam (2008), who demonstrated that the "complex multicomponent recognition cues can be perceived and discriminated by ants at close range," and concluded that contact chemo-sensilla are not cru-cial for this recognition and discrimination process.

While these and several other studies strongly indicate that discrimination of foe from nestmate is based on specific blends of hydrocarbons on the cuticle's surface, the precise neurobiological mechanisms responsible for the detection and coding of that information within the olfactory system remained ambiguous. Ozaki et al. (2005) conducted the first electrophysiological studies in *Camponotus japonicus* and identified multi-porous sensilla (most likely sensilla basiconica) involved in this recognition or discrimination process. They propose that these sensilla respond only to hydrocarbon blends of non-nestmates, whereas colony members have become desensitized to their own colony's specific hydrocarbon blends. In a way, Ozaki et al. (2005) propose that nestmate recognition is in fact non-nestmate discrimination. In subsequent papers that report different experimental approaches, Guerrieri et al. (2009), van Zweden and d'Ettorre (2010), and Ferguson et al. (2020), came to similar conclusions.

Kavita Sharma and colleagues (2015), employing electrophysiological recordings, identified sensilla basiconica on the antennae of workers of *Camponotus floridanus* as major olfactory receptors for hydrocarbons. Each of these sensilla "contains multiple olfactory receptor neurons that are differentially sensitive to cuticular hydrocarbons and allow them to be classified into three broad groups that collectively detect every hydrocarbon tested, including queen and worker-enriched cuticular hydrocarbons." Presumably, this broad-spectrum sensitivity enables the ants to detect cuticular hydrocarbons from both nestmates and non-nestmates, which are in most cases identical, yet mixed in colony-specific quantities. Therefore, the authors also recorded the summations of action potentials of single sensilla. This made it possible to test dose-dependent relationships between stimulations with mixtures of different quantities of hydrocarbons and electrophysiological responses. The results of these studies do not support the hypothesis that the odorant receptors of *C. floridanus* workers do not respond to their own colony's hydrocarbon blends and react only to the hydrocarbon profiles of foreign colonies. In fact, the analysis of the electrophysiological recordings strongly suggests that the worker olfactory sensilla detect both nestmate and non-nestmate cuticular hydrocarbon blends.

These results are consistent with whole-antenna recordings (antennograms) and calcium-based imaging of olfactory glomeruli in the antennal lobes of *C. floridanus* conducted by Christoph Kleineidam, Andreas Brandstaetter, and their colleagues (Brandstaetter and Kleineidam 2011; Brandstaetter et al. 2011). These authors demonstrated that neuronal information about odors "is represented in spatial activity patterns in the primary olfactory neuropile of the insect brain, the antennal lobe, which is analogous to the vertebrate olfactory bulb." Colony odors of nestmates and non-nestmates elicited partly spatially overlapping activity patterns, distributed across different antennal lobe compartments. The authors also propose, based on their results, that "information about colony odors is processed in parallel in different neuroanatomical compartments, using the computational power of the whole antennal lobe network. Parallel processing might be advantageous, allowing reliable discrimination of highly complex social odors" (Brandstaetter and Kleineidam 2011, 2437).

Since the previous work by Kelber et al. (2010) has demonstrated to where in the antennal lobe the sensilla basiconica project, Brandstaetter and Kleineidam could confirm the significance of these sensilla for the hydrocarbon perception. However, the authors convincingly argued that besides the sensilla basiconica, the sensilla trichodea curvata are also involved in the perception of cuticular hydrocarbons. They measured neuronal responses to colony odors in glomeruli clusters, which are innervated not by sensilla basiconica but by sensilla trichodea curvata.

Kleineidam, Brandstaetter, and their colleagues confirmed previous suggestions that colony odors change over nest space and time and "the nervous system has to constantly adjust for this template reformation." In fact, Neupert et al. (2018) found that workers adjust their nestmate recognition by learning novel, manipulated cuticular hydrocarbon profiles but still accept workers carrying the previous profile. We agree with the authors' assessment that nestmate recognition in ants is considerably more complex than previously assumed. It "is a partitioned (multiple template) process of the olfactory system that allows discrimination and categorization of nestmates by differences in their cuticular hydrocarbon profiles." And, in a follow-up study by Andreas Brandstaetter, Wolfgang Rössler, and Christoph Kleineidam (2011, 1), the authors conclude, "Ants are not anosmic [smell blind] to nestmate colony odors. However, spatial activity patterns in the antennal lobes alone do not provide sufficient information for colony odor discrimination and this finding challenges the current notion of how odor quality is coded. Our result illustrates the enormous challenge for the nervous system to classify multi-component odors and indicates that other neuronal parameters, e.g., precise timing of neuronal activity, are likely necessary for attribution of odor quality to multi-component odors."

In conclusion, we have to understand that although these specific hydrocarbon blends are certainly the most significant parameters in colony discrimination, several other studies show that additional intrinsic colony parameters and extrinsic factors can affect or modulate the nature of colony discriminators (e.g., Liang and Silverman 2000; for further discussion see Hölldobler and Wilson 2009).

Identity Theft and Other Means of Intrusion

After the pioneering discovery by Howard, McDaniel, and Blomquist (1980) of cuticular hydrocarbon mimicry in the termitophile staphylinid beetle *Trichopsenius frosti* and the subsequent discovery by Vander Meer and Wojcik (1982) on the acquisition of cuticular hydrocarbons by the scarabaeid beetle *Martineziana dutertrei* (formerly *Myrmecaphodius excavaticollis*) from its *Solenopsis* host ants, many papers have been published on so-called chemical mimicry in social insect colonies (for reviews see Dettner and Liepert 1994; Lenoir et al. 2001; Akino 2008; Guillem et al. 2014).

Scarabaeid, Coccinellid, and Histerid Myrmecophiles

Robert Vander Meer and D. P. Wojcik (1982) report that the cuticular hydrocarbons of *Martineziana dutertrei* differ conspicuously with each of its four *Solenopsis* fire ant hosts, and the beetles can easily shed the hydrocarbons of one *Solenopsis* species and acquire the pattern of another (Figure 3-1). In addition to species-specific characteristics, the hydrocarbon mixtures probably also contain

colony-specific patterns. This, in part, explains how *Martineziana* is able to infest a variety of *Solenopsis* species and colonies. Vander Meer and Wojcik demonstrated this effect with the following experiment. Beetles from *S. richteri* colonies were isolated for two weeks and then introduced into colonies of *S. invicta*. After five days, the beetles were removed and analyzed for cuticular hydrocarbons. It was found that *M. dutertrei* had acquired the cuticular hydrocarbons of its new hosts. Vander Meer and Wojcik observed,

> The same phenomenon occurred when previously isolated beetles were introduced into *Solenopsis geminata* and *S. xyloni* colonies. Although the switching of hydrocarbon patterns from one host to another weakens the likelihood that they are synthesized by the beetle, we also found that freshly killed isolated beetles had acquired *S. invicta* hydrocarbons within two days after exposure to the ant colony. These data eliminate biosynthesis as a possibility and support a passive mechanism of hydrocarbon acquisition. When initially introduced into a host colony, the *M. dutertrei* were imme-

diately attacked. The response of the beetles was to play dead and wait for the attacks to cease, or they moved to an area less accessible to the ants. Within two hours, after introduction into a host colony, the beetles' cuticle contained 15 percent of host hydrocarbons. The accumulation of hydrocarbons continued up to four days until the beetles' cuticle contained about 50 percent host hydrocarbons. Beetles surviving this long were generally no longer attacked. (Vander Meer and Wojcik 1982, 807)

Vander Meer et al. (1989) obtained similar results with the eucharitid parasitoid wasp, *Orasema* sp., of the fire ant *Solenopsis invicta,* and Akino and Yamaoka (1998) report corresponding findings with the aphidiid parasitoid wasp *Paralipsis eikoae,* which lives in nests of *Lasius sakagamii* and oviposits into the honeydew-producing aphids tended by the ants. The wasp is not only tolerated by the ants, but it even succeeds in releasing regurgitation of food in the ants (Figure 3-2). Toshiharu Akino and Ryohei Yamaoka were able to demonstrate that the wasps, by having close bodily contact with and being groomed by the ants, acquire the colony-specific cuticular hydrocarbon profiles. Whereas they are treated like nestmates by the workers of their host colony, they are fiercely attacked by conspecific ants of a foreign colony.

These examples clearly demonstrate that the acquisition of the colony-specific hydrocarbon profile is a simple mechanism that enables the parasitoids or myrmecophiles for better or worse to be tolerated by the host ants. It is a strategy that empowers the myrmecophilous *Martineziana* beetles to avoid being recognized as foreign. The more complete the acquired host hydrocarbon profile, the more effectively the foreign inhabitant goes undetected by its host ants. The beetles seem to make their living primarily as scavengers in fire ant nests, eating dead ants and larvae and booty brought into the nest by ant foragers (Wojcik et al. 1991).

Another intriguing example are the myrmecophilous larvae of the ladybird beetles *Thalassa saginata* (Coccinellidae), which live in the brood chambers of the dolichoderine ant *Dolichoderus bidens.* These ants and their ladybird

3-2   The aphidiid parasitoid wasp *Paralipsis eikoae* lives in nests of *Lasius sakagamii* and succeeds in releasing regurgitation of food in the host ants. (Courtesy of Taku Shimada).

myrmecophiles were studied by Jérôme Orivel and colleagues (2004) in the forests of French Guinea. The ants build polydomous carton nests under leaves in trees. Usually, the colonies consist of one queen and hundreds to several thousand workers. The authors noted that 26 out of the 103 *D. bidens* colonies inspected (25.2%) sheltered one of the developmental stages of *T. saginata* or exuviae (shed exoskeleton of pupae). The larvae were attended by the ants, frequently licked, and when a brood nest was relocated the ants carried the beetle larvae just like their own brood to the new location. The beetle larvae and pupae always remained in the brood pile of the host ants, but it could not be determined whether they feed on the ants' brood. At emergence, the adult ladybirds remained inside the burst pupal exuvia until their exoskeleton was fully hardened. They then exited the nest hurriedly, because the adults would be attacked by the host ants. A comparative analysis of the surface hydrocarbon blends of beetle larvae and ant larvae revealed a close match between them. The authors hypothesize that the myrmecophilous larvae may biosynthesize mimics of surface hydrocarbon mixtures of their host larvae, yet there is no proof of that, and the authors concede this. Orivel and his colleagues also report that the beetle larvae secrete substances from their hairs and anal gland that are eagerly licked by the ants. We propose that during the ants' licking and the myrmecophiles' close contact to ant larvae and nurse ants, they can easily acquire the host's surface hydrocarbons. Most likely these beetle larvae produce a mimic of the ants' brood pheromone, which triggers in the nurse ants the adoption and grooming behavior, as we know it from other myrmecophilous immatures (Hölldobler 1967; Hölldobler and Wilson 1990; Hölldobler et al. 2018).

Alan Lenoir and his colleagues (2012) investigated the chemical integration of the myrmecophilous beetles *Sternocoelis hispanus* (Histeridae, Haeterinae), which live in nests of the myrmicine ant *Aphaenogaster senilis* (Figure 3-3). The genus *Sternocoelis* contains twenty-eight species, distributed in the Mediterranean area (Morocco, Algiers, Central and Southern Portugal, and Spain) (Lackner and Yélamos 2001; Lackner and Hlaváč 2012).

RECOGNITION, IDENTITY THEFT, AND CAMOUFLAGE

All species have been collected in ant nests, and they appear to be quite host-specific. Different species have been found in nests of the myrmicine genera *Aphaenogaster* and *Messor,* and of the formicine genera *Cataglyphis* and *Formica.* They have been reported to prey on the brood of their host ants and feed on ant cadavers (Yélamos 1995; Lenoir et al. 2012). *Sternocoelis hispanus* beetles were occasionally observed riding or clinching onto the host ant's body, especially in the process of gaining access to the inside of the ant colony. This very close contact with host ant workers intuitively suggests that, similar to the above discussed *Martineziana* beetles, these myrmecophiles also acquire the ants' chemical colony signature, and indeed, Lenoir et al. (2012) found a close match of the species-specific and colony-specific cuticular hydrocarbon profiles on the beetles' cuticle. However, the beetles' hydrocarbon blends contain additional components, different from the ants' blends. What is particularly interesting is that the beetles maintain most of these hydrocarbon components of the cuticle surface even after being isolated from the host colony for up to one month, suggesting that the beetles synthesize most of them themselves. Though the profiles of these isolated beetles differ somewhat from those of the host colony, when readopted by the host colony, the beetles eventually gain a cuticular hydrocarbon profile that is not identical to but resembles that of their hosts in many aspects.

The histerid beetle *Sternocoelis hispanus* lives in nests of the ant *Aphaenogaster senilis*. It prefers to reside in the brood nest and appears to feed on the ant larvae. (Courtesy of Pavel Krásenský).

The adoption process in *Sternocoelis* beetles is not an active behavioral act of the ants, but is rather "initiated" by the beetles, which often clinch to or ride on the bodies of the ants. The ants may even treat the beetles aggressively at first, by grasping them on the leg and dragging them around, before eventually carrying them into the nest. Once inside the nest, the beetles are mostly tolerated even when they are sitting on the ant larvae and feeding on them. Lenoir et al. (2012) argue that the relatively close match of the cuticular hydrocarbon blends of *Sternocoelis* and their particular host ant species might be a coevolved trait, because *S. hispanus* is easily adopted also by *Aphaenogaster iberica,* which is considered a sister species of *A. senilis,* and both ant species feature very similar cuticular hydrocarbon profiles. In contrast, *A. simonelli* and *A. subterranea* have very different cuticular hydrocarbon profiles, and *S. hispanus* did not find access to colonies of these ant species. Although the sample sizes of these tests were rather low, the reported results are suggestive. One fact, however, comes out very clearly from these studies. The match of the cuticular hydrocarbon blends between host ants and guests is not the releasing key stimulus for the adoption, but it rather facilitates toleration inside the host colony.

The other histerid myrmecophilous beetle genus is *Haeterius* (Haeterinae), which contains thirty species in the Palearctic and Nearctic regions, all of which are associated with ants (Figure 3-4). We must make a brief comment concerning the spelling of the genus name. Over decades, dating back to books and papers by Wasmann, Donisthorpe, Wheeler, and Hölldobler and Wilson, this genus has been wrongly spelled *Hetaerius*. However, as Bousquet and Laplante (2006) correctly pointed out, the genus should be credited to Dejean (1833) with the spelling *Haeterius,* and not to Erichson (1834) with the spelling *Hetaerius*.

Haeterius beetles have been found throughout Central and North America as well as in Europe, Africa, and Asia. Only a few species have been studied, but from what is known, their behavioral interactions with host ants are quite different from those of their phylogenetically closely related myrmecophilous genus *Sternocoelis*. In most of the recorded finds, the host species belong to the genus *Formica,*

3-4 The histerid beetle *Haeterius ferrugineus* is also often found among the brood of its *Formica* hosts, and it has been reported that they prey on the larvae. (Courtesy of Pavel Krásenský).

3-5 *Haeterius* beetles (in this picture *H. brunneipennis*) often straighten up and wave with their forelegs. (Courtesy of Gary Alpert).

but occasionally the genera *Lasius,* and *Aphaenogaster* are also listed (Wasmann 1920; Wheeler 1908a, 1910; Martin 1922; Maruyama et al. 2013). According to Wasmann (1905; cited in Wasmann 1920), the European species *Haeterius ferrugineus* is most frequently found in nests of *Formica fusca,* where it feeds on dead and wounded ants. Occasionally it also consumes ant larvae. Most of the time the ants pay no attention to the beetles. When they do attack *Haeterius,* the beetle exhibits death-feigning behavior by holding perfectly still with the legs closely appressed to the body. The ants frequently react to this nonviolent resistance by carrying the beetle around, licking it, and finally releasing it. It has been suggested that *Haeterius* has special trichome glands opening at the margins of the thorax (Wasmann 1903, cited in Wasmann 1920; Wheeler 1908a), but no histological investigations have been undertaken to confirm this assumption. Based on our own observations (B. H.) we can state that the ants preferably lick the frontal part of the beetle's body, particularly the region at or close to the head. This behavior very much suggests the presence of some exocrine glands. Wheeler (1908a) observed that adults of the North American species *Haeterius brunneipennis* solicit regurgitated food from the host ants. The beetle sometimes waves its forelegs toward the passing ants and, by this action, appears to attract the ants' attention (Figure 3-5). A very similar behavior was observed in the *H. ferrugineus* by one of us (B. H.). The soliciting beetle takes an upright position, stretching the forelegs widely apart and waving them slightly toward the

RECOGNITION, IDENTITY THEFT, AND CAMOUFLAGE

3-6 *Haeterius ferrugineus* waves with its forelegs toward an approaching *Formica* host ant. Such behavior can lead to trophallactic food exchange between the ant and the beetle. (Bert Hölldobler).

approaching ant (Figure 3-6). The ant antennates and licks the beetle with her mandibles held nearly closed. Tracer experiments have revealed that the beetles receive small amount of regurgitated food (Hölldobler and Wilson 1990).

Although we do not know whether the cuticular hydrocarbon profiles of the beetles resemble those of the host ants, because the beetles have relatively close contact to their hosts and are often licked by them, we assume they may

3-7   The scarabaeid beetle *Cremastocheilus opaculus,* one of approximately forty-five *Cremasto-cheilus* species that are associated with ants. (Courtesy of Gary Alpert).

have acquired some of their hosts' chemical signatures. However, the observational data suggest that other means, such as secretions from exocrine glands, are more important for their integration into their hosts' societies.

Finally, let us consider the scarabaeid myrmecophilous genus *Cremastocheilus,* of approximately forty-five species, which all appear to be associated with ants (Figure 3-7). Several subgenera have been proposed, and a new phylogenetic analysis, based on morphological characters, has been presented by Glené Mynhardt and John Wenzel (2010). *Cremastocheilus* species are reported to prey on host ant larvae (Cazier and Mortenson 1965; Alpert and Ritcher 1975; Alpert 1994). But in contrast to the scarabaeid *Martineziana dutertrei* (which we discussed above), *Cremastocheilus* beetles are endowed with several exocrine trichome glands, which reportedly are licked by the host ants and may elicit adoption behavior in their hosts (Cazier and Mortenson 1965).

In fact, Gary Alpert (1994), in the hitherto most comprehensive study of the genus *Cremastocheilus,* reports that some species of *Cremastocheilus* are carried into the nests by foraging ants; however, he could not find any evidence that this is elicited by the glandular secretions. Gary Alpert provided a detailed morphological and histological study of the trichome glands in males and females of at

RECOGNITION, IDENTITY THEFT, AND CAMOUFLAGE

least one species of each of the five subgenera of *Cremastocheilus*. Alpert covers many aspects of the natural history of several species, but in the context of our current topic, the following excerpt from Alpert's monograph is of special interest. Using two examples, Alpert writes,

> *Cremastocheilus stathamae* are carried into the ants' nests from as far away as 25 ft. (7,62 meters). The beetles appear to land spontaneously after flying randomly over *Myrmecocystus depilis* nesting areas. Then they wander about waiting for the ants to carry them into the nests. *Cremastocheilus hirsutus* fly low over the ground searching for *Pogonomyrmex barbatus* nests, land, and move straight for the nest entrances which they enter unhindered. Among all species, the ants frequently eject beetles, but the net movement is in. (Alpert 1994, 228)

Alpert further noted that *Cremastocheilus* beetles were frequently attacked in an observation nest, but these attacks seldom resulted in the death of the beetles. It appears that often, when the beetles were initially attacked by the host ants, these attacks waned and were converted into licking bouts, as if the glandular secretions serve as appeasement substances. For example, *Cremastocheilus harrisii* that were introduced to *Formica schaufussi* were licked by the ants primarily at the "anterior pronotal angles, the mentum area where the frontal glands empty and a carina over the eye with a dense pad of short setae. These are areas of concentration of gland cells."

Alpert (1994) confirmed that *Cremastocheilus* species feed primarily on ant larvae, and in choice tests they preferred the larvae of their respective host ants. However, when they had no choice, they also fed on larvae of other ant species. He also refuted the claim by Kloft et al. (1979), who reported adults of *Cremastocheilus castaneus* receive regurgitated liquid food from host ants (*Formica integra*). Based on their work with radioactively labeled food, they concluded that the beetles elicited food regurgitation in the ants, which were initially attracted by the beetles' trichome gland secretions. Alpert repeated these tracer experiments with *C. castaneus* and *F. integra* with a larger sample of test specimens and better controls. He concluded that *Cremastocheilus* myrmecophiles do not participate in the host ants' social food flow.

From what we know through the work by Cazier and Mortenson (1965) and Alpert (1994), the development of various *Cremastocheilus* species is quite uniform. In spring or early summer all species deposit eggs "in the soil mixed with vegetable matter," usually in or close to the nests of the host species. The development through the three larval instars takes about four to six weeks. The fully grown last instar constructs a pupal case out of soil and fecal material. Adults eclose by the end of summer. They leave the nest for mating, which occurs in the American Southwest usually after the rainfalls. Different *Cremastocheilus* species choose different mating sites. For example, in *C. beameri* mating takes place in rodent burrows (for more details see Alpert 1994). The beetles overwinter in the host ants' nests.

Although, to our knowledge, the cuticular hydrocarbons in *Cremastocheilus* beetles have not been investigated, from the behavioral studies presented by Alpert, it seems that these myrmecophiles move into their host ant nests without hesitation, or they may even be picked up by the ants (often such beetles fake death) and carried into the nest. For their toleration inside the nest, the secretions from the trichome glands seem to play a key role.

The Slug-Like Larvae of Microdontine Syrphid Flies

As we discussed above, cuticular hydrocarbons have been postulated as a major chemical cue involved in social recognition systems, and considerable indirect, and some direct, evidence from studies with ants and their myrmecophiles support this assumption. The investigation of the myrmecophilous syrphid genus *Microdon* is an exemplar case study.

There exists conflicting information about the number of species within this genus that belong to the Microdontinae, the smallest of the three subfamilies of the Syrphidae (Reemer 2013). According to Reemer and Ståhls (2013a,b), 388

3-8   The larva of a syrphid *Microdon* species in the nest of the dolichoderine ant *Linepithema oblongum.* (Courtesy of Alex Wild / alexanderwild.com).

species names were previously classified in a single genus, *Microdon* Meigen 1803, but a new generic classification of the Microdontinae resulted in sixty-two species for the genus *Microdon* Meigen sensu stricto (s.s.), forty-four species for the six subgenera of *Microdon,* and 126 species for the species groups and "unplaced species" of *Microdon* sensu lato (s.l.).

The immatures of all known species of the hover fly genus *Microdon* live in ant nests (Reemer 2013). Deducing from the literature, it appears that *Microdon* species exhibit a relatively high host specificity, that is, each *Microdon* species has only one, or very few, closely related host ant species; yet, as we discuss later, their host species specificity is considerably more complex. Though the adult flies are treated with hostility by the ants, and after eclosion the flies hurriedly vacate the ants' nest, they are frequently observed hovering around the nest area

3-9 The pupa of *Microdon ocellaris* (upper), and an eclosed *Microdon* fly on top of the empty pupal case. (Courtesy of Gary Alpert).

of the host ant nest from where they emerged. In contrast, the fly larvae and pupae, which live inside the ants' nests, often even in the brood chambers, are usually ignored and tolerated by the ants.

The appearances of the *Microdon* larvae and pupae are rather unusual, and at first glance they look like slugs (Figure 3-8). In fact, they were once described as mollusks or as coccids. They have an elliptical shape with a convex dorsal

RECOGNITION, IDENTITY THEFT, AND CAMOUFLAGE

surface and a "flat creeping-sole," and they move very slowly. When ready to pupate, their whitish-beige color turns to light brown (Figure 3-9).

The integument becomes stiff and brittle and now serves as pupal case inside which the pupa metamorphoses. After completion, the "fly emerges by breaking off the anterior dorsal third of the puparium" (Figure 3-10). William Morton Wheeler (1910), who observed several *Microdon* species himself, cites Verhoeff (1892), who observed a *Microdon* fly ovipositing in the host ant nest. Apparently, the fly had to enter the nest but was repeatedly attacked by resident ants and driven away, but it kept returning until the eggs were deposited.

Anyone who has watched *Microdon* larvae in the formicarium realizes that it is difficult to observe the very slow-moving larvae feeding, but the cumulative published records seem to establish that they prey on the ants' brood (Andries

3-10   The freshly eclosed *Microdon ocellaris* fly displays shriveled wings (upper). Once the wings are fully expanded (lower), the fly departs from the host ants' nest. (Courtesy of Gary Alpert).

1912; Van Pelt and Van Pelt 1972; Akre et al. 1973, 1988; Duffield 1981; Barr 1995). Richard Duffield reported that third-instar *Microdon* larvae could consume eight to ten ant larvae in thirty minutes, and Boyd Barr (1995) stated that a *Microdon* larva may consume up to 125 ant larvae during its life. Menno Reemer (2013) postulates, "With an average number of five or six *Microdon* larvae per nest, over 700 ant larvae would be consumed per nest."

Graham Elmes and his collaborators (1999) made a most interesting, yet puzzling, discovery when studying *Microdon mutabilis* in "extremely localized, small populations that typically persist for many generations on the same small (often 50.1ha) isolated patches." In this area, *M. mutabilis* lives exclusively with polygynous colonies of *Formica lemani,* the nests of which are subdivided into several smaller subnests. Elmes et al. (1999) observed that *M. mutabilis* is extremely localized even though the *F. lemani* population is "widespread and abundant." As previously reported, the *Microdon* females appear quite sedentary and deposit their eggs inside or close to the nest entrances of the *F. lemani* colony from which they had emerged previously. In order to find out how the eggs survive putative ant predation, Elmes et al. (1999) conducted comparative laboratory experiments, presenting *M. mutabilis* eggs of different females to several *F. lemani* test colonies. The authors discovered a strong maternal effect: "New-laid eggs had a more than 95% survival when introduced to the individual ant colony that reared each mother fly or to its close neighbors, but survival declined as a sigmoidal logistic function of distance from the mother nest, with *F. lemani* colonies from two and 30 km away killing 80 and 99% of eggs, respectively, within 24 hours." The authors suggest that the eggs "may be coated with a mimetic chemical disguise that lasts for three to four days after oviposition" (Elmes et al. 1999, 447). The authors propose that there exists an extreme local adaptation by a *M. mutabilis* population, not simply to one species of host, but to an individual host population and possibly to local strains or family groups within an *F. lemani* population.

From a behavioral point of view, it is hard to imagine with what kind of host signal the *Microdon* eggs might be coated. To our knowledge no colony-specific brood pheromones are known, and usually eggs and larvae can easily be transferred from one colony to another. As far as we could see from the published literature, no equivalent comparative tests were conducted with ant eggs of the host ant colonies. On the other hand, Graham Elmes and colleagues point out that the *Microdon* eggs "never appeared to be attractive" to the ants. "The best outcome for the egg was to be superficially examined and then ignored." No question, these discoveries are extremely interesting but also very puzzling. In this connection we want to report, at least briefly, another intriguing observation by these researchers.

Formica lemani colonies that are infested with *Microdon* myrmecophiles seem to produce significantly more reproductive winged females (gynes) than colonies that are free of *Microdon* larvae. At first glance this seems paradoxical. However, it could be seen as a colony-level reaction to escape the parasitism, because alates leave the colony and, after mating, either get adopted into a more distant *F. lemani* colony without *Microdon* parasites, or independently find their own colony, which subsequently may develop secondary polygyny without *Microdon* parasitism. Thomas Hovestadt and coworkers (2012) developed a model "for resource allocation within polygynous ant colonies, which assumes that whether an ant larva switches development into a worker or a gyne depends on the quantity of food received randomly from workers. Accordingly, *Microdon* predation promotes gyne development by increasing resource availability for surviving broods." Indeed, because of the considerable predation on ant larvae the "brood: worker ratio" becomes very low, so workers will feed higher quantities of vitellogenin-rich larval food to individual larvae, which leads to an increase of gyne production. To make this work perfect, it would have been nice to conduct a series of experiments with artificial predation. Instead of *Microdon* larvae preying on ant larvae, the experimenter could have removed larvae from the colonies, thus artificially affecting the "brood / worker ratio," and subsequently determine whether these manipulated colonies raised more alate females.

Let us now consider how the *Microdon* larvae manage to survive inside the brood nests preying on the host ants' immatures. Garnett et al. (1985) have studied three *Microdon* species: *M. albicomatus, M. cothurnatus,* and *M. piperi.* They report that the host ants occasionally carry the *Microdon* larvae, during which the fly larvae assume a curled, cylindrical shape, and the authors postulate that the host ants apparently do not distinguish the *Microdon* larvae from the ant immatures. They write, "In nests of their host ants (species of *Camponotus* and *Formica*), the 1st and 2nd larval instars resemble the ant cocoons upon which they prey and are transported with the cocoons by workers. Immatures are not attacked by their host ants and appear to possess chemical as well as physical attributes providing integration with their host" (Garnett et al. 1985, 615).

On the other hand, Wheeler describes observations of the host ants killing a young *Microdon* larva "that failed to get hold of a surface with its vulnerable creeping-sole." But in general, Wheeler states, the ants do not seem to notice the *Microdon* larvae, and he observed that the fly larvae were left behind when the ants moved to a new nest. One of us (B. H.) has observed *Microdon* larvae in nests of *Formica sanguinea* and *Formica fusca,* and came to the same conclusion, that the fly larvae are rather ignored or tolerated by the ants, but not tended and treated like ant brood.

Nevertheless, the question has to be asked: Why are the *Microdon* larvae tolerated or ignored by the host ants, despite the fact that they are preying on the ants' brood? The answer was most likely discovered by Ralph Howard, Roger Akre and William Garnett (1990). They compared the cuticular hydrocarbon profiles of the host ant species *Camponotus modoc* (adults and immatures), with the surface hydrocarbon blends of the *Microdon piper* and found striking correlations. Both *Camponotus* adults and larvae contained the same hydrocarbons, in instar-specific proportions, and the hydrocarbon profiles of *Microdon* larvae were identical with those of the host ants. However, the cuticular hydrocarbon profiles of the adult *M. piperi* flies were quite different from those of the host ants, and indeed, the adult flies were recognized as intruders and attacked by the ants.

This study is an excellent example of hydrocarbon profiles functioning as discriminators. Though the *Microdon* larvae are not nursed by host ants like ant larvae are, they obviously have frequent contact with ants throughout their larval stages and most likely will be coated with the ants' hydrocarbons. In a second paper on *Microdon albicomatus,* which lives in nests of *Myrmica incompleta,* Howard et al. (1990) found similar hydrocarbon profiles in the myrmecophiles and host pupae, and in the host workers, although in the latter the substances were found in different relative abundances. Radio-labeling experiments using $1\text{-}^{14}C$-acetate with the *Microdon* larvae seem to indicate that the fly larvae biosynthesize their hydrocarbons rather than procuring them from their hosts. Surprisingly, the authors investigated only the hydrocarbons of the ant pupae and not those of the ant larvae, yet the main prey of *Microdon* larvae are the host larvae. From our interpretation of the table listing the hydrocarbons found in *Microdon* larvae, host ants, and host ant pupae, it seems that the hydrocarbons of the *Microdon* larvae consist of a mixture of their own and acquired compounds from the hosts, as is mostly the case, even among the ant nestmates and their immatures.

Whether we can call this chemical mimicry is a semantic question. We would rather consider it a "sneaky usurpation" of the host colony's identity code, or we may call it "identity theft."

Other Syrphid Flies in Ant Nests

All these studies dealt with microdontine species of Europe and North America. Although there are many species known from the tropics that are myrmecophiles (see Duffield 1981; Reemer and Ståhls 2013a; Pérez-Lachaud et al. 2014; Schmid et al. 2014), not much is known about their natural history. In only two instances has predation of ant larvae been documented: *Microdon tigrinus,* the larvae of which live in the nests of the fungus-growing ant *Acromyrmex coronatus* (Forti et al. 2007) and *Pseudomicrodon biluminiferus,* of which the third-instar larvae were reported to feed on the larvae of the host ants *Crematogaster limata* (Schmid et al. 2014). Perhaps there is a third case: Hölldobler discovered in 1980 hitherto unknown slug-like syrphid larvae in nests of an Australian *Polyrhachis* species (possibly *P. australis*) (Hölldobler and Wilson 1990). Only recently has this species been described as *Trichopsomyia formiciphila* (Downes et al. 2017).

The most surprising, more recent, discovery was made by Gabriella Pérez-Lachaud and her colleagues (2014), who described the first parasitoid hover fly species in ants. The species *Hypselosyrphus trigonus* was reared from cocoons of the arboreal ponerine ant *Pachycondyla villosa,* nesting in *Aechmea bracteata* bromeliads in southern Quintana Roo, Mexico. This is a new record of parasitoid flies. Previously this was known only from phorid and tachinid flies (see Chapter 2).

Finally, one more case of a myrmecophilous syrphid fly, with an entirely different life history, deserves to be briefly reported. The larvae of the syrphine *Xanthogramma citrofasciatum* live in nests of the formicine ants *Lasius niger* and *L. alienus* (K. Hölldobler 1929). They are supposed to feed on root aphids (Aphididae) cultured by the host ants inside their subterranean nests (K. Hölldobler 1929; Speight 2017), though the early instar larval stages are unknown for most *Xanthogramma* species (Nedeljković et al. 2018). The later stages, when they reached the size of ant queen larvae, were usually found in the brood chambers of the host ants, where they were attended by the ants and treated by the hosts like ant larvae. Karl Hölldobler, who first observed *X. citrofasciatum* in formicaria, reported that he noticed several behavioral instances that suggested trophallaxis between the syrphid larva and nurse ants, but only in one case was the mouth-to-mouth contact long enough for justifying this suggestion. The ants lost interest in and ignored the myrmecophile larva once it began preparing itself for pupation inside the peripheral nest chambers without ant brood. After eclosion the adults quickly left the ant nest. According to Karl Hölldobler's observations in formicaria, the *X. citrofasciatum* guests overwinter in the host ants' nest twice, once as larvae and a second time as prepupae or pupae. The adults leave the nests in April to early May (Figure 3-11).

Although no studies exist concerning the mechanisms that make the *Xanthogramma* larvae so attractive to the host ants, and why they are still tolerated once

3-11 The syrphid fly *Xanthogramma citrofasciatum,* whose larvae develop inside ant nests. (Photo: Aleksandrs Balodis / Wikimedia Commons).

the guest larvae have lost the attractiveness for the ants, from what we know today from other similar cases, we dare to suggest that the syrphid larval stages that are tended by the host ants inside the brood nest may mimic a brood pheromone of the ant larvae. During the tending and grooming by the ant nurses, the guest larvae also acquire the ants' surface hydrocarbons, and these assure that the hosts will tolerate or ignore the immature guests once they have reached the prepupal stages.

Camouflage in Chrysomelid Beetle Larvae

Identity theft is not the only mechanism underlying survival for a predator of ant larvae inside the brood nest of the host ants. The larvae of leaf beetles (Chrysomelidae) belonging to the Camptosomata construct larval cases within which they develop. These cases consist of fecal material, soil, and rotting vegetable material. The architecture of the portable cases or quivers is taxon-specific, and so are the behavioral, temporal, and spatial aspects of larval case construction (Erber 1968, 1969, 1988; Wallace 1970; Brown and Funk 2005). The primary function of these larval quivers is defense-protection and camouflage (Eisner et al. 1967; Olmstead and Denno 1993; Vencl et al. 1999; Eisner and Eisner 2000; Nogueira-de-Sá and Trigo 2002).

The taxon Camptosomata consists of two subfamilies, the Lamprosomatinae and the Cryptocephalinae. The latter include the three tribes Fulcidacini, Cryptocephalini, and Clytrini, but only in the latter two tribes have myrmecophily of larvae been documented (Agrain et al. 2015), yet information on their myrmecophilous interactions remained fragmentary. The best-studied example is the

3-12 The larva of the Japanese chrysomelid species *Clytra arida* inside the larval case, with the host ant *Formica yessensis*. (Courtesy of Taku Shimada).

genus *Clytra* of the tribe Clytrini. The larvae of all *Clytra* species live inside the nests of ants, although they are not particularly host-specific.

For example, Jong Eun Lee and Katsura Morimoto (1991) report that *Clytra arida*, the only *Clytra* species known from Japan (Figure 3-12), has been found in nests of the ant genera *Formica, Camponotus, Lasius,* and *Cataglyphis* (Medvedev 1962 cited in Lee and Morimoto 1991).

Horace Donisthorpe (1902) was first to reveal the complete life history of the myrmecophilous species *Clytra quadripunctata,* which commonly occurs in nests of species of the *Formica rufa* group, including *F. pratensis* and *F. sanguinea.* Occasionally it also occurs in nests of *Formica fusca* and *F. rufibarbis* (for further data and review of host specificity, see Skwarra 1927; Erber 1988; Agrain et al. 2015) (Figure 3-13).

According to Donisthorpe, the adult beetles eclose from their pupal case inside their host ants' nests and shortly afterward leave the nest. When attacked by the ants they often react by "feigning death." The beetles mate on bushes and trees. Females ready to oviposit seek shrubs close to a nest of host ants and drop the eggs to the ground beneath. However, before they do this, they compress small, flat pieces (shingles) of excrement around each egg using their

RECOGNITION, IDENTITY THEFT, AND CAMOUFLAGE

3-13 The adult beetle of *Clytra quadri-punctata*. (Courtesy of Marion Friedrich, arthropodafotos). The lower picture depicts a fully grown larva of *C. quadri-punctata* (inside its larval case, with only the sclerotized head visible). (Courtesy of Thomas Parmentier).

metathoracic tarsi. According to Vernon Stiefel and David Margolies (1998) this is a unique behavior of females of most species in the tribe Clytrini. The excrement surrounding the eggs is apparently attractive to the ants. They transport the coated eggs into their nest, where the larvae hatch. The empty egg case with its fecal coat "give[s] the larvae a foundation on which to start the larval case. When this case grows larger the egg covering may be found embedded in it or may be broken off. If this should happen the larva fills up the hole" (Donisthorpe

3-14　Larvae of *Clytra quadripunctata* with host ant *Formica polyctena*. Lower: Different developmental larval instars of *C. quadripunctata*. (Courtesy of Thomas Parmentier).

1902, 1927). As the larva grows through several molts, the cylinder-shaped case increases incrementally. It is open at one end, but it can be closed by the hard, sclerotized head of the larva (Figure 3-14).

Regarding the feeding habits, the literature is controversial. Whereas Donisthorpe reports that the *C. quadripunctata* larvae feed partly on vegetable detritus in the ant nest, but also on "droppings and pellets of the ants," others, such as Escherich (1906) or Skwarra (1927), report observations of *Clytra* larvae feeding on ant brood, and Schöller (2011) observed *Clytra laeviuscula* preying on larvae of their host ants *Lasius emarginatus*. Finally, in a quantitative study of predation on ant larvae by myrmecophiles of *Formica rufa,* Thomas Parmentier and his coworkers (2016a,b) found that 45% of the *Clytra* larvae were seen in the brood chambers, and 67% were recorded preying on ant brood, though aggressive interactions with the host ants were relatively low. These authors also confirmed that the beetle larvae feed on booty brought in by the ants. Interestingly, most of the time the beetles are not treated with hostility by the ants; in fact, Parmentier (2019) observed beetle larvae being picked up and dragged or carried by the ants for short distances along the trail when the colony was moving to a new nest site. This confirms similar observations previously reported by Donisthorpe (1902). In considering the behavioral mechanisms that elicit such behavior in the ants, Parmentier offers two possibilities. Perhaps the beetle larvae release some kind of appeasement substance that dampens the ants' aggression. We doubt this, because the appeasement process in most cases known is a short-lasting gentle chemical defense strategy (see Hölldobler 1971; Hölldobler et al. 1981; Hölldobler and Wilson 1990; Hölldobler and Kwapich 2019). The alternative suggestion is a more intriguing one: "The ants might mistake the larval case for building material, which is constantly brought into the nest." However, to our knowledge, the wood ants do not carry nest-building material into the brood chambers, but perhaps the beetle larvae, once inside the ant nest, move on their own into the brood chambers. It is also possible that *Clytra* larvae that feed on ant larvae acquire the nonvolatile brood pheromone of the ant larvae, so that the ants do not instantly detect them as intruders. In general, all reports indicate that aggressive interactions with the host ants are relatively low and, when attacked, the larvae retreat into the larval case, blocking any attempts of intrusion by the ants with their hard heads. Similar observations were reported from the clitrine genera *Lachnia* and *Tituboea* (Erber 1969, 1988).

The larvae of the clitrine species *Anomoea flavokansiensis,* which is common in the grasslands of northeastern Kansas (USA), live in nests of the myrmicine ant *Crematogaster lineolata.* The ant workers are attracted to the *Anomoea* eggs and carry them into their nests, where the larvae live as myrmecophiles until they pupate (Erber 1969, 1988). Vernon Stiefel and David Margolies (1998) made the interesting discovery that, in their study area, *C. lineolata* workers exhibited strong preference for eggs of *A. flavokansiensis* females that fed exclusively on Illinois bundle flowers (*Desmanthus illinoensis,* Fabaceae) compared to eggs from females that fed exclusively on honey locust (*Gleditsia triacanthos,* Fabaceae). In fact, the focus colony of *C. lineolata* foraged mainly on extrafloral nectaries of bundle flowers, and the authors suggest that the fecal material of beetle females, with which they coat their eggs, may contain some chemical compounds of the extrafloral nectaries that are attractive to the ants. In contrast, honey locust does not feature extrafloral nectaries on which the ants forage, and therefore eggs deposited by beetle females that feed on honey locust are mostly ignored by the ants (Stiefel and Margolies 1998).

Although *Anomoea flavokansiensis* is not an obligate myrmecophile, Stiefel and Margolies demonstrated that those *A. flavokansiensis* larvae that hatched and developed inside the host ant nest were less vulnerable to predation or other environmental hazards. Generally, the *A. flavokansiensis* larvae feed on decaying plant material whether they live inside or outside the ants' nest. Nevertheless, inside an ant nest they enjoy indirect protection by the ants. The ants largely ignore the beetle larvae, and should they occasionally be handled by the ants, the larvae retreat into their case. The ants seem to treat the larval cases like nest material, which is from time to time handled and rearranged by the ants.

Stiefel et al. (1995) provide circumstantial evidence suggesting that *A. flavokansiensis* larvae spend the latter part of their life cycle (especially the phase before pupation) in neighboring nests of *Formica pallidefulva,* where they overwinter and pupate. But not much is known regarding how the beetle larvae change hosts, or whether this is a coincidental or regular event. Most likely it is coincidental, because as far as we understand these findings, even the myrmecophily appears to be only facultative, and beetle larvae may leave the ants' nest and feed on detritus outside. In any case, the larvae of *A. flavokansiensis* are exclusively detritivores (obtaining nutrients by consuming detritus), and much of their larval development they spend under the "protective umbrella" provided by the interior of an ant nest.

This example most likely represents an evolutionary grade from which detritivores could evolve to become increasingly predatory on ant brood inside the nest, like *Clytra quadripunctata,* the larvae of which are obligate myrmecophiles. However, in both cases, the protective disguise provided by the larval case, which the ants seem to treat like nest material, is the main mechanism to assure survival inside the ant nest.

Camouflage in Tineid and Other Moth Larvae

We find a fascinating convergence of this kind of myrmecophily in the tineid moths also known as "fungus moths" or "true moths." They include more than a dozen subfamilies, one of which is of particular interest to us. The subfamily Myrmecozelinae includes the genus *Myrmecozela,* in which the larvae of some species construct larval cases like those of clitrine beetle larvae. Larvae of *M. ochraceella* have been found near or on the nest of mound-building *Formica* species, feeding on detritus. The adults have been reported resting on grass blades near *Formica* nests (references in Gaedike 2019). The larvae of this and other *Myrmecozela* species seem to be detritovores, which are often loosely associated with ants, and like some previously discussed clitrine species, some myrmecozeline species also spend their larval phase in ant nests, where they prey on the ant brood and even sometimes adult ants. A striking example is the genus *Ippa* (formerly called *Hypophrictoides* or *Hypophrictis*), which was first described by W. Roepke, who studied its myrmecophilous habits in Java (Roepke 1925). He proposed the genus name *Hypophrictoides.* About a decade earlier, E. Meyrick (1916,

cited in Roepke 1925) had described a very similar tineid genus from India, which he had named *Hypophrictis*. Subsequently G. S. Robinson considered the taxonomic differences between both genera too insignificant to justify a separate genus status, and therefore *Hypophrictis* (the senior name) replaced *Hypophrictoides*. But in a renewed revision, Gaden Robinson (2001, references in Gaedike 2019) included the type species *Hypophrictis* in the genus *Graphara,* which caused considerable confusion, at least in us, because *Graphara* is a genus within the Noctuidae. Konrad Fiedler (personal communication) resolved this puzzle for us. He pointed out that *Graphara* was previously used for a genus within the Noctuidae; therefore, because of homonymy, this genus name cannot be used within the Tineidae. The final and valid genus name for this interesting myrmecophilous tineid moth is *Ippa* Walker 1864.

Meyrick (1916, cited in Roepke 1925, 175) provided a succinct description of the *Ippa* larvae and the elaborate case they construct: "Larvae found in nests of

3-16 The larva of *Ippa conspersa* attacks a worker of the host ant, probably *Lasius nipponensis*. (Courtesy of Kyoichi Kinomura).

Crematogaster (Formicidae) at Ambulangoda, Ceylon [reside] in a singular nearly flat case composed of two dark grey sections of stout silk joined together round the edges, shaped somewhat between an ellipse and an hourglass, or like two coalescing circles, length 15 mm, greatest width 8 mm, contracted in middle to 6 mm" (Figure 3-15).

Other *Ippa* species were found with the formicines *Polyrhachis* and *Lasius,* and in nests of myrmicine ants. The species described by Roepke (1925) lived in nests of *Dolichoderus bituberculatus.* Roepke's report is, to our knowledge, the most detailed description to date of the predatory behavior of *Ippa* larvae inside the host ants' nest.

The ants seem not to notice the larvae residing in their elaborately disguised cases. In formicaria, where Roepke observed the interactions between *Ippa* larvae

3-17 The larva of *Ippa conspersa* attacks a young alate queen of *Lasius* sp. The lower picture depicts a worker ant that attempts to rescue the captured alate nestmate but to no avail. (Courtesy of Taku Shimada).

with the *D. bituberculatus* ants, the ants completely ignored the larval cases. The *Ippa* larva, when reaching out of its case, performed with the frontal part of its body circling searching movements. When incidentally touching an ant, it retreated into the protective case in a split second. Nevertheless, soon after, the larva emerged again and again with searching movements, and it did not take long until it captured an ant larva or pupa and rapidly retrieved the booty into her case. Shortly afterward, the larva's head appeared again, and other ant larvae and pupae were captured and retrieved. It appears as if the parasitic larva first gathers enough prey inside its case before it devours the booty. After these most interesting observations in artificial nests, Roepke investigated whether these parasites can also be found and observed inside natural *Dolichoderus* nests, and indeed *Ippa* larvae were found in larger numbers, occasionally twenty to thirty specimens coexisting in one nest. Observations in natural nest and in formicaria confirmed that these tineid larvae live exclusively as disguised or camouflaged predators in ant nests. Other *Ippa* species from Japan choose as host *Lasius nipponensis* (sometimes wrongly called *fuliginosus*), and apparently prey also on adult ants (Figures 3-16, 3-17).

Naomi Pierce (1995) reports that predation by moth larvae in brood nests of ants occurs in other moth families too. Some species of the pyralid subfamily Wurthiinae live parasitically in ant nests. For example, *Niphopyralis myrmecophila* prey on the brood of the weaver ants *Oecophylla smaragdina,* and larvae of *N. aurivilli* live as brood parasites in nests of *Polyrhachis bicolor.* It has been suggested that the *Niphopyralis* larvae mimic ant recognition signals, which enable them to be adopted by the host ants (Kemner 1923), but no experimental evidence has been provided. According to Pierce, the most specialized moth species preying on ants are found among the Australian family Cyclotornidae (also called ant worms) exemplified by *Cyclotorna monocentra.* Here is the succinct description by Naomi Pierce:

> The larvae of these moths begin life as parasites of leafhoppers in the Cicadellidae, and then move to the nests of meat ants, *Iridomyrmex purpureus* Smith, where they complete their development by feeding on the brood. Dodd (1912) observed that females of this species lay large numbers of eggs near the trails of ants attending the leafhoppers. The first instar larva spins a pad of silk on the abdomen of the host beneath the wings, with a small sac at the anterior end to protect the larval head. Once the larva leaves the leafhopper, it builds an oval, flat cocoon where it molts into a broad, dorsoventrally flattened larva with a small head that can retract into the prothorax. It adopts a particular posture when encountered by a meat ant, raising the anterior half of the body and curling its posterior over its back to expose the anus. Following inspection, a meat ant will carry the larva into the nest, where it becomes a predator on the ant brood. In the nest, the

larva continues to produce an anal secretion that is attractive to the ants. Once the larva has completed development, which may take weeks or possibly months, it emerges from the ant nest, and spins its cocoon in a protected spot nearby (Common 1990). (Pierce 1995, 418–419)

It would certainly be of great interest to know whether these parasitic moth larvae mimic ant brood pheromones, but to our knowledge neither behavioral nor chemical evidence exists that would support this hypothesis. For additional examples for myrmecophagy in Lepidoptera, see Pierce (1995) and Chapter 4.

Cuticular Hydrocarbons Revisited

These examples illustrate the many avenues into ant nests, one of which might indeed be chemical identity theft, although these findings are based mostly on correlations, such as the myrmecophiles' cuticular hydrocarbon blends matching those of their host ant species to various degrees. We already learned about the case of the syrphid fly *Microdon piperi* and its host ants *Camponotus modoc,* where the myrmecophilous predator *M. piperi* acquires at least part of its recognition cues from its host ants, or the scarabaeid beetle *Martineziana dutertrei* that picks up cuticular hydrocarbons from its *Solenopsis* hosts. Other examples have been reported for lycaenid caterpillars of the genus *Phengaris,* which we discuss in Chapter 4, and in *Myrmecophilus* ant crickets, which we discuss in Chapter 7.

In general, relatively few experimental studies that demonstrate the behavioral significance of these chemical resemblances exist. One of them was conducted by Christoph von Beeren and his coworkers (2011), who discovered that the kleptoparasitic silverfish *Malayatelura ponerophila* most likely acquires cuticular hydrocarbons directly from its host ant, *Leptogenys distinguenda*. This inference is based on the measurements of transfer of a stable-isotope label from the cuticle of workers to the silverfish. In a second experiment, the silverfish were separated from their host for six or nine days. Separated individuals showed reduced chemical host resemblance, and they received significantly more aggressive rejection behavior than did unmanipulated individuals.

Slightly different results were previously reported by Volker Witte and his colleagues (2009) studying the spider *Sicariomorpha* (formerly *Gamasomorpha*) *maschwitzi* (Oonopidae) and the silverfish *Malayatelura ponerophila*. Although the cuticular hydrocarbon profiles of the spider were less similar to those of the host ants than the profiles of the silverfish were, the spiders received significantly fewer attacks from host ants and survived longer in laboratory colonies than the silverfishes did. The latter were frequently attacked and often even killed, whereas spiders remained in contact with the confronting host ant and aggression eventually ceased. The authors concluded that the spiders relied less on chemical mimicry but nevertheless managed to be accepted. How they achieved this is not quite clear. The authors also point out that the high mortality of the silverfishes in laboratory nests could be due to the simplified nest architecture. Indeed, silverfishes thrive in natural host colonies, where they occur in high numbers. The nest structure of mature colonies of *Leptogenys distinguenda* is very distinct, providing many escape and hiding niches for the myrmecophiles (Witte 2001). Volker Witte and his colleagues (2009) argue that "encounters with aggressive mature workers were frequent, which constantly disturbed the silverfish and probably hindered the updating of their cuticular hydrocarbon profiles by physical contact." But nevertheless, the hydrocarbon profiles of the silverfish species resembled those of host ants better than did those of the myrmecophilous spider species. Rettenmeyer (1963) also observed high mortality due to worker aggression in myrmecophilous silverfishes associated with ecitonine army ants. Considering all these observations, it is tempting to conclude that the similarity of hydrocarbon profiles between hosts and guests is not the most important parameter that makes coexistence of myrmecophiles with their host ants possible.

Other examples of hydrocarbon acquisition and the putative behavioral significance are discussed in separate chapters. An interesting discussion concerning the semantics of the terms used in putative adaptive chemical resemblance has been provided by von Beeren et al. (2012b).

Although a considerable number of studies have revealed various degrees of correlations of the cuticular hydrocarbon profiles between myrmecophiles and their hosts—for example, in the Asian army ant genus *Aenictus* (Maruyama et al. 2009), or in leafcutter ants and myrmecophilous cockroaches (see Chapter 5)—instances where the cuticular hydrocarbon profiles of well-integrated myrmecophiles do not match those of their host species also exist. Parmentier et al. (2017a) compared the cuticular hydrocarbons of a large number of myrmecophilous arthropods associated with the mound-building red wood ants (*Formica rufa* group) and detected no or very marginal resemblances with the host ants. Though we do not deny that the resemblance of the cuticular hydrocarbon profiles with those of the host colony can be one means of assuring survival inside the ant colony, we have to guard against a hasty conclusion that the hydrocarbon match represents the effective mechanism for the acceptance of myrmecophiles by their hosts. For example, von Beeren et al. (2012a) describe the behavior of a myrmecophilous spider, *Sicariomorpha maschwitzi,* which acquires hydrocarbons from its hosts, *Leptogenys distinguenda.* Although specimens that were isolated from the host species for several days lost these hydrocarbons, when reintroduced to the host ants, they showed no change in their integration with their host colony. Recently Parmentier et al. (2018, 2021) presented additional examples that demonstrate that resemblance of the cuticular hydrocarbons between myrmecophiles and hosts is not the key to survival in host colonies. Other behavioral mechanisms, such as risk avoidance and quiet defense strategies, are more important (see also Hölldobler and Kwapich 2019; von Beeren et al. 2021b).

Although recognition is an essential part of group cohesion and defense, there is only limited experimental evidence suggesting that myrmecophiles benefit by sharing the recognition cues of their adult host ants. Likewise, while some myrmecophiles can be exchanged between colonies without prejudice, the ability

RECOGNITION, IDENTITY THEFT, AND CAMOUFLAGE

of myrmecophiles to mimic ant brood remains largely untested, and the identity of important brood pheromones remains unknown. The possibility that some myrmecophiles camouflage themselves as detritus, prey, or benign objects invites enticing directions for future research, which may further clarify how myrmecophiles overcome the strict recognition systems of their hosts. In Chapter 4, we visit the lycaenid butterflies, the larvae of which are endowed with special organs that aid in mutualistic and parasitic transactions with ants. The function of glandular appeasement substances and use of other types of communication in this group have been well studied across a full spectrum of mutualistic, parasitic, and predatory caterpillars.

4 The Lycaenidae: Mutualists, Predators, and Parasites

THE LYCAENIDAE IS among the largest families of butterflies, with almost 5,400 species worldwide (Fiedler 1991a, 2012; Pierce et al. 2002; Fiedler 2021). They are commonly called the "blues, coppers, and hairstreaks" because of the brilliant structural coloration and markings on the wings of many species. Adults are small, with a wingspan of rarely more than 3 to 4 cm. Most lycaenid caterpillars feed on plants; however, some species are carnivores, preying on aphids and scale insects. A few live as predators and kleptoparasites in ant nests, where they prey on ant larvae and eggs or even receive regurgitated food from the ants. However, the majority of myrmecophilous lycaenid species live on plants, where they engage in relationships with ants that appear to be mutualistic, inasmuch as the caterpillars provide secretions for the ants in exchange for protection against parasites and predators, and each symbiotic partner appears to gain a net benefit from the other (Hinton 1951; Atsatt 1981; Henning 1983a; see reviews in Pierce 1987; Pierce et al. 2002; Fiedler 2006, 2012). It seems likely that the parasitic life pattern seen in some species (discussed later) evolved from such ancestral, mutualistic interactions with ants (Devries 1991a, 1997; Campbell and Pierce 2003).

The Mutualists

Several morphological structures in lycaenid larvae are important in maintaining associations with ants. Scattered over the surface of the larvae and pupae are many innervated sensory structures. These are the so-called pore cupolas, the

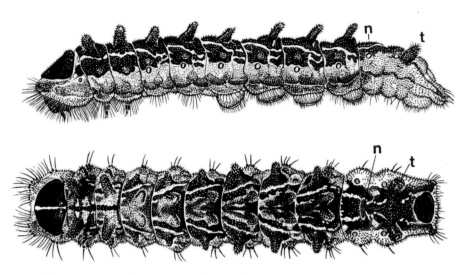

4-1   Fifth-instar larva of the Australian lycaenid *Jalmenus evagoras.*
n = Newcomer's gland (dorsal nectar organ); t = tentacular organ. (Courtesy of Roger
Kitching).

significance of which was noted by Hinton (1951) and subsequently described in detail by Malicky (1969), Downey and Allyn (1979), Kitching (1983), Pierce (1983), Wright (1983), Franzl et al. (1984), and Kitching and Luke (1985) (Figures 4-1, 4-2). These perforated structures have been found in the larvae of all lycaenids and are present in the pupae of some species. They are derived from innervated glandular setae that are believed to exude either ant attractants or appeasement substances (Henning 1983b; Pierce 1983).

The larvae of many lycaenid species also possess the so-called dorsal nectary organ or honey gland, also known as Newcomer's gland because it was first described by the lepidopterist E. J. Newcomer (1912). It is located on the dorsum of the seventh abdominal segment (Figures 4-1, 4-2, 4-3). This organ occurs in most species of the subfamilies Polyommatinae and Theclinae, but is absent in other subfamilies, including the Lycaeninae, Miletinae, Liphyrinae, and Poritiinae. The dorsal nectar organ presumably developed from dorsal epidermal pores (Malicky 1969, 1970; Kitching and Luke 1985). Similar honey-gland structures have been found in other groups and include the "perforated chambers" of the Curetinae (Devries et al. 1986), the "pseudo-Newcomer's gland" of the Miletinae (Kitching 1987), and the nectar organs in the Riodinidae (Ross 1966; Schremmer 1978; Cottrell 1984).

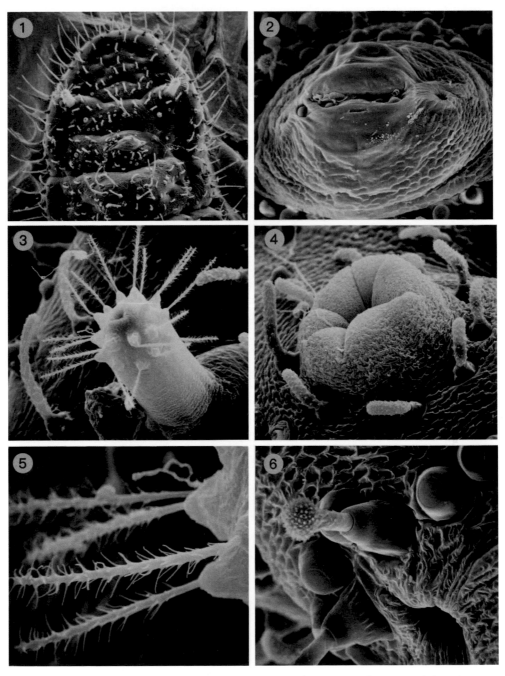

4-2 Scanning electron microscopic images of the external morphology of myrmecophilous organs in an older larva of the lycaenid *Lysandra coridon*: (1) posterior and dorsal view; (2) Newcomer's gland opening; (3) everted tentacular organ; (4) inverted tentacular organ; (5) spinose hair on tentacular organ; (6) pore cupolas. (Courtesy of Roger Kitching).

The nectar organ of most species of Curetinae, Polyommatinae, and Theclinae is flanked on either side by a pair of eversible tentacles, which are called lateral organs, or tentacle organs (Figures 4-1, 4-2). They are absent in the Lycaeninae. Histological investigations revealed that they are also secretory organs (Hinton 1951; Claassens and Dickson 1977). Though the function of the tentacles and their secretions is unclear, it has variously been supposed that they produce substances that are attractive to the ants, or that alert or alarm ants when the nectar organ is depleted. Some experimental evidence indicates that the tentacular gland secretions serve at least in part to evoke alarm behavior in the attending ants (Henning 1983b; DeVries 1984; DeVries et al. 1986; Fiedler and Maschwitz 1987;

4-3   A worker of the ant *Oecophylla smaragdina* collects a droplet of nectar secreted from the Newcomer's gland of the caterpillar of the lycaenid butterfly *Hypolycaena erylus*. The lower picture depicts an adult butterfly of *Hypolycaena*, with eye spots and tails that simulate a head. The illusion helps to divert predators from the true head and body while the butterfly escapes. (Courtesy of Konrad Fiedler).

Fiedler et al. 1996). The tentacles are particularly well developed in species of *Curetis* and some Aphnaeinae and may be used themselves for defense by lashing predators or parasitoids like a cat-o'-nine-tails (Boyle et al. 2015; Pierce and Dankowicz in press).

Structures similar to the dorsal nectary organ and tentacle organs exist in two tribes of the Riodinidae (metalmark butterflies), but they are located in different anatomical positions than those in the Lycaenidae. The nectar organ of the Riodinidae is in the eighth abdominal segment and consists of paired glandular tentacles (Ross 1964) that secrete honeydew. Riodinid caterpillars also have an additional pair of eversible tentacles located in the third thoracic segment that appear to have a defensive function (Kitching 1983; Kitching and Luke 1985; DeVries et al. 1986; DeVries and Baker 1989; DeVries 1991a,b, 1997).

It has been suggested that ant association in the Lycaenidae and Riodinidae evolved independently because their myrmecophilous organs are located on different anatomical segments (DeVries 1991a,b, 1997), but this was initially questioned by Dana Campbell and Naomi Pierce (2003), who argued, based on their preliminary phylogenetic work (Campbell et al. 2000), that the tentacular organs might be more parsimoniously explained as homologous structures, with shifts in function and position over time. However, ancestral-state reconstruction of ant association later carried out by Naomi Pierce and her colleagues as part of a recent tribe-level phylogeny of all butterflies indicated that ant association has indeed evolved independently in Lycaenidae and Riodinidae, and that, in the riodinids, where ant association has been documented from only two tribes, it may have arisen twice (Espeland et al. 2018; for recent phylogenies of the Riodinidae, see Espeland et al. 2015; Seraphim et al. 2018).

In the following sections, we focus mainly on the myrmecophilous lycaenids. From its first description by Newcomer (1912), the lycaenid dorsal nectary organ was thought to secrete a sweet liquid that was imbibed by the ants. Ulrich Maschwitz and his collaborators (1975) provided the first analytical proof for this hypothesis. They found that the secretions of the nectar organ of the lycaenid species *Lysandra hispana* consists of a solution of sugars (fructose, sucrose, trehalose, glucose) concentrated seven to ten times higher than that found in the larva's hemolymph. In addition, minor quantities of protein and the amino acid methionine were found. Thus, the lycaenid caterpillar provides the ants rewards in the form of a highly enriched sugar solution.

A more detailed analysis of the contents of nectar organs in lycaenids was later provided by Holger Daniels, Gerhard Gottsberger, and Konrad Fiedler (2005). They analyzed larval nectar secretions and hemolymph of three facultatively ant-attended lycaenid species in Europe (*Polyommatus coridon, P. icarus,* and *Zizeeria knysna*), among which *P. coridon* is the most consistently found in the company of ants. It occurs in central and southern Europe, where the larvae feed on plant species of the genera *Hippocrepis* and *Coronilla* in the Fabaceae. *Polyommatus icarus* can be found in the Palearctic region; it also feeds on various species of Fabaceae and is irregularly found in association with ants. *Zizeeria knysna* occurs in Africa and Southern Europe. The larvae of this species use various food plants, and they are found only occasionally in the company of ants.

The main component of the nectar organ secretions of all three lycaenid species was sucrose, and about half of the *P. coridon* samples also contained glucose, whereas in *P. icarus* and *Z. knysna,* melezitose was found to be the second most common sugar, followed by fructose and glucose. The mean sugar concentrations in the nectar were 4.3% for *P. coridon,* 7.4% for *P. icarus,* and 6.8% for *Z. knysna.* Up to fourteen identified amino acids were found in the nectar-gland secretion of *P. coridon,* with leucine constituting about 50% of all amino acids. Other amino acids were tyrosine, proline, arginine, and phenylalanine. The mean concentration of all amino acids in the secretion of *P. coridon* was 0.97%. In *P. icarus,* the larval nectar contained six amino acids (mostly tyrosine and phenylalanine; mean concentration 12%), and in *Z. knysna,* alanine and proline (mean concentration 0.03%).

These comprehensive results support a hypothesis proposed by Naomi Pierce (1987) that lycaenid species with no, or only sporadic, interactions with ants produce nectar secretions consisting mostly of carbohydrates, whereas the nectar of species that entertain regular or even permanent ant attendance contains a higher concentration of nitrogenous compounds. In fact, those species that have close mutualistic relationships with ants also appear to prefer nitrogen-fixing plants or other protein-rich food plants (Pierce 1985). Together with the previous findings by DeVries and Baker (1989), who found an even higher percentage of amino acids in the secretions of the larvae of the riodinid *Thisbe irenea*, the hypothesis that amino acids play an important role in mutualistic riodinid- and lycaenid-ant associations is strongly supported. Not only the quantities but also the varieties of amino acids seem to be correlated with the intimacy of associations with attendant ants. More strongly myrmecophilous species may have richer and more diverse mixtures of amino acids (Daniels et al. 2005). In addition, different species often vary in their composition of main nectar amino acids.

The functional significance of these variations is unknown, although Yao and Akimoto (2002), working on the composition of amino acids in aphid-produced honeydew, suggested that serine and glycine play significant roles in all myrmecophilous insect-ant interactions. Daniels et al. (2005) correctly point out that although one or the other of the two amino acids is present in some obligate and parasitic lycaenid myrmecophiles, in all facultative lycaenid myrmecophiles, serine and glycine were negligible.

However, a potentially more manipulative explanation for these results cannot be ruled out. In more recent work, Masaru Hojo, Naomi Pierce, and Kazuki Tsuji (2015) observed that secretions from the dorsal nectary organ of the caterpillar of the lycaenid *Arhopala (= Narathura) japonica* have a behavior-changing effect on the attending ants (*Pristomyrmex punctatus*). Ants that imbibed secretions from the nectar organ exhibited markedly reduced locomotor activity and a significantly lowered response threshold to reacting aggressively to the eversion of the tentacle organs. These observations alone would perhaps not yet cause too much excitement; however, the authors were able to demonstrate that "neurogenic amines in the brains of workers that consumed caterpillar secretions showed a significant decrease in the levels of dopamine compared with controls." When ant workers were treated with reserpine, a compound that has been identified as a dopamine inhibitor in the *Drosophila* flies, the ants also exhibited significantly reduced locomotor activity. These results suggested that the secretions from the dorsal nectary organ were involved in altering host ant behavior in a way that would enhance fidelity to the caterpillar and increase aggression toward possible threats or disturbances. Given the strong correlation between ant association in the Lycaenidae and preference for nitrogen-rich host plants and parts of plants, it's possible that nitrogen is required to produce compounds capable of manipulating ant behavior via the dopaminergic pathway. These could conceivably involve amino acids such as glycine, which, in earlier work, Hojo had shown to elicit responses from the taste-enhancing "umami" receptors of their host ants such that the combination of glycine and sucrose generated a significantly stronger response than either glycine or sucrose on its own (Hojo et al. 2008). Additional work is needed before further conclusions can be drawn regarding whether these kinds of trophobiotic lycaenid-ant associations are truly mutualistic, or are simply sophisticated forms of parasitism involving the exchange of secretions that only appear superficially to be beneficial for attendant ants.

Nevertheless, the nutritive contributions of carbohydrates from these secretions suggest that, in many cases, they might function as nutritive rewards. For example, Daniels et al. (2005) conclude from the published studies and their own work that, in general, the nectar organ secretions contain 5%, 10%, sometimes even 15% sugar, "with the possible exception of the Australian obligate ant mutualist *Paralucia aurifera*," whose secreted nectar has been reported to contain 34% sugar. The main sugar components in all lycaenid and riodinid species studied are sucrose and glucose.

A series of elegant behavioral experiments by Konrad Fiedler and Ulrich Maschwitz (1989a) further support the nutritive role of these secretions. They placed fourth-instar larvae of *Polyommatus coridon* in a small arena that was connected by a narrow bridge made of two wooden sticks to a nest of myrmicine *Tetramorium caespitum*. In nature, *P. coridon* is associated with several ant species belonging to three subfamilies. As a control they used the lycaenid species *Lycaena tityrus,* which belongs to the subfamily Lycaeninae and, as we learned, does not have nectar organs. The results were unequivocal: larvae lacking a nectar organ did not release a recruitment response in ants that encountered them, whereas ants that discovered *P. coridon* larvae laid chemical recruitment trails along which nest mates moved over the bridge into the little arena where *P. coridon* caterpillars were placed (Fiedler and Maschwitz 1989a). The mature larvae of *P. coridon* produces "an average of 30.9 droplets of nectar per hour." Considering that, in the study area, the population density is about twenty caterpillars per square meter, and each individual produces over the entire larval period an estimated 22–44 µl of nectar, which equals an energy content of 55–110 joules (J), one can estimate that the *P. coridon* larvae "produce carbohydrate secretions with an energy equivalent of 1.1–2.2 kJ per square meter" (Fiedler and Maschwitz 1988, 205). Thus, we can conclude that these carbohydrate secretions are a significant source of nutrition for the attending ants.

Let us now consider again the function of the tentacle organs. That caterpillar-attending ants react aggressively to the eversion of the tentacles should not be surprising; it has been reported by various authors that during the eversions of the tentacles putative chemical alarming stimuli are released by the caterpillar, which trigger, as Fiedler and Maschwitz (1987) expressed it, "excited runs" in the attending ants. As we already stated, caterpillars of *Polyommatus coridon* can be found with several ant species. The strongest reaction to the tentacle eversion was observed in *Plagiolepis pygmaea* and other formicines such as *Lasius alienus,* while ant species of other subfamilies that have also been seen attending *P. coridon* caterpillars had less of a reaction. Many formicines employ the hydrocarbon undecane from the Dufour's gland as an alarm pheromone (Hölldobler and Wilson 1990), and it is suggestive that the caterpillar secretion might contain undecane components.

   In fact, Henning (1983a,b), investigating an African lycaenid, *Aloeides dentatis,* reports that the attending formicine *Lepisiota (Acantholepis) capensis* ants responded with alarm behavior to the caterpillar's tentacle eversions. Henning was also able to release such alarm response in the ants with extracts made from tentacle organs. Apparently, these alerting stimuli emitted by the myrmecophilous caterpillars are quite specifically tailored to match the "host" ant's alarm pheromones. Similar results were reported by DeVries (1984) for a lycaenid *Curetis* species and its attending ant, *Anoplolepis longipes*.

The specificity of such putative alarm pheromone mimics (also called allomones) is likely most noticeable in species in which lycaenid caterpillars are obligate myrmecophiles, often associated with one or a small guild of phylogenetically closely related attending ant species. On the other hand, as Fiedler and Maschwitz have demonstrated in their study of a facultative myrmecophile with a variety of attending ant species, *Polyommatus coridon,* even here, interspecific communication works best with only a few ant species belonging to the same subfamily.

Now that we have discussed the response of ants to the tentacle organ in several lycaenid species, let us return to the exciting findings by Hojo et al. (2015) that suggest the nectar-gland secretions of the lycaenid *Narathura japonica* larvae contain a neuromodulating component that affects the ant's brain and causes reduced locomotor activity. The authors also observed that such ants show significantly higher aggressive responses to the eversions of the caterpillar's tentacle organs. Apparently, only "experienced" ants—that is, ants previously in contact with lycaenid larvae—exhibit an enhanced aggressive reaction, whereas "inexperienced" ants, which did not yet imbibe nectar gland secretions, "did not respond aggressively and simply ignored the eversion" of the tentacles. If the hypothesis is true that *N. japonica* caterpillars also emit compounds that mimic the alarm pheromones of the attending ants, we wonder why "inexperienced" *Pristomyrmex* ants do not show any reaction to these mimics. We speculate that the alarm-allomones are released in such small quantities that, under normal conditions, the number of molecules is below the typical response-threshold concentration. However, in "experienced" ants that have imbibed the neuromodulator secreted by the nectar organ gland, the ant's response threshold is modulated in such a way that they react to much lower concentrations of the allomones than they normally would.

Thus, the neuromodulating effect caused by the nectar imbibed by the ants most likely does not trigger the ant's aggressive response toward tentacle eversion but may just lower the response threshold in the ants to the alarm pheromone mimic released from the tentacle organ glands. Of course, we must keep in mind that these interesting discoveries concern only one lycaenid species, *Narathura japonica*, and one attending myrmicine ant species, *Pristomyrmex punctatus*. Future work will determine whether these findings also apply more generally to other myrmecophilous lycaenid species and their associated ants.

For decades, the relationship between lycaenid and riodinid caterpillars feeding on plants and the attending ants has been considered mutualistic. In the previous paragraphs we discussed the benefits the ants gain from these interactions, but for a long time the caterpillars' return was more or less taken for granted.

In the late 1970s and early 1980s Naomi Pierce, at the time a graduate student at Harvard University and now Hessel Professor of Biology and curator of Lepidoptera at the same university, wanted to experimentally test whether lycaenid larvae that feed on specific plant species and are attended by certain ant species really benefit from the presence of ants. As we have learned, the ants attend the caterpillars that have reached their third or fourth (final) instar. This is the larval developmental stage wherein the caterpillars are most prone to attack by parasitoids. It is also the time when the dorsal nectar organ has reached its secretory peak. Remember, the caterpillars of many lycaenid butterflies (and also of many other species of the closely related family Riodinidae, the metal mark butterflies) eat plant tissue but use some of the nutrients and energy to manufacture honeydew and amino acids in the nectar organ; and as we have seen, these secretions provide a sizeable amount of valuable nutrition to the attending ants. Obviously, this entails a cost for the caterpillars, but what do the caterpillars get from the ants in return?

Naomi Pierce and her student assistant, Paul Mead, who is now a well-known bacteriologist, set out to experimentally test whether the ants' attendance is really

4-4   The adult of the lycaenid butterfly *Glaucopsyche lygdamus*. (Courtesy of Naomi Pierce).

of any benefit for butterfly larvae. They conducted their studies in the Colorado mountains, and their work focused on the "silvery blue" (*Glaucopsyche lygdamus*), the caterpillars of which feed on lupine plants (Figure 4-4). For the experiment, they chose two different habitats, one in a dry region where *G. lygdamus* caterpillars feed on inflorescences of *Lupinus floribundus* and are attended primarily by ant workers of the species *Formica altipetens*. In the other habitat, a wet meadow, *G. lygdamus* caterpillars feed on flowers, seedpods, and leaves of *Lupinus barberi*, and the attending ants belong to the species *Formica fusca*. In both populations the ants were observed attending the caterpillars once they had reached the third larval instar (Figure 4-5).

In a series of well-designed experiments, the ants were either prevented from accessing or allowed to access the caterpillar's host plants. Pierce and Mead achieved this by arranging one set of *Lupine* plants with viscous barricades (Tree Tanglefoot), which made it impossible for the ants to climb up the plant. A second set of plants (controls) was also treated with Tree Tanglefoot. However, in this treatment, one side of the plant stem remained free of the barricades so the ants could still access the upper leaves and florescence. The results were unequivocal: a significantly higher proportion of caterpillars without ant attendance suffered attacks by parasitoids (such as the braconid wasps or tachinid flies) than did the

4-5 The caterpillar of *Glaucopsyche lygdamus* is tended by a worker of *Formica fusca*. The picture below shows the *Formica* host ant attacking a parasitic wasp that attempted to oviposit into the caterpillar. (Courtesy of Naomi Pierce).

control groups that continued to have ant attendance. These and other experiments clearly demonstrate the benefits *Glaucopsyche* caterpillars obtain from ant attendance (Pierce and Mead 1981). In a follow-up study, Pierce and Easteal (1986) fully confirmed the previous findings and substantially extended the database.

4-6   Ventral and dorsal view of the Australian lycaenid butterfly *Jalmenus evagoras*.
(Upper: Courtesy of Benjamin Twist; Lower: Courtesy of Tobias Westmeier).

4-7 Caterpillars of *Jalmenus evagoras*. They are often attended by many host ants of *Iridomyrmex anceps*. (Courtesy of Tobias Westmeier).

Subsequently, Naomi Pierce teamed up with Roger Kitching and his collaborators in Australia, and together they conducted a thorough analysis of the costs and benefits of cooperation between the Australian lycaenid *Jalmenus evagoras* and its attendant ants (Pierce et al. 1987) (Figure 4-6).

This lycaenid species is an especially interesting case. *Jalmenus evagoras* larvae and pupae can be found in aggregations on the *Acacia* plants, several species of which serve as food plants (Figures 4-7, 4-8). The primary attendant ant species is the dolichoderine *Iridomyrmex mayri,* but in some other populations, *I. rufoniger* has also been observed in association with *J. evagoras.*

THE LYCAENIDAE

4-8   Larvae and pupae of *Jalmenus evagoras* typically occur in groups, and both developmental stages are attended by host ants. This picture depicts a group of *Jalmenus* pupae. (Courtesy of Benjamin Twist).

Pierce and Elgar (1985) noted that females of *J. evagoras* show a significant preference for food plants for oviposition on which *Iridomyrmex* ants are present, and that the butterflies appear to sense the presence of the ants when still in flight. The butterflies also seem to distinguish whether the ants are just foraging on the tree or tending homopterans, and appear to prefer the latter situation when deciding to land on the tree and oviposit. As Pierce and Elgar (1985) point out, it is not unusual that lycaenid butterfly females use the presence of prospective attendant ant species as a cue for suitable oviposition sites. They tabulated a total of forty-six lycaenid species from twenty-nine genera and five subfamilies of the Lycaenidae of which observations and data have been published concerning ants on food plants serving as cues for choosing oviposition sites. This appears to be a widespread phenomenon, yet not universal. In some lycaenid species, the female butterflies seem not to use the presence of ants as a cue for oviposition. For example, Pierce and Easteal (1986) observed that in *Glaucopsyche lygdamus,* ants do not play a significant role in selecting oviposition sites.

In addition to influencing oviposition site in some species, the aggregation of larvae and pupae appears to be influenced by the presence of ants, because young and last-instar larvae are found significantly more often on plants with ants than on plants where ants are experimentally excluded (Pierce et al. 1987). The exclusion experiments further demonstrate that "larvae and pupae disappeared significantly more often from plants without ants than from plants

4-9 Occasionally the defenses of host ants fail. Here, the predatory ant *Myrmecia nigrocincta* succeeds in catching a *Jalmenus* caterpillar. (Courtesy of Naomi Pierce).

with ants." Observational evidence indicates that dominant predators include arthropods such as spiders, reduviid and pentatomid bugs, and, most of all, the ant *Myrmecia nigrocincta* (Figure 4-9) and the vespid wasp *Polistes variabilis*. In addition, unattended *Jalmenus* larvae are prone to attack by parasitic wasps of the genera *Trichogramma* and *Brachymeria*.

The fact that *J. evagoras* larvae aggregate enables small larvae (first and second instar) to benefit from ant protection, because older instars ensure the ants' presence by providing the valuable nectar-organ secretions to the ants. Pierce et al. (1987) calculated that the secretions "can amount to as much as 409 mg dry biomass from a single host plant containing 63 larvae and pupae of *J. evagoras* over a 24-hour period."

All these experimental field studies, complemented by investigations in the laboratory, provide ample proof that these interrelationships between ants and lycaenid caterpillars are truly mutualistic. Nevertheless, scientists next wondered whether the continuous production of high-grade nectar entails a measurable burden that is reflected in the caterpillars' developmental process. Indeed, in some species it has been found that caterpillars and pupae suffer developmental costs. The most profound effects have been reported for *J. evagoras* (Pierce et al. 1987; Baylis and Pierce 1992). As we have learned, in this species, the larvae provide substantial amounts of energy-rich secretions to assure ant attendance. Production of these valuable exudates is obviously costly, and several studies have shown that larvae and pupae tended by ants have significantly lower weight and smaller size than do lycaenid immatures experimentally cultured on food plants without ants in the laboratory (Pierce et al. 1987; additional studies reviewed in Pierce et al. 2002). However, under natural conditions *J. evagoras* larvae and pupae without ant attendance would not survive and would inevitably fall victim to predation and parasitism.

Yet, we have to point out that, although the experimental studies we have discussed above clearly demonstrate that ants play a significant role in protecting the lycaenid caterpillars and pupae, there are cases where this effect is not so clear, especially in some species with sporadic or facultative associations with ants (Pierce and Easteal 1986; Fiedler and Hölldobler 1992; Wagner 1993, 1995; Wagner and Del Rio 1997). In some of these species, the secretions from the nectar organ might just function as appeasement substances, muting the ants' aggressive behavior, or as Pierce et al. (2002) expressed it, "The food rewards they provide attendant ants can be regarded as a kind of bribery." This was proposed by Malicky (1969, 1970) and in fact might have been the original function of the nectar organ in facultative lycaenid-ant interactions. Over the course of the evolution of ever-closer mutualistic symbiotic bonds, the nectar organ developed into a true trophobiotic organ, producing not just tiny appeasing "taste samples," but copious amounts of honeydew, as exemplified in the case we discuss next.

Let us now consider the remarkable mutualistic symbiosis between the obligate myrmecophilous lycaenid, *Anthene emolus,* and its host ants, the weaver ant

Oecophylla smaragdina, which was studied in detail by Konrad Fiedler and Ulrich Maschwitz (1989b). First, a few sentences about the remarkable weaver ants.

There exist two species of the formicine genus *Oecophylla.* One is *O. longinoda* of tropical and subtropical Africa, and the other is *O. smaragdina* of tropical and subtropical regions of Asia and Australia. Both species build leaf-tent nests, and a mature colony usually lives in hundreds of such leaf-tents, spread over the canopies of several trees (Figure 4-10). The *Oecophylla* are called weaver ants because, in order to hold the leaves that form the tent in place, the adults use the silk produced by the last instar of their own larvae. Normally, in formicine ants (*Oecophylla* belongs to the subfamily Formiciniae), the last-instar larvae use the silk to spin a cocoon in which they pupate. However, in *Oecophylla,* the larval silk is used for binding together leaf-tents, and the pupae are left "naked" (not inside a silken cocoon).

During nest construction, *Oecophylla* workers hold last-instar larvae between their mandibles, and the almost motionless larva, apparently responding to the tactile signals given by the ant worker with its antennae, releases silk from the labial glands, which the workers attach to the leaves by moving the larvae like a weaving shuttle, back and forth (see Figure 2-19). However, *Oecophylla* not only build leaf-silk nests to house their queen, brood, and nestmates, they also build special leaf and silk pavilions for their honeydew-producing Hemiptera (Figure 4-11) and for some of their lycaenid symbionts. One of those is *Anthene emolus* (Figure 4-12).

The females of *A. emolus* use the presence of *Oecophylla smaragdina* as a cue for oviposition. Interestingly, female butterflies that alight on an *Oecophylla*-occupied tree branch or leaf oviposit without being attacked by the resident ants. As soon as the first-instar larvae, freshly hatched from their eggs, are discovered by the ants, they are carried to the pavilions, where the young *Anthene* larvae feed on plant tissues. These young larvae do not yet have a functional nectar organ, but nevertheless, the ants take care of them. The older larvae leave the pavilion on their own and are carried by the ants to the appropriate sites on the tree where the larvae can feed. Sometimes they are brought back by the ants

4-10 A leaf-tent nest of *Oecophylla longinoda*. The opened nest shows the many ant larvae and pupae housed inside these nests. (Bert Hölldobler).

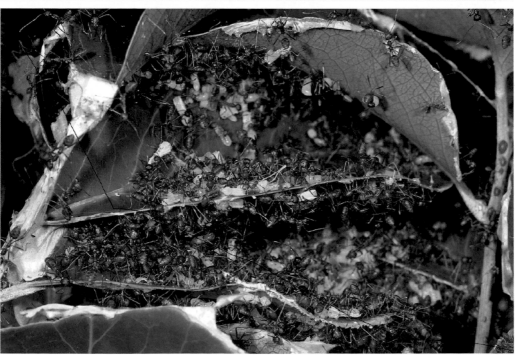

4-11 Special "stable" nest of *Oecophylla longinoda*, where the ants culture their honeydew-producing coccids. (Bert Hölldobler).

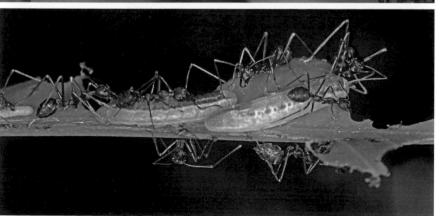

4-12 Young caterpillars of the lycaenid *Anthene emolus* are housed in special nests built by their host ants *Oecophylla smaragdina*. The upper picture depicts such an opened nest, with several small *Anthene* caterpillars grouped together. The lower picture shows later instar caterpillars feeding on plants while being guarded by the attendant weaver ants, *O. smaragdina*. (Courtesy of Konrad Fiedler).

to the pavilion shelters. Once they have reached the third and fourth (final) larval instar, the caterpillars' nectar organs are fully active. Each nectar organ secretes droplets of a clear fluid that is immediately imbibed by attendant ants. The droplets are relatively large and can be seen with the naked eye. The frequency of droplets appearing on the nectar organ is also remarkably high: "Full-grown larvae produce on average more than 50 droplets in ten minutes," and in third-instar and young fourth-instar larvae, the frequency is about 30% lower (Fiedler and Maschwitz 1989b). A major component of the secretion is glucose. Concerning the gain for the myrmecophilous lycaenids, the authors present the following calculation: "Under the assumption (a) that one fourth instar larva [L4] produces on the first day 200 droplets per hour (the value obtained with young L4) and on the following two days 300 droplets per hour (the value of older L4) and (b) that the larva is milked by *O. smaragdina* worker ants the whole time, the complete secretion amount of the fourth instar within three days is at least 80 µl of nectar gland contents." Assuming that 15% of the secretion consists of carbohydrates (like in *Polyommatus hispanus,* Maschwitz et al. 1975) the authors make a rough estimate of the energy gain the ants enjoy: "The whole larval population of a single young shoot [where the caterpillars graze] is often more than 50 individuals which . . . account for an energy production of 10 kJ [kilojoules] or more" (Fiedler and Maschwitz 1989b, 842).

This tight interrelationship is obviously beneficial for the ants, but considering the energy the caterpillars invest to produce these copious amounts of high-value nectar gland secretions, one wonders whether the cost / benefit ratio is in balance for *Anthene emolus.* Fiedler and Maschwitz (1989b, 844) list all the pluses and minuses for both symbionts and finally conclude "in summary the benefits *A. emolus* derives from its tight relationship towards *O. smaragdina* outweigh the costs. *Anthene emolus* is an abundant species at disturbed, open-habitats in Malaysia where *O. smaragdina* is one of the most dominant species." The fact that *Oecophylla* workers already take good care of the first- and second-instar larvae of the lycaenid myrmecophile, which do not produce nectar-organ secretions, must be considered a major benefit for *A. emolus.* This is comparable to *Jalmenus evagoras,* discussed before, that has not evolved the same tight interrelationship with ants, but also forms aggregations comprised of larvae from more than one instar. In this way, young larvae indirectly enjoy protection by attendant ants, although only the third- and fourth-instar larvae deliver the nectar-organ secretions.

Finally, it is worthwhile mentioning here a discovery by Alain Dejean and his coworkers (Dejean et al. 2016), although the myrmecophile is not a lycaenid but the African noctuid moth *Eublemma albifascia,* the caterpillars of which develop inside the nests of the African weaver ants, *Oecophylla longinoda.* The adult females of this moth deposit their eggs onto the outsides of the leaves forming the leaf-tent nests. *Oecophylla* workers retrieve the freshly hatched caterpillars into the nests, which are usually packed with ant brood and, in addition, one especially well-guarded nest that houses the queen. According to the authors, the moth larvae are better fed by the host ants with regurgitated food and trophic eggs than are the ant larvae and queen, and the infestation with the myrmecophilous moth can reach such magnitude that the queen's fertility rapidly wanes, and the colony perishes. Though such dramatic effects caused by myrmecophiles are rare, specialized predators are common in the Lycaenidae, which we consider in the following section.

Predators and Socially Parasitic Myrmecophiles in the Genus *Phengaris*

Most myrmecophilous lycaenid species appear to live in a more or less mutualistic relationship with ants, but about 5% of them are predators or myrmecophilous social parasites inside the host ants' nests. Phylogenetic analyses have revealed that socially parasitic relationships between lycaenids and ants have arisen independently at least fourteen times (Fiedler 1998; Pierce et al. 2002). We focus here on the genus *Phengaris* and choose only a few species that best represent the different evolutionary grades of adaptation as myrmecophiles. In fact, several species of the genus *Phengaris* are among the most thoroughly studied myrmecophilous parasites, mainly due to the research efforts of Jeremy A.

Thomas and his collaborators, in particular Graham W. Elmes, Karsten Schön-rogge and J. C. Wardlaw, and to the work of David Nash and his colleagues. The following descriptions of several case studies are based primarily on their work.

First, we should briefly consider some taxonomic issues concerning these lycae-nids. In most publications of the Thomas group, the genus name *Phengaris* is absent and *Maculinea* is instead used to refer to this European ant-associated species group. Already in 1998, Konrad Fiedler addressed the possibility that the East Asian taxon *Phengaris* is the senior genus name of the *Phengaris-Maculinea* group, and eventually *Phengaris* should replace *Maculinea,* provided a molecular phylogenetic analysis were to support this notion. Subsequently, Als et al. (2004) published such an analysis and concluded that *Phengaris* and *Maculinea* together form a monophyletic group. However, only the species from the so-called *Maculinea* group are intertwined with the ant genus *Myrmica* and, phylogenetically, form an undisputed monophyletic group, whereas the three East Asian *Phengaris* species do not.

In an ensuing extended phylogenetic revision, Ugelvig et al. (2011) confirmed that *Maculinea* is indeed monophyletic and nested within the *Phengaris-Maculinea* group, whereas the *Phengaris* group alone is paraphyletic. Independently, Fric et al. (2007) arrived at the same conclusion and suggested the genus name *Maculinea* von Eecke, 1915, should be replaced by the senior synonym *Phengaris* Doherty, 1891. Obviously, this would require a more inclusive definition of the genus, which would subsume all species of *Maculinea* and two of the three *Phengaris* species, and because of priority reasons, the entire taxon would then be called *Phengaris* sensu lato. Nevertheless, the new inclusive taxon *Phengaris* is a relatively small clade comprising hardly more than ten species. Finally, the International Commission on Zoological Nomenclature decided in favor of this change, despite the fact that this decision caused some concern among ecologists and conservation biologists who have studied the endangered lycaenid genus in Europe to find feasible ways to protect the genus and its habitats. These scientists argued that, in most publications on this genus, the junior name

Maculinea has been used and that replacing it with the senior name *Phengaris* would cause much confusion. We sympathize, because we experienced similar taxonomic problems with several of our study objects, where genus names, used for more than a century, were recently changed based on new phylogenetic studies. However, we also understand the scientific mission of systematists and taxonomists and recognize that dealing with several names for the same genus would cause even more confusion. In fact, in two recent publications, the authors use the genus name *Phengaris* in one paper (Casacci et al. 2019b), and in the other *Maculinea* (Tartally et al. 2019).

After a very instructive discussion with Konrad Fiedler, one of the world's leading scholars of lycaenids, we decided to use the genus name *Phengaris*.

Before we proceed, we must clarify a second taxonomic issue. Most of the studies of the Jeremy Thomas group deal with what were once believed to be two species, *Phengaris (Maculinea) alcon* and *P. rebeli*. However, taxonomic and phylogenetic studies (Fric et al. 2007; Ugelvig et al. 2011; K. Fiedler personal communication) suggest that separating these two "ecotypes" into two species is not supported, even though they show different host preferences, live in different habitats, and use different host plants. The fact that one "ecotype" lives in xerophytic habitats and the other in wet habitats, and each use different ant hosts, led Jeremy Thomas and his coworkers to maintain the separate species names, albeit both "ecotypes" might taxonomically belong to one species. After much consideration, we decided to retain the two names *P. alcon* and *P. rebeli*, because this is the most parsimonious and least confusing way to present the extensive behavioral research on these lycaenids, but to indicate the ecotype status of *P. 'rebeli,'* we frame it in quotation marks.

Socially Parasitic Myrmecophiles: The Cuckoo Strategists

As we discussed previously, the best ecological niche for a myrmecophilous species is the area of the ant nest where the larvae are housed and nursed by the ants. Although the adult butterfly of *Phengaris alcon* and *P. 'rebeli'* has no or only

4-13 The adult butterfly of *Phengaris alcon* laying eggs on the gentian plant, the food plant of the first three instars of the *P. alcon* larvae. (Courtesy of Izabela Dziekańska).

marginal interactions with ants, their caterpillars (larvae) undertake their major growth inside the host ants' (*Myrmica* species) brood chambers. But let us start from the beginning.

Much of the work by Jeremy Thomas and his group focused on the two species or ecotypes *Phengaris alcon* and *P. 'rebeli.'* *Phengaris 'rebeli'* prefers dry, xerophytic habitats, where the butterflies choose mainly *Gentiana cruciata* for oviposition. In contrast, *P. alcon* lives in wet grass or heathland and lays its eggs preferably on *Gentiana pneumonanthe* (Figure 4-13). After the first-instar larvae hatch from the eggs, they remain on the gentian food plant until they have reached the end of the third or beginning of the fourth and final larval instar. At

THE GUESTS OF ANTS

4-14 A fourth-instar caterpillar of *Phengaris alcon,* which dropped to the ground and was discovered by a host ant worker of *Myrmica scabrinodis.* The lower picture shows how the *Myrmica* worker carries the *Phengaris* larva into the ant nest. (Courtesy of Izabela Dziekańska).

this stage, they drop to the ground and, as soon as a *Myrmica* worker encounters them, they are picked up and carried to the ants' nest, where they are housed together with the ant larvae in the brood chambers (Figure 4-14).

It has been suggested that, to a certain degree, the caterpillar might even be able to follow its host ants' trail pheromone, which may enable it to reach the host ants' nest entrance independently, should it have been dropped to the ground close to a *Myrmica* nest. Ulrich Maschwitz and his student Martin Schroth (Schroth and Maschwitz 1984) thought they had demonstrated in the laboratory that larvae of the lycaenid species *Phengaris teleius* follow artificial trails laid

4-15 Inside the *Myrmica* nest, the freshly adopted *Phengaris alcon* larva is groomed and fed (lower picture) by the host ants. (Courtesy of Darlyne A. Murawski).

with the trail pheromone of the *Myrmica* host species. However, in a follow-up study, Konrad Fiedler (1990) was unable to confirm these findings, concluding that *P. teleius* larvae do not, in fact, follow the trail pheromone of *Myrmica* ants.

Interestingly, although the caterpillars reach the third larval instar on the food plant within about three weeks, they do not gain much weight. It is during the

4-16 The *Phengaris alcon* larvae are continuously surrounded by nurse ants, which groom and feed the myrmecophile brood more than their own larvae. This picture shows how *Myrmica scabrinodis* workers feed a caterpillar by regurgitation. (Courtesy of Marcin Sielezniew).

fourth instar, inside the ants' brood chambers, that the caterpillars acquire most of their biomass. This delayed larval growth is most remarkable and obviously an adaptation for maximal resource exploitation in the host colony. The *Phengaris* larvae are treated like ant larvae; they are groomed by the nurse ants and they exhibit ant-larva-like begging behavior by moving the frontal part of their body upward, seeking contact with "the mouth" of the nurse ants (Figure 4-15). This behavior, possibly in combination with mechanical and chemical releasers, triggers feeding by trophallaxis in the nurse ants. In addition, the caterpillars receive trophic eggs and pieces of booty brought in by ant foragers and may even eat some of the ant eggs and larvae. As they grow, they are continuously surrounded by nurse ants that groom and feed them (Figure 4-16).

The time span for reaching the size at which the caterpillar is ready to undergo pupal metamorphosis can take ten to eleven months, but in about two-thirds of the individuals, approximately twenty-two months are needed before pupation (Figure 4-17). Such a polymorphism in larval growth has been found in both ecotypes of *Phengaris alcon*—that is, *P.* 'rebeli' (Thomas et al. 1998, 2005) and *P. alcon* (Schönrogge et al. 2000; Als et al. 2002; Thomas et al. 2005). This unusually long larval development phase inside the host ants' nests appears to

4-17   Once the *Phengaris alcon* caterpillars reach the right size, they get ready for pupation. *Myrmica* workers (in this case *M. sabuleti*) keep tending and grooming them, and even the pupae (lower picture) are attended by ants, although to a lesser degree. (Courtesy of Marcin Sielezniew).

be a special trait in these socially parasitic myrmecophiles, because Henning (1984a,b) describes the same pattern for South African Lycaenidae. In species not dependent on ants, the larval phase lasts about twenty to forty days, whereas for species that are dependent on ants, the larval stages last ten to eleven months. Pupation also occurs in the nest close to the brood chambers, and the freshly eclosed adult butterfly must crawl out of the nest before it can unfold its wings and fly away.

Although all *Myrmica* species in the respective habitats of the two *Phengaris* ecotypes pick up caterpillars lying beneath the *Gentiana* plants, Thomas et al. (1989) have identified *Myrmica schencki* as the main host species in their study population of *P. 'rebeli.'* However, the caterpillars are also accepted by other *Myrmica* species, and they can complete their development in these alternative host ant nests, yet do not do so well as when raised by the "proper" *Myrmica* species. Why is that? Jeremy Thomas and coworkers (2005) summarize their extensive studies as follows: "On leaving its gentian [food plant] *M. rebeli* [*P. 'rebeli'*] secretes a simple mixture of surface hydrocarbons that weakly mimic those of *Myrmica* (Akino et al. 1999; Elmes et al. 2002). Its profile is closest to *M. schencki,* and vice versa, but it is sufficiently similar to all *Myrmica* species for the caterpillar to be quickly retrieved by the first worker to touch it (Akino et al. 1999). At this stage we doubt whether it is possible for a *Maculinea* [*Phengaris*] larva to evolve a species-specific retrieval cue, since *Myrmica* foragers will gently retrieve the larva of any congener that is artificially placed in its territory" (Thomas et al. 2005, 498).

We do not think the current evidence supports the assumption that the caterpillar's cuticular hydrocarbon profiles are the functional releasers for such adoption and tending behavior. They might be important for colony membership and cohesion, and perhaps have a modulating effect in the interactions between caterpillars and host ants. Likewise, the bioassays reported by Akino et al. (1999) do not, in our view, justify a conclusion that the hydrocarbon profiles elicit the particular brood care behavior in the ants. When larvae of different *Myrmica* species and dead fourth-instar caterpillars of *Phengaris 'rebeli'* were placed outside a laboratory nest of *M. schencki,* Akino et al. (1999) noted that there were significant

interspecific differences in the time intervals in discovering and retrieving the larvae of the different *Myrmica* species and *M. 'rebeli'* caterpillars, whereby their own larvae were usually picked up first. But ultimately all larvae were retrieved into the nest of *M. schencki*. When the authors presented glass rod dummies (1 mm in diameter and 2–3 mm long), both dummies, treated with hexane extract from either *M. schencki* larvae or *P. 'rebeli'* caterpillars, were picked up within about the same time frame. However, in contrast to the larvae, these dummies were deposited on the ants' garbage site.

The cuticular hydrocarbon patterns exhibited by *Phengaris* caterpillars (Akino et al. 1999) weakly resemble those of the host ant species, and this might have a modulating effect—for example, by lowering the response threshold in ants to the putative mimic of the nonvolatile brood pheromone presented to potential host ants. They might be the reason for a relatively high population-specific host ant fidelity. In this context, it is important to point out that Guillem et al. (2016) demonstrated that the species-specific cuticular hydrocarbon profiles of twelve *Myrmica* species, collected in four countries, remained qualitatively stable across Europe.

According to Thomas et al. (1989), *Phengaris 'rebeli'* is found in dry habitats in the Pyrenees, mainly in nests of *Myrmica schencki,* while in the Southern Alps it also occasionally occurs in nests of *M. ruginodis* and *M. scabrinodis,* and in the Northern Alps and in East Westphalia (Germany), it is found mainly with *M. sabuleti* (Meyer-Hozak 2000; Steiner et al. 2003). Similar variation in host specificity was discovered in *Phengaris alcon,* which prefers more humid habitats. Here too, the preference for particular host species varies significantly according to geographic locality and habitat. In Sweden, *P. alcon* occurs exclusively in nests of *M. rubra,* in Southern Europe only with *M. scabrinodis,* in the Netherlands with *M. ruginodis* and to a lesser degree with *M. rubra* (Gadeberg and Boomsma 1997; Als et al. 2001, 2002).

In this context, it is important to pay attention to the work of Marcin Sielezniew and colleagues (Sielezniew et al. 2012). They investigated the genetic variability of the two ecotypes of *Phengaris alcon* (*P. 'alcon'* and *P. 'rebeli'*) in Poland. They found that the two ecotypes differed significantly in their genetic variability; *P. 'rebeli'* was less polymorphic and the populations were much more differentiated than those of *P. 'alcon.'* The work suggests that all populations of *P. 'alcon'* "form a single clade but that *P. 'rebeli'* can split into either six or two clades. The former model would indicate many independent origins, especially in the mountainous area of Southern Poland. The latter, not mutually exclusive, grouping clearly reflects the use of different host ants."

The perplexing records concerning the diversity of host specificities were critically viewed by Pavel Pech and his collaborators (2007), who conducted a statistical analysis of all published records of host specificity in several *Phengaris* (*Maculinea*) species; they concluded that the "discoveries of known ant hosts increase with the number of ants surveyed in three European species (*alcon, rebeli, teleius*)" and state that there might be some local association with a particular frequently occurring *Myrmica* species, but this does not mean that a true host specialization exists. Indeed, it was long assumed that any *Myrmica* species can raise any *Phengaris* species (Malicky 1969), and according to Wardlaw et al. (1998), all *Phengaris* species can survive in laboratory nests of many *Myrmica* species, though Elmes et al. (2004) demonstrated that under food stress *Phengaris* species do better in nests of their "primary host."

In a recent comprehensive review and new analysis of all published data, Tartally et al. (2019, 2) conclude, "While all but one of the *Myrmica* species found on *Maculinea* [*Phengaris*] sites have been recorded as hosts, the most common is often disproportionately highly exploited. Host sharing and host switching are both relatively common, but there is evidence of specialization at many sites, which varies among *Maculinea* [*Phengaris*] species."

Apparently each *Phengaris* species in a population prefers one "primary host ant species" of *Myrmica,* and one or a few other *Myrmica* species might serve as secondary hosts. This may also explain the at-first-sight puzzling, but very interesting, discovery by Karsten Schönrogge and coworkers (2004), who introduced *Phengaris* 'rebeli' larvae into *Myrmica* colonies cultured in the laboratory. They found that, after adoption of the caterpillars into the ants' brood nest, 90% of the myrmecophiles survived in the first forty-eight hours in a *M. schencki* colony; however, only 45%–60% survived in nests of other *Myrmica* species. Yet, once the hurdle of the first forty-eight hours had been overcome, these same caterpillars did well in any *Myrmica* nest. The authors propose that this might be due to the closer resemblance of the initial surface hydrocarbon pattern of *P. 'rebeli'* with that of the primary host *M. schencki*. The authors also observed that about six days after the caterpillar's adoption in a *M. schencki* nest, *M. 'rebeli'* possess a more complex hydrocarbon profile that closely resembles that of the host ants, and they assume that the caterpillars are able to produce these specific hydrocarbon profiles themselves, because similar rapid adjustments were not observed when the caterpillars were placed into nests of other *Myrmica* species (Schönrogge et al. 2004; Elmes et al. 2004; Thomas et al. 2005; see also Thomas et al. 2013).

An alternative explanation could be that the host ants' specific hydrocarbon blends are transferred from host ants to the *Phengaris* caterpillars during the ants' grooming and feeding the myrmecophile larvae. In fact, as we discussed in Chapter 3, in ants, like in all insects, hydrocarbons are manufactured by the oenocytes within the body and transported into the hemolymph (blood), to the cuticle, and to the ants' postpharyngeal gland for storage and later distribution. This organ is an important source not only of hydrocarbons, but also of liquid food fed to the larvae and most likely also to the myrmecophile larvae (Hölldobler and Wilson 1990). In addition, the hydrocarbons are spread over the body of each ant as she grooms herself and her nestmates by licking motions of her tongue or lower lip. Besides having a social hygienic function, this action is also a means of homogenizing and distributing the colony's unique cuticular hydrocarbon blends, which are believed to constitute the colony's chemical signature (Lahav et al. 1999; Lenoir et al. 2001; d'Ettorre et al. 2002; Hölldobler and Wilson 2009). Conceivably, the *Phengaris* caterpillars, which are licked and groomed by the host nurse ants, will most likely also be coated with the colony's hydrocarbon blends. This process might be delayed in *P. 'rebeli'* caterpillars that were introduced to other *Myrmica* species, because their primary (preadoption) hydrocarbon pattern is somewhat closer to that of *Myrmica schencki* than to other *Myrmica* species. In fact, some of the results reported by Schönrogge et al. (2004) appear to support this hypothesis, while others do not.

Here is a summary of their puzzling results: When fourth-instar caterpillars of *P. 'rebeli'* were introduced to *Myrmica schencki, M. sabuleti,* and *M. rubra,* and after three weeks the cuticular hydrocarbon blends of the caterpillar were compared with those of the three ant species, the authors found that "the hydrocarbons of postadoption caterpillars were more similar (78%, 73%) to the ant colony profiles of the non-host species than were caterpillars reared in colonies of *M. schencki* (42% similarity)." When caterpillars taken from the *M. schencki* nest were isolated for four days, their cuticular hydrocarbon profile did not change. In contrast, the similarity of the hydrocarbon profiles of caterpillars isolated from the *M. sabuleti* and *M. rubra* colonies fell to 52% and 56%, respectively. However, "six compounds, presumably newly synthesized, were detected on the isolated caterpillar that could not have been acquired from *M. sabuleti* and *M. rubra* (nor occurred in preadoption caterpillars), five of which were found on the natural host *M. schencki*." The authors speculate that "these new compounds may relate to the high rank caterpillars attain within the hierarchy of *M. schencki* societies." Since five of these six compounds are also present on members of the *M. schencki* host colony, one wonders how these six compounds can be responsible for the caterpillars' "high rank." In addition, there appears to be a contradiction, because in a later paper (Barbero et al. 2009), the authors assert "that neither begging nor chemical mimicry explains the high rank achieved by *Phengaris 'rebeli'* within its host's social hierarchy."

In any case, in a number of studies, including a recently published extensive population-biological investigation of the host specificity and "chemical mimicry" of *Phengaris 'rebeli,'* Casacci et al. (2019b) found local variations in host specificities "that are consistent with similarities found in chemical profiles of hosts and parasites on different sites." This is, in fact, the first field study that measured the correlation between the similarities of the cuticular hydrocarbon profiles of *P. 'rebeli'* larvae and those of the host workers of *Myrmica* species. These field studies confirmed that the preadoption surface hydrocarbon profiles of *P. 'rebeli'* caterpillars from most population studies were closer to *M. schencki* when compared to those of other *Myrmica* species. However, caterpillars from one *P. 'rebeli'* population exhibited preadoption hydrocarbon profiles that resembled somewhat those of *M. lobicornis*. In this population, a large number of *P. 'rebeli'* larvae and pupae were found in *M. lobicornis* colonies. In addition, Casacci et al. (2019b) found one population where the preadoption hydrocarbon profiles of *P. 'rebeli'* were closest to *M. sabuleti* workers, which turned out to be the primary host of the caterpillars in this population. Thus, this comprehensive study confirms that host specificity can vary in different populations, and that one species of *Phengaris* can consist of several ecotypes adapted to parasitize one particular *Myrmica* species. However, this host preference is not strictly exclusive.

We have paid special attention to *Phengaris 'rebeli'* because, for this group, the most extensive data have been assembled in the literature, but David Nash, Koos Boomsma, and their collaborators studying *P. alcon* in Denmark have found similar results for this species (ecotype) (Gadeberg and Boomsma 1997; Als et al. 2001, 2002; Nash et al. 2008).

All these comparative studies of the cuticular hydrocarbon profiles of the myrmecophiles and host ant species reveal certain correlations, but experimental evidence demonstrating their function is still scarce. The mechanism underlying the acquisition of a host-specific cuticular hydrocarbon profile by *Phengaris* caterpillars after adoption is far from understood. A recent study of endosymbiotic microbes indicates that they are involved in the biosynthesis of lipids and "could potentially be involved in the *Phengaris* chemical mimicry" (Di Salvo et al. 2019; Szenteczki et al. 2019). These new studies might open up entirely new perspectives for understanding these complex, interspecific chemical interactions in ant societies.

To summarize, the previously held and passionately defended assumption that each *Phengaris* species has one specific *Myrmica* host species does not hold across wider geographic ranges. Casacci et al. (2019b) point out that different *Phengaris* species have different propensities for using multiple *Myrmica* hosts or shifting hosts. In fact, there might even be networks of hosts, and myrmecophilous parasites may switch between them in several subsequent generations because of an evolutionary arms race.

David Nash and his collaborators (Nash et al. 2008) present just such an interesting case. They studied *Phengaris alcon* in Denmark and confirmed a certain host specificity with *M. rubra*. However, in this geographic area, *P. alcon* can be found with both *M. rubra* and occasionally with the sympatric *M. ruginodis*. They discovered geographic variations in the cuticular hydrocarbon profiles of *M. rubra*, which also reflects significant genetic differentiation among *M. rubra* populations. The authors hypothesize that these variations are due to coevolutionary "arms race" between hosts and parasites. By producing aberrant hydrocarbon profiles, the host species might stifle parasitism by *P. alcon*, that is, the parasite might be more easily identified as an intruder and rejected. In contrast, according to Nash et al. (2008), *M. ruginodis* has much more panmictic populations and therefore hydrocarbon profiles among geographic populations do not vary significantly. Therefore, although the *Phengaris* caterpillars' surface hydrocarbon profile are less close to those of *M. ruginodis* than to some population groups of *M. rubra*, the match is good enough to be adopted by *M. ruginodis*. Nash et al. (2008) conclude the *M. ruginodis* "may provide an evolutionary refuge for a parasite during periods of counteradaptation [to] their preferred hosts."

This is a clever interpretation of these described facts, but many questions remain to be answered, which should not be surprising to those studying such complex parasitic symbioses.

As we indicated above, in our view it has not been demonstrated that the adoption of these lycaenid larvae into the host ant's brood chambers is due to certain similarities in the surface hydrocarbon profiles. Hydrocarbon matching might be important for colony membership and cohesion, and perhaps have a modulating effect in the interactions between caterpillars and host ants, but the key chemical stimuli provided by the myrmecophilous larvae that release adoption behavior in the host ants are most likely different compounds. We propose that these lycaenid larvae most likely produce mimics of the ants' brood phero-mones. (For a succinct discussion of these issues see also Pierce et al. 2002.)

Here are our arguments: It is a common phenomenon in myrmecology that ant larvae and pupae can be easily transferred not only between conspecific ant colonies, but also between different colonies of congeneric species within a cer-tain guild of species. Such exchange is even possible with callow workers not older than a few days. In fact, Norman Carlin and Bert Hölldobler (1986, 1987) were able to create artificial carpenter ant (*Camponotus* spp.) colonies consisting of two to four species. Such cross-fostered workers from mixed colonies, when confronted with genetic sisters from their mother colony, fiercely attacked their true siblings, but would readily adopt larvae from that or other congeneric col-onies. We made similar observations with *Formica* species and previous myrme-cologists; in particular, Adele M. Fielde (1903, 1904) and Luc Plateaux (1960a,b,c), conducted studies with such artificially mixed ant colonies. Thus, within a guild of congeneric or phylogenetically closely related species, larvae and pupae are readily adopted by colonies of other species, and even exploited by dulotic slave-raiding ant species that pilfer the brood from nests of phylogenetically closely related ant species. Once these "stolen" pupae eclose in the foreign nest, the young worker ants will slowly adopt and learn the colony-specific nest odor and will function as workers in a foreign colony. They suffer a total loss of inclusive fitness but enhance the inclusive fitness of their unrelated nestmates, and there-fore they are called "slave ants." Bert Hölldobler discovered that larvae of the myrmecophiles *Lomechusa* and *Lomechusoides* can be as easily transferred as ant larvae to colonies of several *Formica* species. This suggests that there exists a brood pheromone that is mimicked by the myrmecophile larvae. Indeed, there is good evidence for the presence of brood pheromones in ants. For example, consider the detailed experimental studies by John Walsh and Walter Tschinkel (1974), who demonstrated that, in the fire ants (*Solenopsis invicta*), a nonvolatile contact brood pheromone is distributed over the whole cuticle of the larvae and pupae. Although it was not possible to isolate the pheromones, the brood could be deprived of their attractiveness by extraction. The authors concluded, "There exists substantial evidence for a brood pheromone. The retrieval of skin and larval body extracts on blotter, the persistence of the signal for such long periods after death (72 hours), despite disfigurement of larval cuticle and the ability of organic solvents to destroy the signal without visibly altering the cuticle are compelling evidence for a pheromone" (Walsh and Tschinkel 1974, 702). Similarly, Murray V. Brian (1975) reports results from *Myrmica rubra* and *M. scabrinodis.* He was able to extract a substance from the *Myrmica* larvae that was "insoluble in hexane, water, methanol, and 70 percent ethanol, but soluble in acetone, ether and chloroform," and he suggests that the compound or compounds function as brood pheromones. Finally, in a very straightforward behavioral analysis of the adoption of lycaenid social parasites into their host ants' brood chambers, Stephen F. Henning (1983b) was able to demonstrate a putative mimic of the ant brood pheromone in the South African lycaenid *Aloeides dentatis,* whose final-instar larvae live in nests of the ant *Lepisiota (Acantholepis) capensis,* and the lycaenid *Lepidochrysops ignota,* the larvae of which live in the brood nest of the ant *Camponotus niveosetosus.* Henning employed the solvent dichloromethane to extract compounds from the host ant larvae, as well as from the larvae of the myrmecophiles, and applied the extracts to "corn crushed small enough for the ants to pick up." Corn "grits" that were "bathed" in larval extracts were significantly better adopted into the brood nests than were grits contaminated only with the solvent (controls), with no difference between the larval extracts of host ants or lycaenid myrmecophiles. Henning tested the larval extract of a third lycaenid species, *Euchrysops dolorosa,* which is not dependent on ants. *Lepisiota capensis* workers ignored grits bathed in the larval extract and so did *Camponotus niveosetosus. Camponotus maculatus* exhibited initially some "interest," inspecting the grits with their antennae, but eventually carried them to the refuse pile.

The very detailed analysis of the brood pheromone in *Solenopsis invicta* by Walsh and Tschinkel (1974) clearly demonstrates that this pheromone is likely nonvolatile and almost impossible to extract intact. Hölldobler though (1967, and unpublished data), using acetone (dimethyl ketone), was able to extract adopted as member of a host society it mimics adult ant acoustic (particularly queens) to advance its seniority toward the highest attainable position in the colony hierarchy" (Barbero et al. 2009, 785).

The authors also state that in previous experiments by Akino et al. (1999), using inert dummies coated with extracts of cuticular hydrocarbons taken from ant larvae or *Phengaris 'rebeli'* caterpillars, respectively, the former were always retrieved in preference to the latter. However, in a rescue situation, living *P. rebeli* larvae are always picked up first by the host ant workers and carried to safety (see Thomas et al. 1998; Thomas et al. 2005), and in other emergency situations, *P. 'rebeli'* caterpillars are also treated preferentially by the host ant workers. Barbero et al. (2009) hypothesize that this is due to the sounds produced by *Phengaris* larvae and pupae. Most remarkably, according to the interpretation by the authors, the *M. schencki* queens seem to mistake the pupae of *P. rebeli* for those produced by competing queens and react with fierce rivalry and hostility.

Although we regard these findings as intriguing, we think it would be desirable to relate these data gathered in a very artificial setting to the natural social surroundings of the host ants. For example, the authors describe using a laboratory culture consisting of twenty-five workers and four queens and some brood, to which they introduced four *Phengaris* pupae, and they observed "each of the four test pupae were violently attacked" by the queens. The authors interpret this behavior as the queens perceiving the sound the pupae produce as queen sounds, and therefore, they attack a putative rival. We wonder why the ant queens do not show hostility to their nestmate queens, whose stridulation "sounds" more closely resemble their own. Comparing the published sonograms, we cannot recognize a striking similarity between the stridulation sounds of *Myrmica* queens and those of *Phengaris* pupae and larvae, except that the frequencies of all three are lower than those of the ant workers. Nevertheless, the queen's sound frequency differs markedly from that of *Phengaris,* and the pulse length and pulse repetition rate (two parameters that are important in most insects using sound or substrate-borne vibrations for communication) are strikingly dissimilar. Barbero et al. (2009) played back the recorded stridulation sounds with a "miniature speaker" and recorded the behavioral response of ten workers that were placed in a box measuring 7 cm by 7 cm by 5 cm. In addition, a small piece of wet sponge, on which three ant larvae of the workers' colony were placed, was positioned in the middle of the box. According to the author, the ants aggregated around the speaker and exhibited increased antennation when recorded stridulation sounds were played, and the "sounds of workers and queens elicited similar amounts of antennation." Queen sounds induced significantly higher occurrence of what the authors called "on-guard attendance" than did worker sounds. Similarly, the sounds of *Phengaris* pupae played back to the worker group elicited six times more "royal on-guard attendance" responses than the sounds of adult workers.

Neither in the publication nor in the "supporting online material" has it been stated whether these behavioral tests were made "double-blind," or video-recorded and subsequently "double-blind" evaluated. The behavioral response patterns described by the authors—antennation, standing still with raised head, gaping mandibles—is usually observed in several ant species that perceive substrate-borne vibrational signals caused by stridulation or knocking of nestmates (see for example Markl and Fuchs 1972; Fuchs 1976a,b). We assume that these vibrations serve as an "attention" signal that lowers the response threshold to chemical releaser signals, such as alarm pheromones. Such signals are called modulatory signals and are typically employed in alarm or recruitment communication (see Markl and Fuchs 1972; Markl 1985; Markl and Hölldobler 1978; Baroni Urbani et al. 1988; Hölldobler and Wilson 1990). As the queen- or the *Phengaris*-produced sounds are more energetic, they may elicit greater response effect in the workers. Finally, we want to point out, Barbero played back airborne sounds. Although Autrum (1936) suggested that *Myrmica rubra* workers could theoretically respond to sound caused by wind stimulation, he found no evidence that they do. To our knowledge, no verified evidence exists that ants perceive airborne sounds, though, as we stated above, they are extremely sensitive to substrate-borne vibrations (Autrum 1936; Markl 1967, 1970, 1973, 1985; Masters et al. 1983; Roces et al. 1993; for review see Hunt and Richard 2013).

4-18　The lycaenid butterfly *Phengaris arion*. (Courtesy of Marcin Sielezniew).

no secretions and may not have a significant function during the adoption process and inside the ants' brood nest.

Incidentally, both the cuckoo *Phengaris* and the predatory *Phengaris* species have no tentacular organs. It has been suggested that, in the course of the evolution of an increasingly tight association with ants inside the host's nests, these organs became dispensable. As they serve for alarming attendant ants outside the nest, there is no need for that inside the host ant nest. Indeed, several lycaenid species lost their tentacular organs as a consequential adaption to specialized ecological niches such as endophytic species and species living as kleptoparasites or predators inside ant nests (Fiedler and Maschwitz 1987).

Anyway, during the cumbersome adoption process of *P. arion* and *P. teleius*, *Myrmica* workers frequently lick and imbibe sweet secretions from the nectar organ, which might have an appeasement function. In addition, scattered across the surface of the caterpillar are perforated pore cupolas, which apparently secrete substances attractive to ant workers. Jeremy Thomas (2002) suggested that

4-19 *Phengaris arion* female depositing eggs on a *Thymus* plant. The lower picture shows an egg of *P. arion* on *Thymus* flowers. (Courtesy of Marcin Sielezniew).

4-20 *Phengaris arion* third-instar caterpillar dropped from the food plant to the ground. (Courtesy of Peter Eeles). Once it has been discovered by a *Myrmica* worker, in this case *M. rugulosa,* the ant is enticed to lick on the dorsal nectar organ, though it is not very productive. Eventually the ant will carry the caterpillar into the nest. (Courtesy of Marcin Sielezniew).

4-21   Inside the ants' nest, caterpillars are tolerated by ants as they feast on ant brood. In this case *Myrmica lobicornis* (upper picture) and *Myrmica constricta* (lower picture) hosts are pictured. (Courtesy of Marcin Sielezniew).

during the licking process the *P. arion* larvae might acquire some of the ants' colony "gestalt odor," which facilitates the adoption process. Konrad Fiedler (1990) describes a similar adoption process for *P. teleius,* the other predatory species.

Once retrieved into the nest, the caterpillar settles in an outer chamber near the brood chambers, where it stays most of the time during the next ten months or so. For feeding, it moves to the brood chambers, where the larger ant larvae are housed, and preys extensively on the ants' brood almost unopposed by the ants (Figure 4-21). It then returns to its resting place, where it may remain for up to ten days before it again feasts on the ant brood. Jeremy Thomas and his coauthors (Thomas et al. 2005) make an interesting observation concerning the ergonomics of the *Phengaris* predation: "By eating the largest available prey, they [*Phengaris* larvae] initially kill only those [ant] larvae that will soon pupate and be lost as food. At the same time, the fixed number of larvae in the second (over-wintering) cohort of *Myrmica* brood is let to grow larger before it is killed. When this occurs, large individuals are again selected, leaving small ones to grow on." For further ergonomic calculations, see Thomas and Wardlaw (1992).

Also, in this predatory *Phengaris* group the host specificity is somewhat controversial. According to Thomas et al. (2005), *P. arion* caterpillars will be adopted "with equal facility by any of up to five *Myrmica* species"; however, "the mean survival of the caterpillars was 6.4 times higher in colonies of *M. sabuleti* than with any other *Myrmica* species." Likewise, *P. teleius* will survive 2.9 times better with *M. scabrinodis* than with any other *Myrmica* species that adopted it in the field. *Myrmica scabrinodis,* which appears to be the primary host of *P. teleius,* would readily kill *P. arion*. Whether this specificity is due to the similarity of hydrocarbon profiles or other chemical cues, or whether acoustical (vibration) signals used by the host ants are imitated by the caterpillar, is not yet clear. Thomas et al. (2005) write, "The churring sounds produced by all *Maculinea* [*Phengaris*] species differ greatly from those of typical lycaenids and mimic the adult stridulations of the genus *Myrmica,* but not any particular *Myrmica* species (De Vries et al. 1993)," so the vibrational signals would hardly be the reason for the fine-tuning

to particular host species. However, more recent studies (Sala et al. 2014; Riva et al. 2017; Schönrogge 2017) propose that lycaenid sounds are more specific and may even have host ant–specific parameters that may play a role, perhaps in combination with chemical signals, in multimodal communication between myrmecophiles and their hosts (Casacci et al. 2019a). In our view, no hard evidence is yet available.

As we already discussed, the claim of host specificity of *Phengaris* species is not accepted by all scholars, yet, from our interpretation of the literature, there exists substantial direct and indirect evidence that suggests *Phengaris* species exhibit a population-based host preference; that is, optimal development and survival is usually possible in nests of one particular *Myrmica* species. This preferred host species might vary in different populations. Whether this is due to a certain phenotypic plasticity, or genetic population-level variation (see also the population genetical studies by Sielezniew and Rutkowski 2012), or to other ecological and behavioral parameters is currently not known. We doubt, however, based on the current evidence, that the cuticular hydrocarbon blends or the stridulation stimuli play a significant role in the adoption process, and, if at all, they may only marginally affect certain host species preferences.

Miletine Predators, Tripartite Symbiosis, and Indirect Parasitism

As we pointed out previously, the more common and probably ancestral type of association of lycaenids with ants is a facultative mutualistic relationship, though we have also insinuated that the mutualistic association is rarely a balanced one that benefits each partner equally. Instead, these relationships are prone to evolve into full parasitism. This is strikingly exhibited in *Phengaris (Maculinea)* species, which were covered in the above disquisitions. These species most likely evolved from herbivorous ancestors to become carnivorous, preying on the ants' brood and in some species parasitizing the social food flow from nurse ants to immature ants. These highly socially adapted myrmecophilous parasites are nursed and tended by the ants as much as or even more than the ant brood. Most likely they achieve this by imitating the essential key stimuli that elicit not only adoption and protection in the ants, but also false brood-caring behavior.

However, there is at least one other type of predatory lycaenid whose caterpillars live inside the brood chambers of their host ants and prey ferociously on the ants' brood. These lycaenids apparently do not copy social signals, and though the ants attack them continuously, it is to no avail. We are talking about the genus *Liphyra* (Miletinae, Liphyrini), particularly the species *L. brassolis,* one of the largest species of lycaenid butterflies, which is also known as the moth butterfly. The genus occurs in India, Southeast Asia, Indonesia, the Philippines, New Guinea, Solomon Islands, and North Australia (Eastwood et al. 2010). This is the geographic range of the weaver ant species *Oecophylla smaragdina,* which indeed is the most common host species of *L. brassolis.* The aberrant, slug-like butterfly larva has been aptly characterized by the late Densey Clyne (2011, 1) as follows: "Oval in shape, it is covered by a kind of carapace, a tough integument that curves protectively around and partly under the body. Flattened on top with a rim all around, the golden-brown carapace looks for all the world like a well-cooked apple pie. It is totally impervious to attacks by the ants" (Figures 4-22; 4-23). This is of prime importance, because the *Liphyra* caterpillar spends most of its life inside the leaf-tent nests that constitute the ants' nurseries, and feeds on the ant larvae and pupae (Johnson and Valentine 1986).

Why is the carapace of the *Liphyra* caterpillar, which lacks integumental sclerotized cuticular plates, such an effective armor against attacks by *Oecophylla* workers endowed with sharp, pointed mandibles? This question was resolved by Steen Dupont when he was a visiting student in Naomi Pierce's laboratory (Dupont et al. 2016). They conducted histological and scanning electron microscopic investigations of the caterpillar's integument and found several structural novelties that impart "protection from ant attacks whilst maintaining the flexibility necessary to walk." We found especially interesting that there are distinct differences in the integumental structures between early- and late-instar caterpillars of *L. brassolis.* In the late instars, the authors found the cuticle covered "with lanceolate setae, which act as endocuticular struts, and overlapping scale-like sockets, which form a hard, flexible integument." Remarkably, these

4-22 Larva of the miletine *Liphyra* sp. in the brood nest of the weaver ant *Oecophylla smaragdina*. (Courtesy of Taku Shimada).

structures are not yet well developed in early-instar caterpillars, which are instead endowed with many putative secretory pores spread over the integumental surface. They appear to be homologous to the pore copula organs present in larvae of other lycaenid species. The putative secretions may attenuate aggressive behavior in the host ants. The authors hypothesize that "the importance of these pores presumably wanes as structural (setal) cuticular defenses are reinforced in later instars." The *Liphyra* caterpillars pupate inside the last instar larval skin. The puparium retains the protective cuticular structures. It is firmly attached with silk inside the ants' nest. Over several days its shape changes. "The carapace swells up into a dome." It now resembles "a nicely browned loaf of bread rising in the oven" (Figure 4-24) (Clyne 2011). The adult

4-23   The upper picture shows the underside of the *Liphyra* caterpillar with its appendages. The caterpillar is feeding on a prepupa of its host ants. The lower picture is a close-up of the underside of the *Liphyra* caterpillar, with the head, mouthparts, and forelegs clearly visible. (Courtesy of Taku Shimada).

butterfly emerges from the pupal case inside the ant nest. The soft-bodied imago would be easy prey for the ants, but it is endowed with remarkably effective defensive tools. We cannot do better than citing again Densey Clyne (2011, 3), who might be the first to photographically document the pupal emergence of *Liphyra brassolis* inside the weaver ant nests. "Its abdomen is clothed with a mass of wiry black scales, and its wings and legs with slippery white scales. Ants that attack its wings immediately toboggan off. Ants attempting to bite its body find their mandibles so entangled they lose interest in the butterfly." And thus, the butterfly will slowly walk out of the ants' nest, and while extending its wings, it sheds the scales and hairs and, in the process, one presumes, also sheds the remaining "entangled" ants (Figure 4-24).

The moth-like butterfly appears to be unable to feed because its proboscis is completely atrophied. Apparently, it depends entirely on fat and protein reserves acquired in the larval stage and stored within the body (Hoskins 2015). After mating, females are reported to lay their eggs on the undersides of branches of trees with ant nests. Like *Liphyra brassolis,* all other miletine species are aphytophagous, meaning they do not feed on plants.

With several notable exceptions—such as the *Liphyra* caterpillars that exclusively feed on ant brood; their wholly African sister genus, *Euliphyra,* whose larvae are fed via trophallaxis by workers of their host ants, *Oecophylla longinoda;* as well as the remarkable Southern African genus *Thestor,* whose twenty-seven described species are thought to be almost exclusively parasitic on ants, particularly *Anoplolepis custodiens*—the majority of miletine species prey on Hemiptera (aphids, coccids, membracids, and psyllids) (Kaliszewska et al. 2015). For example, in the miletine *Logania malayica,* the caterpillars feed on aphids and the adult butterflies imbibe honeydew produced by the aphids. Ulrich Maschwitz and his collaborators (1988) studied the natural history of these "sternorrhynchophagous miletines" (feeding on species of the suborder Sternorrhyncha) and discovered a remarkable diversity (for further extended studies of the ecological and behavioral diversity of miletine species preying on hemipterans, see Pierce 1995;

4-24   The pupa of *Liphyra brassolis* surrounded by workers of *Oecophylla smaragdina*.
(Auscae / Universal Images Group via Getty Images). The lower picture shows the adult butterfly
of *Liphyra*. (Courtesy of Taku Shimada).

Lohman and Samarita 2009; Kaliszewska et al. 2015). The Maschwitz study, for example, revealed that *Miletus biggsii* feeds on several species of aphids as well as on coccids. All these honeydew-producing hemipterans are attended by ants, most often species of *Dolichoderus*. Circumstantial evidence suggests that butterfly females of *Miletus biggsii* and *Logania malayica* use the ants as cue for locating oviposition sites, a phenomenon also found in other lycaenid species (e.g., Pierce and Elgar 1985).

Another particularly interesting case is the miletine species *Allotinus apries,* wherein the young caterpillars feed on coccids and the older caterpillars are carried by ants of the myrmicine *Myrmicaria lutea* into the ants' nest, where the caterpillars probably prey on the ants' brood. Indeed, pupae of *Allotinus apries* were found inside the carton nest of *M. lutea,* and young instar lycaenid larvae were collected from inside the ant nest, but it was not possible to determine whether these were *Allotinus* larvae. However, this conclusion is suggestive, because the authors also demonstrated in the laboratory that the lycaenid larvae taken from *M. lutea* nests readily fed on *Myrmicaria* brood, and pupae of *Allotinus apries* have been found only in *M. lutea* nests. As we discussed previously, the larvae of the Miletinae all lack a dorsal nectar organ and tentacular organs, but they are richly endowed with pore cupola organs. The latter are supposed to produce secretions that mitigate aggressive behavior in the ants, and indeed, the predatory miletine caterpillars are largely ignored by the ants attending the trophobiotic hemipterans. Nothing is known, however, about how the advanced instars of *Allotinus apries* caterpillars elicit adoption behavior in their host ant, *Myrmicaria lutea.*

A final example we would like to mention is that of *Allotinus subviolaceus,* which was found on the creeping plant *Uncaria* sp. (Rubiaceae). Membracid aggregations on this plant were tended by the formicine ant *Anoplolepis longipes.* Lycaenid eggs and different instars of the caterpillars were found near the membracids, and the caterpillar preyed only on the young membracid larvae.

"Second instars and older caterpillars were observed grasping with their thoracic forelegs first and second instar membracids and feeding on them" (Maschwitz et al. 1988).

Obviously, adults and larvae of many miletine species are somehow associated with ants, although most species feed on Hemiptera. Maschwitz et al. (1988) call this "indirect parasitism" on ants. The miletine caterpillars, by feeding the ant-attended aphids and coccids that provide nutritious honeydew to the ants, negatively affect this trophobiotic ant-hemipteran association. A second mode of indirect parasitism, or parasitism of the tripartite symbiosis (myrmecophyte, ants, aphids, and coccids), also first noticed by Maschwitz and his collaborators, is that of lycaenid larvae parasitizing the symbiotic association of ants with the myrmecophytic plants (Maschwitz et al. 1984). A striking case is the Paleotropical plant genus *Macaranga* (Euphorbiaceae), many species of which live in symbiosis with the myrmicine ant genus *Crematogaster*. The hollow internodes of the tree branches provide ideal nesting sites for the ants, and in addition, the plants produce food bodies in the leaf stipules, which are harvested by the ants. In return, the ants protect the plants from herbivores, ferociously attacking any foreign intruder that attempts to feed on the plants' leaves. This defense, however, fails with the phytophagous lycaenid *Arhopala* (Theclinae, Arhopalini). The caterpillars of several species of this genus overcome the ants' aggression by deploying the secretions of their pore copula, the nectar organ, and possibly also the tentacular organs. This enables the *Arhopala* caterpillars to feed unmolested on *Macaranga* plants and, in addition, to attract the ants, which protect them against predators and parasitoids.

But there are exceptions. Usun Shimizu-kaya and colleagues (2013) demonstrated that *Arhopala zylda,* contrary to other *Arhopala* species, have no firm association with ant species. They are neither attended nor attacked by the ants on their host plants. Even when exposed to other ant species, *A. zylda* was ignored by the ants. Shimizu-kaya (2014) also demonstrated that caterpillars of this lycaenid preferably feed on the food bodies of the *Macaranga* trees, even though these were frequented by the symbiotic ants. The author suggests that particularly the younger caterpillar instars preferably feed on the food bodies. How the parasitic caterpillars evade ant aggression is unknown.

A possible answer to this question was suggested by Yoko Inui and colleagues (2015). They studied three *Arhopala* species, *A. dajagaka, A. amphimuta,* and *A. zylda,* which feed on *Macaranga rufescens, M. trachyphylla,* and *M. beccariana,* respectively (Okubo et al. 2009). As already noted by Maschwitz et al. (1984), the symbiotic partnerships between *Macaranga* myrmecophiles and their ant protectors, mainly the myrmicine *Crematogaster* species (often referred to by their subgenus name, *Decacrema*), are highly species-specific. Of the five *Arhopala* species known to feed on *Macaranga,* each fed on one or two closely related species (Inui et al. 2015).

The research by Inui and colleagues confirmed previous observations that ants on *Macaranga* plants (shorthand called plant-ants) behave quite differently to various *Arhopala* species. *Arhopala zylda* caterpillars were treated indifferently by the ants, independently of whether they were on their host *Macaranga* species or placed on another, non-host *Macaranga* species. In contrast, *A. amphimuta* and *A. dajagaka* were much more frequently attended and less frequently attacked by the plant-ants on their respective host *Macaranga* species than when transplanted to non-host species, though the difference in *A. dajagaka* was not significant. The authors hypothesize that the *Arhopala* larvae employ chemical camouflage, perhaps even mimic the cuticular hydrocarbon blends of the plant-ants on their *Macaranga* trees. The chemical analyses did not reveal clear-cut results: "*A. dajagaka* matched well the host plant-ants, *A. amphimuta* did not match, and unexpectedly, *A. zylda* lacked hydrocarbons," or rather, only small amounts of hydrocarbons were detected. The behavioral tests with Teflon rod dummies coated with the respective hydrocarbon extracts indicated that the specific hydrocarbon blends might play a role in these symbiotic interactions. Rods contaminated with extract of *A. dajagaka* elicited some attraction in host plant-ants and also in plant-ants from non-host *Macaranga* species, whereas dummies coated with extract of *A. amphimuta* were "often attacked by all plant-ants," but significantly less so by ants from their own host plant. Dummies with extract from *A. zylda* "were ignored by all plant-ants" (Inui et al. 2015).

We are fully aware that these experiments are demanding and often incomplete, because it is difficult to collect enough specimens. We assume that this is one of the reasons that additional studies could not be carried out. Especially with *A. dajagaka,* it would be desirable to know whether there is a host-colony difference with the same host plant species, or whether the hydrocarbon blends in larvae of *A. dajagaka* and *A. amphimuta* change when the larvae are exposed to non-host plant-ants for longer periods of time. Almost in passing, the authors report the interesting observation that *A. dajagaka* larvae frequently evert the tentacular organs when attacked by ants, "and the behavior of the ants were converted to attending within a few minutes." They did not observe this in the other two *Arhopala* species. In fact, they observed *A. dajagaka* caterpillars produce larger droplets of nectar from their nectar organs, and they reasoned "the abundant nectar may reward the plant-ants for attending rather than attacking *A. dajagaka* larvae, at least in the short term." Would it not be possible that *A. dajagaka* caterpillars, having more "intimate" interactions with the plant-ants, are more likely to acquire some of the ants' cuticular hydrocarbons than the other two species that have fewer close interactions with the ants? In any case, this is a most fascinating system, from which many new insights about symbiotic species interactions will arise.

Shouhei Ueda and colleagues (2012) conducted a phylogenetic study of *Arhopala* lycaenid butterflies that feed on *Macaranga* plants inhabited by *Crematogaster* ants that attend coccids. They discovered that *Macaranga* and *Crematogaster* species have been coevolving and diversifying over the past sixteen to twenty million years, and that the tripartite symbiosis (*Macaranga, Crematogaster,* coccids) was formed nine to seven million years ago, when the coccids became involved in the plant-ant mutualism. The parasitism of this tripartite symbiosis (or indirect parasitism of ants) by lycaenids dates back about two million years (Ueda et al. 2012).

Nevertheless, during studies in the field, researchers repeatedly collected caterpillars that later turned out to be parasitized by parasitoids (Fiedler et al. 1992). Konrad Fiedler and colleagues studied seventeen lycaenid species, representing thirteen genera and two subfamilies, of which field-collected caterpillars were brought into the laboratory together with their host ants. In many of them, parasitoid larvae emerged from the lycaenid caterpillars to pupate. In contrast to most other parasitoids, caterpillars infested by the braconid genus *Apanteles,* which we discussed above, remain attractive to the ants, and the pore cupola organs and dorsal nectar organs remain functional several days after the parasitoid larvae have emerged from their host caterpillar. As we discussed above, the lycaenid butterfly *Anthene emolus* is an obligatory myrmecophile usually associated with the weaver ant *Oecophylla smaragdina.* Caterpillars of *A. emolus* secrete droplets from the dorsal nectary organ at particularly high rates. They are frequently parasitized by solitary parasitoid species of the *Apanteles ater* group. Konrad Fiedler and his colleagues (1992) observed in the field and laboratory fourteen parasitized caterpillars. They report,

> All caterpillars remained fully attractive to their specific host ant even after the parasitoids had emerged. One or two *O. smaragdina* workers constantly attended and antennated each larval "carcass." These responded with eversions of the DNO (dorsal nectar organ), and droplets of secretions appeared at the nectar organ. *Oecophylla smaragdina* ants eagerly harvested every single droplet. The ability to deliver DNO secretions persisted in *A. emolus* caterpillars up to 3 days after the braconid larva had emerged. Attractiveness of the caterpillar carcasses to *Oecophylla* ants persisted 4–5 days, and the adult braconids eclosed after a pupal period of 5–6 days. (Fiedler et al. 1992, 160) (Figure 4-25)

Similar observations were made with the above-mentioned Palearctic facultative myrmecophiles *Polyommatus bellargus* and *P. icarus.* They are frequently parasitized by *Apanteles* species; in one case fourteen wasp larvae emerged from a

4-25 *Oecophylla smaragdina* worker ants drink secretions from the dorsal nectar organ of a parasitized lycaenid caterpillar *Anthene emolus*. The white cocoon of the wasp *Apanteles* is attached to the ventral side of the *Anthene* caterpillar, which served as the host of the parasitoid. (Courtesy of Konrad Fiedler).

4-27   The parasitoid wasp female of *Ichneumon eumerus* that invades *Myrmica* nests that harbor *Phengaris alcon*. (Courtesy of Marcin Sielezniew). The lower picture shows two *P. alcon* pupae. The lower individual is parasitized by *I. eumerus,* which develops inside the *P. alcon* pupa. (Courtesy of Izabela Dziekańska).

the intruding wasp, which is able to enter the ants' brood chambers with little or no resistance rendered by the ants. A very similar technique was previously discovered by Fred Regnier and Edward Wilson (1971) in the slave-raiding ants *Formica subintegra* invading nests of the so-called slave species *Formica subsericea.* The raider ants release from their hypertrophied Dufour's gland hyperbolized alarm pheromones that cause complete confusion and undirected "panic" reactions in the assaulted *F. subsericea* colony. The raider ants exploit this confusion and, almost unresisted, invade the foreign nest and steal the immatures of the slave-ant species.

The work by Jeremy Thomas, Graham Elmes, and colleagues (2002) suggests that *I. eumerus* wasps employ similar "propaganda allomones," which enable them to invade the brood chambers of the *Myrmica schencki* nests. Although the anatomical origin of these "propaganda allomones" has not been identified, the authors indirectly isolated and subsequently synthetized three longer chained alcohols ($Z\text{-}9\text{-}C_{20}$-ol, $Z\text{-}9\text{-}C_{22}$-ol, $Z\text{-}9\text{-}C_{24}$-ol) and the corresponding three aldehydes. In bioassays, the authors demonstrated that $Z\text{-}9\text{-}C_{20}$-ol attracted *M. schencki* workers, $Z\text{-}9\text{-}C_{24}$-ol amplified aggression, and $Z\text{-}9\text{-}C_{22}$-ol and $Z\text{-}9\text{-}C_{24}$-ol strongly repelled the ants. "Mixed together, these chemicals drew ants to the parasitoid, where, having become aroused to a state of high aggression, they were quickly repelled. This resulted in three to eight times more attacks being made on kin ants than on *I. eumerus,* the stimulator of the aggression" (Thomas et al. 2002, 505). The parasitoid wasps exploit this utter confusion among the ants and intrude the brood chambers of the ants, where they find lycaenid host larvae for oviposition.

This is, indeed, a fascinating story, although several key questions remain open. The adult female wasps seek out *Myrmica schencki* nests that house *Phengaris 'rebeli'* larvae, and it has been suggested that they are able to detect such nests by smell. Although the laboratory experiments are based on only two *Ichneumon eumerus* females, the data support this conclusion (Thomas and Elmes 1993). The authors also propose that the recognition of the most suitable host species with lycaenid host larvae is probably based on hardwired olfactory cognition, because they could not find any evidence for learning, and they suggest that the species-specific "cocktails" of mandibular gland secretions in the host ants (Cammaerts et al. 1982) might serve as recognition cues. In fact, the mandibular gland secretions serve as alarm pheromones; they are quite volatile and possibly unsuited for such specific olfactory recognition. More likely candidates are the cuticular hydrocarbon blends, which in the genus *Myrmica* exhibit a high degree of species specificity (Elmes et al. 2002; Guillem et al. 2016). However, the previously assumed extremely high host ant specificity in *P. alcon,* respectively *P. 'rebeli,'* cannot be supported anymore. Instead, there seem to exist different ecotypes of this lycaenid species that prefer one or another *Myrmica* species as host in particular habitats and geographic zones. It is hard to understand how the *Ichneumon eumerus* populations evolve parallel habitat-specific innate preferences for specific *Myrmica* species that happen to be preferred by the lycaenid larvae in this particular ecosystem. One possibility could be that the young adult wasp that emerges from the pupal case inside the *Myrmica* nest gets imprinted on the species-specific hydrocarbon blends of the *Myrmica* species and *Phengaris alcon,* and when ready to search for host larvae (about ten days after eclosion [Thomas and Elmes 1993]), they will be primed to search first for the *Myrmica* species they became familiar with during their own development. In this context, a more recent discovery is of interest. Natalia Timuş and colleagues (2013) reported that several *Phengaris alcon* pupae were collected in nests of *Myrmica scabrinodis,* and eight *Ichneumon balteatus* wasps emerged from these pupae. This was a very surprising find, because this *Ichneumon* species was known as a parasitoid of the moths *Melitaea cinxia* and *Calliteara pudibunda.* Although this is only one collection record, nevertheless it indicates that the host-finding system in *Ichneumon* species is more plastic than previously postulated.

Another unresolved question remains: How do the wasps preordain the presence of lycaenid larvae in the *M. schencki* nest? According to Thomas and Elmes (1993), the wasps preferably enter *M. schencki* nests that house *P. 'rebeli'* larvae. Yet, if the guest larvae and host ants' larvae smell alike, as has been claimed, how do the wasps distinguish volatile odorants of ant larvae and myrmecophiles? No experimental evidence for this kind of olfactory discrimination or identification is yet available. Furthermore, it would also be interesting to know in which glandular organ *I. eumerus* females produce this remarkable cocktail of alcohols and aldehydes of very low volatility. Equally interesting is the sensory mechanism underlying the ants' confusion. These "propaganda allomones" are not known from *Myrmica* ants. If they would elicit just a repellent reaction, this would be plausible; however, they have a behavior-modulating effect, misdirecting the ants' aggression. As an alternative explanation, we suggest that these repelling allomones only mask the colony recognition cues, and thereby cause *Myrmica* workers to attack nestmates. These are some of the questions that might be answered in the future.

Ant-associated lycaenid caterpillars stand out among myrmecophiles because of the diversity of techniques they use to appease their ant hosts and even persuade them to adopt and feed them. They can be found in nearly every niche created by an ant colony, including in brood chambers at the heart of the nest and in the ants' stockyards on distant food plants. In Chapter 5, we focus on one of these hot spots of myrmecophile diversity, the foraging trails. Along ant trails, one can also find questing roaches and flies that wait to be carried belowground, a collection of thieves that shake down returning ant foragers, and the diverse menageries of nomadic army ants.

5 Foraging Paths and Refuse Sites

THE FORAGING PATHS and trails of ant colonies not only provide recruitment signals and orientation cues for the ants that create them, they also inadvertently create numerous ecological niches within which diverse symbionts specialize. In the following sections we consider several studies of myrmecophiles that live as predators and kleptoparasites—organisms that steal food from another organism—along the foraging or emigration trails of ants.

Ant-Mugging Flies on the Ants' Trails

The dipteran family Milichiidae comprises a group of small flies, several species of which are known to be associated with ants, usually as larvae that develop in ant nests (Donisthorpe 1927; Moser and Neff 1971; Waller 1980; Peeters et al. 1994). The milichiid larvae typically live as commensals, feeding on debris and rotting nest material. Some other milichiid species are kleptoparasites on other arthropods, including ants (Sivinski et al. 1999; Brake 1999; Swann 2016). An especially interesting case of kleptoparasitism has been reported and marvelously documented in photographs by Alex Wild and Irina Brake (2009). They encountered several individuals of the milichiid species *Milichia patrizii* on trails of the myrmicine ant *Crematogaster castanea tricolor,* located on acacia trees "along the edge of a clearing in a coastal forest in the St Lucia Estuary, KwaZulu-Natal, South Africa." The flies individually patrol the ants' trail, where each invades the ants' foraging formation and attempts to isolate a single ant. The fly uses its cup-shaped basal antennal flagellomeres to grasp the terminal antennomeres of the smaller *Crematogaster* worker, which responds by standing still and crouching down. This allows the fly to move its proboscis to the ant's mouthparts, especially the labium (lower lip), to stimulate the ant's regurgitation response (Figure 5-1). This interaction takes about twenty-three seconds, "with the food exchange itself occurring in the final ten seconds." The authors note that 80% to 90% of these kleptoparasitic attempts were unsuccessful. On the same acacia tree branches where *Crematogaster* commute, other ant species such as *Tetraponera* spp., *Cataulacus brevisetosus,* and *Camponotus troglodytes* (the last appears to mimic *Crematogaster*) are also present. *Milichia patrizii* flies never attempt to attack workers of these species. Similar kleptoparasitic behavior is suspected in other *Milichia* species (*M. dectes, M. proectes,* and *M. prosaetes*) (Farquharson 1918, cited in Wild and Brake 2009), but no details have been reported.

Another fascinating but little-investigated highwayman behavior has been reported from the mosquito genus *Malaya* (formerly *Harpagomyia;* Culicinae, Sabethini) first observed by Jacobson (1909, 1911) in Java and subsequently confirmed by Farquharson (1918) in tropical Africa. The larvae have been found in water-filled tree holes near the nests of the myrmicine ant genus *Crematogaster,* which usually build their carton nests on or in tree trunks. The adult flies were observed patrolling the tree trunk along which the foraging *Crematogaster* workers return from the honeydew sources, carrying back to their nests in their crops the nutritious excrement of hemipteran trophobionts. The mosquito confronts the forager head on, and the ant usually responds by aggressively gaping its mandibles, which allows the mosquito to poke with its specialized proboscis at the ant's mouth parts (especially the labium). This stimulation elicits regurgitation of crop contents in fully laden foragers, which contents are readily imbibed by the myrmecophilous *Malaya* species. The proboscis of these mosquitoes is highly specialized, and when not in use it is folded backward under the body. It appears that in these species adult males and females sustain themselves exclusively as kleptoparasites of ants, and Jacobson was able to prove

5-1   The milichiid fly *Milichia patrizii* on the trail of the myrmicine ant *Crematogaster castanea tricolor*. The lower picture shows how the fly elicits regurgitation of food in the *Crematogaster* ant. (Courtesy of Alex Wild / alexanderwild.com).

that the mosquitoes really obtained food from the ants because he demonstrated that, when dyed honey was fed to the ants, this food was rapidly distributed to the mosquitoes (Downes 1958).

Bengalia: Brood and Booty Snatcher Flies

"*Bengalia* Robineau-Desvoidy, 1830 is mainly an Afrotropical and Oriental genus of large yellowish or brownish flies but has recently also been discovered in Australia (Farrow & Dear 1978). It is currently classified in a separate tribe Bengaliini (alongside Auchmeromyiini) within the subfamily Bengaliinae of Calliphoridae (Rognes 1998)" (Rognes 2009, 5). This quotation is from a revisionary work by Knut Rognes (2009). In some publications the genus *Bengalia* has been assigned to a separate family, Bengaliidae (Lehrer 2006a,b, 2008). However, the current consensus is to accept the ascertainment of Knut Rognes (2009, 2011), so we will continue with Rognes's classification.

Bengalia flies feed on ant larvae and pupae and capture the pale, soft-looking booty, which they snatch from ants carrying these items along the foraging and emigration trails. This booty and body snatching have been observed by Jacobson (1910), Bequaert and Wheeler (1922), Alston (1932), Mellor (1922), and Maschwitz and Schönegge (1980). Our account is based mainly on the findings by Maschwitz and Schönegge, who provided the best documented description. They conducted their studies in Sri Lanka near Anuradhapura, where they mainly observed *Bengalia emarginata* but found similar results in *B. jejuna* and *B. latro*. They first noticed the *B. emarginata* flies when they excavated a nest of *Camponotus rufoglaucus*. When the five- to ten-millimeter-long ant workers rushed out of the disturbed brood chambers, many of them with larvae and pupae between their mandibles, suddenly flies about 12 millimeters long began circling close above the ant nest or resting on elevated spots near the brood-carrying ants. The flies seemed to recognize brood-carrying ants from about a 50 cm distance, making jerking turns toward the brood carriers and suddenly attacking them with a kind of flight-jump. During the attack, a *Bengalia* fly would grasp the transported larva or pupa with its front tarsi, often also employing its proboscis and simultaneously pushing its middle legs forward. These attacks lasted only a few seconds, and every second or third attack was successful. After having captured the ant brood, the fly carried the booty in flight to an undisturbed place nearby, where it sucked the larva or pupa dry; this took about thirty seconds. Shortly afterward, it returned to the lookout and waited for more brood-transporting ants.

Maschwitz and Schönegge were able to demonstrate that brood or termites that were just placed onto the ground were mostly ignored by the flies, but as soon these items were picked up by the ants, they became targets for fly attacks. The flies did not attack workers carrying dark-colored booty, like insect cadavers, and only attacked ants carrying pale-colored objects. Furthermore, it was demonstrated that the *B. emarginata* does not exhibit ant species specificity and will attack a diverse variety of ant species. Among the targets were several formicines, including the aforementioned *Camponotus rufoglaucus, C. sericeus, Oecophylla smaragdina, Polyrhachis* sp., *Plagiolepis* sp.; the dolichoderine *Technomyrmex albipes;* the myrmicines *Myrmicaria brunnea, Solenopsis* sp., and *Meranoplus bicolor;* and the ponerines *Bothroponera tesseronoda* and *Leptogenys chinensis.* Not surprisingly, the flies' success rates varied with these different target species. Attacks on brood or booty carriers of the only three-millimeter-long *Technomyrmex* workers or "clumsy" *Bothroponera* foragers were usually successful, whereas booty and brood snatching from the ten-millimeter-long and agile *Leptogenys* workers was a struggle and often failed. The authors never observed *Bengalia* attacks on army ants; however, Taku Shimada provided photographic evidence of *Bengalia* sp. attacking emigration and predatory columns of army ants belonging to the *Aenictus laeviceps* species group (Figures 5-2, 5-3). In these events, the fly robs the booty or ant brood from ants on the wing. Like a gigantic helicopter, it hovers over the much smaller *Aenictus* workers and snatches immatures from their mandibles.

5-2   A species of the bengaline fly *Bengalia*. Several species of this genus were observed lingering around trails of ants. In the lower picture, a *Bengalia* fly hovers just above the trail of the army ant *Aenictus laeviceps*. (Courtesy of Taku Shimada).

5-3   The *Bengalia* fly snatches a larva from an *Aenictus laeviceps* worker and sucks the captured larva dry (lower picture). (Courtesy of Taku Shimada).

The Beetle *Amphotis marginata:* Highwayman of *Lasius fuliginosus*

One of the most striking examples of foraging trails are those of the Central European "shining black ant"—the formicine *Lasius fuliginosus*—along which huge numbers of forager ants commute all day and night, transporting honeydew collected on nearby trees in their crops (also called social stomachs). Honeydew is the "defecation" of a diversity of tree-sap-sucking insects (most of them Hemiptera), which is rich in sugars and amino acids. Depending on the temperature, foraging activity of the ants commences in the middle to end of March and continues almost uninterrupted until October. Foraging trails of

5-4   The nitidulid beetles *Amphotis marginata* resting in the ground litter near the nest of *Lasius* fuliginosus. (Courtesy of Konrad Fiedler).

L. fuliginosus consist of a network of trunk routes, running as far as 20 to 30 m away from the nest (Dobrzańska 1966; Quinet and Pasteels 1991). The trails are marked with secretions from the foragers' hindguts, which contain a specific blend of trail pheromone compounds. These trails are among the busiest traffic routes known in ants, because foragers not only collect honeydew and other food items as nourishment for adult nestmates and brood, but also use the collected honeydew as "glue" during the construction of the sophisticated carton nests, and as food for the symbiotic fungal mycelia, which is an integral stabilizing part of the carton structure (Maschwitz and Hölldobler 1970).

The paths of *L. fuliginosus* are frequented by a diversity of myrmecophiles, some of which we will visit later. First, let us consider the nitidulid beetle *Amphotis marginata,* which we call a "highwayman" on the trails of *Lasius fuliginosus,* because these beetles are kleptoparasites that successfully entice the food-laden ants to regurgitate liquid food from their filled crops (Hölldobler 1968; Hölldobler and Kwapich 2017). In a survey of several *L. fuliginosus* nests, we consistently found most of the *Amphotis* beetles outside near the nest entrance, but occasionally discovered beetles under shelters along the foraging trail up to 28 m away from the nest. In one case, we noticed a second accumulation of beetles on the base of a tree where *L. fuliginosus* workers were moving up and down the trunk. In this case, it was not clear to us whether the colony had a second nest inside this tree, because both the main nest tree and this second tree were connected by a busily frequented path along which we also found *Amphotis* under shelters (Figure 5-4).

During the day the beetles usually occupy hideaways near the nest entrance or (although much less frequently) along the foraging trails. At dusk and during the night the beetles successfully intercept and obtain food from the food-laden foragers (Figures 5-5, 5-6). Although we saw food solicitation by beetles in the field only after dusk, in the laboratory food begging by *Amphotis* was observed frequently in the late afternoon, evening, and night, and occasionally during the daytime.

The *Amphotis* beetle induces an ant to regurgitate a food droplet by thrusting its head and thorax upward while approaching with outstretched antennae. The ant briefly antennates the beetle's head and may then continue her trip home, or briefly lick the beetle's head with her extended labium. The beetle simultaneously rapidly drums with its antennae on the ant's head. This exchange takes about one to two seconds and often does not lead to trophallaxis because the ant hectically proceeds with her journey to the nest. However, on average, every fourth or fifth attempt succeeds in keeping the ant's "attention" long enough for the beetle to stimulate the ant's extended labium with its mandibles and maxillary palps, accompanied by rapid drumming with its antennae on the sides of the ant's head. The beetle's stimulation of the fully extended labium of the ant triggers the food flow from the ant to the beetle. Occasionally, a relatively large droplet is spilled over the head of the *Amphotis* beetle. Usually, the ant quickly moves on and the beetle imbibes the food droplet. Radioactively labeled food enabled us to trace the food flow from the ants to the beetles, and not surprisingly, begging time and amount of food transferred were positively correlated. Although there was considerable variation, on average, beetles obtained 24% of food carried by the ants in their crops during a single feeding event. On average, ants shared 1.8 times more food with the beetles than with their own nestmates during individual feeding bouts. In contrast, no food was transferred from the beetles to the ants (Hölldobler and Kwapich 2017).

During interactions, we observed that host ants briefly licked the head of the myrmecophilous *Amphotis* beetle. Histological and scanning electron microscopic studies revealed several well-developed exocrine glands associated with

5-5   The *Amphotis* beetles solicit food that *Lasius fuliginosus* foragers carry inside their crops (social stomachs) to the nest. (Bert Hölldobler).

5-6   If the *Amphotis* beetles succeed in eliciting regurgitation from *Lasius fuliginosus* foragers, the regurgitated food droplet is spilled over the beetle's head. (Bert Hölldobler).

the front parts of the beetle's head, labium, mandibles, and lateral head regions. It is possible that ants are enticed to lick the secretions of these glands, which in turn enables the beetles to gain access to and stimulate the extended labium of the ant. On the other hand, these glands do not protect the beetle from attack by the ants. Ants without fully laden crops are less prone to regurgitate their crop contents, and often attack approaching beetles. Yet even when they are under attack, it seems that the beetle's external body shape is well adapted to its "highwayman" lifestyle in the ant world. During attacks, the beetle protects it-

5-7 Ants that notice that they were tricked by the *Amphotis* beetle occasionally attack the beetle, but usually the beetles are well protected by their turtle-like carapace and their firm attachment to the ground, aided by powerful claws and tarsal setae. (Bert Hölldobler).

self by retracting its appendages and head under its sturdy carapace and flattening itself to the ground (Figure 5-7). Apparently with the aid of its powerful claws and special setae on its tarsi, it firmly attaches its lower body surface to the ground. Most of the time the ant is unable to lift the beetle. Indeed, the "turtle defense strategy" of *Amphotis* is most effective; only rarely have we found injured specimens.

The beetles' begging signal is not ant-species-specific. Although *A. marginata* usually occurs only with *L. fuliginosus,* when placed with non-host species including *Camponotus ligniperdus, Formica pratensis, F. sanguinea, F. fusca,* or *Myrmica rubra,* the beetle succeeded in soliciting regurgitation (Figure 5-8). However, the success rate is much lower than with their host species *L. fuliginosus.*

The question of why *Amphotis* beetles solicit food from the ants mostly during the night can be answered only hypothetically. From observations in a well-lit laboratory, we have the impression that ants visually perceive the beetles and sometimes attack them, even before the beetles have approached them. We hypothesize that in darkness the beetles are not as easily detected by the ants. In this context, it is noteworthy that the last three segments of the beetle's antennae

FORAGING PATHS AND REFUSE SITES

5-8   The food-begging behavior in *Amphotis marginata* is not very specific. The beetle is able to solicit food from a variety of ant species, in this case *Myrmica rubra,* even though the beetle has never been found associated with these ants. (Bert Hölldobler).

are densely bestowed with olfactory sensilla and some extremely long whisker-like setae, which appear to be mechano-sensory hairs (Figure 5-9). The behavior of the beetle suggests that it perceives the ants mechanically and by olfaction. As mentioned above, the ant often stops briefly and licks the head of the beetle, which is richly endowed with exocrine glands. The ant's licking rarely lasts longer than one second because the beetle immediately stimulates the ant's extended lower lip (labium), employing its mandibles, the inner rims of which are endowed with broom-like bristle structure and its "brushy" maxillae and labium. All this is accompanied by rapid antennal strokes, during which the long mechano-sensilla touch both sides of the ant's head.

The genus *Amphotis* belongs to the beetle family Nitidulidae, also called sap beetles or sap-feeding beetles. A total of five *Amphotis* species have been described, and all have been collected near ant nests. A detailed analysis of their interactions with ants has been made only for *A. marginata,* but presumably the other species, which have been found with different ant species, may have a similar lifestyle. To our knowledge *Amphotis* is the only myrmecophilous nitidulid genus. It is reasonable to suppose that they evolved from nitidulid species that were tree-sap eaters, or mycophagous (feeding on mycelia mats of fungi), or fed on rotting wood or the defecations of lachnids (honeydew) and other hemipterans tended by ants in the trees. Beetles might have encountered ants at these sites frequently and eventually "discovered" the easy stimulus that elicits regurgitation in ants with an engorged crop. Over evolutionary time, the beetles specialized on a new, very rich ecological food niche (Hölldobler and Kwapich 2017).

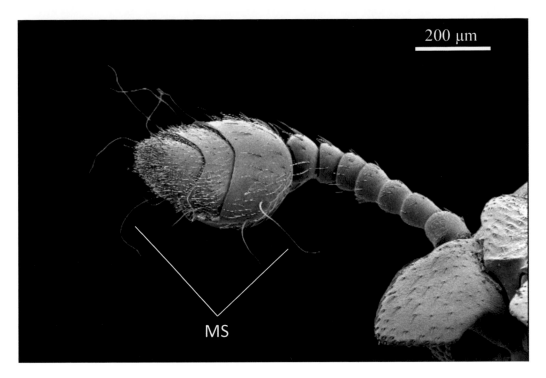

5-9 Club-shaped antenna of *Amphotis marginata*. The long mechano-sensory setae (MS) are a striking feature of these antennae. Most likely they serve in the beetle's close-range orientation when soliciting food from ants under low-light conditions. (Bert Hölldobler).

Predators and Scavengers: The Story of Some *Pella* Species

Besides highwayman beetles, many other myrmecophiles can be found in the garbage dumps and along the trails of *Lasius fuliginosus*. The most abundant are a variety of staphylinid aleocharine beetles. In the study areas in Bavaria and Hesse (Germany), Hölldobler, along with collaborators Michael Möglich and Ulrich Maschwitz (1981), recorded twelve staphylinid species belonging to seven genera associated with *L. fuliginosus*. They paid special attention to the genus *Pella*, many species of which are myrmecophiles (Maruyama 2006). In this study the focus species were Pella *funesta* and *Pella laticollis* (Figure 5-10). The reconstruction of beetles' life cycle is mainly based on *P. funesta,* for which the most comprehensive information is available, because this species was cultured in laboratory nests. However, comparative circumstantial data indicate that the life cycle of *P. laticollis* is similar to that of *P. funesta*. Most of these observations were made with large laboratory colonies in which carton nest containers were connected to large foraging arenas. However, regular observations of the beetles' interactions with ants were conducted also in field colonies.

5-10   Two *Pella* species that are frequently found in the surroundings of *Lasius fuliginosus* nests: upper, *Pella funesta;* lower, *Pella laticollis.* (Courtesy of Pavel Krásenský).

In early spring the female beetles deposit eggs near the kitchen middens of the formicine ant *Lasius fuliginosus,* the shiny black ant that lives in elaborate carton nests the ants construct in cavities of tree trunks or other hidden localities. The beetle larvae develop in the refuse pile and pupate sometime during the period from May to July. In July or August, the adult beetles eclose. The young beetles then apparently emigrate, as indicated by a short phase of high diurnal locomotory and flight activity. After this period, the beetles hunt injured, but also occasionally healthy, ants or feed on dead ants deposited on the refuse pile. They are primarily active during the night and remain mostly hidden in shelters during daytime. *Pella* adults overwinter in dormancy in the peripheral zones of *L. fuliginosus* nests. At the end of winter, they exhibit a second diurnal activity phase, during which mating takes place. After reproduction, the beetles die, normally a few weeks before the new adult beetle generation ecloses.

There is conflicting information in the literature concerning nutrition in different *Pella* species. Do they live entirely as scavengers, eating deceased ants and other dead insects left over by the hosts, or do they also prey on live ants, especially disabled individuals? Erich Wasmann (1920, 1925) reported the latter, though other authors could not confirm predatory behavior for all *Pella* species. For example, Stoeffler et al. (2011) reported behavioral data suggesting that *P. laticollis* is a strict scavenger, feeding on dead ants and other insect cadavers in the ants' garbage dumps, whereas *P. funesta* and *P. cognata* are mainly predators. This does not entirely match observations by Hölldobler et al. (1981). They observed that, as long as enough dead ants were available at the ants' nest middens, the beetles of all species rarely exhibited predatory behavior, limiting themselves to a diet of dead ants or of insect cadavers left over by the ants. However, when the beetles were starved, predation by *Pella* became strikingly obvious. The time of onset was often unpredictable. They saw the beetles hunting during the daytime, but most frequently observed predatory behavior in the evening or at night.

How can we explain the different results obtained by Stoeffler et al. (2011) and the observation made by Hölldobler et al. (1981)? It has been suggested that researchers might have accidentally wrongly identified the *Pella* species. Indeed, such confusion can easily happen in the field. Although Hölldobler and colleagues observed hunting behavior of *Pella* during dusk in a field colony, most of their observations were made with large laboratory colonies, where they could easily confirm the taxonomic identification. It is possible that there might be species-specific differences between *Pella* in their proneness to switch from scavenging to predatory behavior. However, Hölldobler can say with certainty that both *P. funesta* and *P. laticollis* scavenge and exhibit predatory behavior. Although he has no comparative data concerning niche preferences of the different *Pella* species studied, he can state that *P. funesta, P. laticollis, P. cognata,* and *P. humeralis* were observed in the refuse area, as well as on the trails of *L. fuliginosus,* both in nature and in laboratory nests. However, despite these contradictory observations, Hölldobler does not argue against a possible preference of certain *Pella* species, such as *P. laticollis* for the host refuse areas and *P. funesta* and *P. cognata* for the foraging trails. Hölldobler would only ascertain that these putative preferences are not exclusive. His and his collaborators' observations confirm Wasmann's findings of predatory behavior in six *Pella* species, all found with *L. fuliginosus,* including *Pella humeralis,* which also occurs with ants of the *Formica rufa* group (Hölldobler et al. 1981). Donisthorpe (1927) made similar observations. Worldwide, more than sixty *Pella* species are known, and all appear to be associated with ants.

Ant-hunting *Pella* beetles were observed chasing individual ants and pursuing them for distances up to six centimeters. When a beetle made a rear approach, it attempted to mount the ant and insert its head between the victim's head and thorax (Figures 5-11, 5-12). The attacked ant usually stopped abruptly and pressed its legs tightly against its body (Figure 5-13).

Often, this rapid reaction threw the beetle off the ant's back, allowing the ant to escape. In most cases, the ant's defensive response was successful. However,

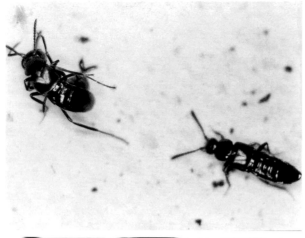

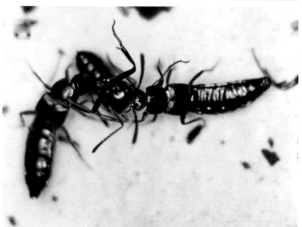

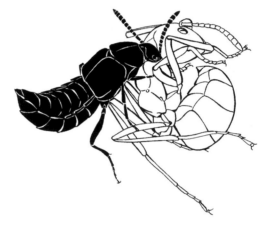

5-11   *Pella* beetles chase a *Lasius fuliginosus* worker. The beetles attack the worker, usually biting it between head and alitrunk (lower). (Bert Hölldobler).

5-12   The upper picture shows a *Pella* beetle attacking a *Lasius fuliginosus* worker. The drawing (lower) renders this behavior with greater clarity. (Turid Hölldobler-Forsyth; ©Bert Hölldobler).

5-13   Often, the attacked *Lasius fuliginosus* worker stopped abruptly and pressed its legs tightly against its body. This reaction frequently threw the *Pella* beetle off the back of the ant. (Bert Hölldobler).

in some instances the beetle was able to partially decapitate the ant, severing the nerve cords. Occasionally two or three beetles were seen chasing an ant. Once the ant was caught by a beetle, other beetles joined in subduing and killing it (Figure 5-14).

Although individual *Pella* often tried to drag the prey away from the others, all members of the pack usually ended up feeding on it simultaneously. No aggression among the beetles was observed on such occasions (Hölldobler et al. 1981). Interestingly, the aleocharine *Drusilla (Santhota) sparsa,* which lives near the nests of *Crematogaster osakensis* in Japan, seems to have a lifestyle similar to the *Pella* species discussed above (Ikeshita et al. 2017). *Drusilla sparsa* mainly feed on dead *C. osakensis* workers but also prey on live host ants, biting them between the head and thorax, as a striking photograph by Taku Shimada suggests (Figure 5-15; for similar photographic documentation see Maruyama et al. 2013).

When foraging on the ants' garbage dumps at the base of the carton nest or running along the ants' trail, *Pella* beetles frequently encounter ants that behave aggressively toward them. How do the beetles manage to escape their hosts? Usually, they run with their abdomens curved slightly upward. When encountering ants, the beetles flex their abdomens over their bodies with greater intensity. This is a frequently observed behavior of many staphylinid myrmecophiles, and it is commonly considered a defensive response. It has been suggested that during abdominal flexing the beetles discharge secretions from their tergal gland, first described by Karl Jordan in 1913 and later characterized in greater detail by Jacques M. Pasteels (1968) and by Hölldobler and collaborators (Hölldobler 1970b; Hölldobler et al. 1981, 2018). The gland is located between the sixth and seventh abdominal tergites. In most cases it consists of a glandular reservoir, which is an invagination of the intersegmental membrane between the two tergites. The intersegmental membrane is hypertrophied and partly transformed to a glandular epithelium to various degrees in different species. In addition, paired clusters of large glandular cells with long duct cells drain their secretions into the tergal reservoir.

5-14 Occasionally two or three *Pella* beetles were seen chasing a *Lasius* worker, and after the ant was captured, other beetles joined in feeding on the killed ant. (Bert Hölldobler).

According to Jordan (1913), Pasteels (1968), and Steidle and Dettner (1993), this tergal gland seems to be present in all aleocharine species, whether they are associated with ants or not. The chemical composition of the tergal gland secretions of several species has been investigated and found to be extraordinarily diverse (Hölldobler et al. 1981; Steidle and Dettner 1993; Stoeffler et al. 2007, 2011, 2013). We will refer here only to those studies, which also consider the behavioral function of the tergal gland contents.

Kistner and Blum (1971) suggested that *Pella japonicus* and possibly also *P. comes,* both of which live with *Lasius spathepus,* produce citronellal in their tergal glands. This substance is also a major compound of the mandibular gland secretions of their host ants, for which it may function as an alarm pheromone. Although no *Pella* tergal gland contents were available for chemical analysis, because irritated

5-15   The staphylinid beetle *Drusilla (Santhota) sparsa* attacking a worker of the host ant *Crematogaster osakensis*. (Courtesy of Taku Shimada).

beetles seemed to smell like the ants' mandibular gland secretions, Kistner and Blum speculated that *Pella* produce citronellal in their tergal glands and thereby mimic the alarm pheromone of their host ants. They suggested that, in this way, the beetles can "cause the ants to reverse their direction; a reaction which allows the myrmecophiles to escape."

The investigations by Hölldobler et al. (1981) of the defense strategy employed by the myrmecophilous *Pella* species led to a different picture. *Pella laticollis,* when irritated mechanically, discharges a pungent-smelling secretion from its tergal gland. Only when the beetles were severely attacked and firmly grasped on their appendages by the ants could we smell the tergal gland secretion. Hölldobler and his colleagues never observed the beetles employing tergal gland contents when they initiated attacks on ants. Ants contaminated with tergal gland secretions usually exhibited a repellent reaction, releasing the grip on the beetles and grooming and wiping their mouth parts and antennae on the substrate. But the beetles had to escape quickly, because other ants close by became alerted and often rapidly approached the scene, apparently alarmed by the ants' alarm pheromone, which in *Lasius* workers derives from the Dufour's gland opening at the gaster tip. The preliminary analysis of the tergal gland secretions of *P. laticollis* did not reveal a resemblance to the mandibular gland secretions of *L. fuliginosus;* instead Hölldobler et al. (1981) identified benzo-quinone and toluquinone as the main compounds. In addition, saturated hydrocarbons and short-chained fatty acids were detected in the secretions of *P. laticollis*. No citronellal could be detected in any of the *Pella* species investigated.

Although Hölldobler et al. (1981) did not find any resemblance between the *Pella* tergal gland secretions and the mandibular gland secretions of *Lasius fuliginosus,* it is noteworthy that the *Pella* secretions contained undecane, a hydrocarbon commonly found in the Dufour's glands of formicine ants and considered to be an alarm pheromone in *L. fuliginosus* (Dumpert 1972). However, in behavioral tests, the tergal gland secretions of *P. laticollis* elicited a repellent reaction rather than an alarm response in *L. fuliginosus* workers. Although it is possible that the subsequent alarm response often observed in ant nestmates in close vicinity might be triggered by the undecane in the tergal gland, we are nonetheless inclined to interpret this as a secondary response originally triggered by the release of alarm pheromones by the repelled ants.

This brings us to an interesting study by Michael Stoeffler, Johannes Steidle, and their collaborators (Stoeffler et al. 2007). This group tested the idea, first proposed by Kistner and Blum (1971) that the myrmecophilous genus *Pella* use alarm pheromone compounds of *Lasius fuliginosus* to avert attacks by their host ants. They confirmed the presence of quinones and undecane in tergal gland secretions of *P. funesta* and *Pella humeralis,* and besides other aliphatic compounds, they also identified 6-methyl-5-hepten-2-one (sulcatone). This latter substance was previously found to be part of the mandibular gland secretions of *L. fuliginosus,* together with the two terpenoid furans perillene and dendrolasin (Bernardi et al. 1967). Sulcatone is a volatile component of citronella oil, and most likely part of the very particular and persistent odor of *L. fuliginosus* workers and their entire nest structure.

Stoeffler et al. (2007) hypothesize that the undecane component in the beetle's gland secretions elicits alarm behavior in host ants, while the addition of sulcatone "blocks the aggression-inducing effect" elicited by undecane and quinones. They consider this to be "a rare example of chemical mimicry in myrmecophilous insects." In support of this hypothesis, Stoeffler et al. (2011) compared *Pella cognata* and *P. funesta,* which live (according to their observations) primarily along the ants' trails, with *Pella laticollis,* a beetle that resides (according to their observations) mainly in the refuse area of the ant nest. *Pella cognata* and *P. funesta* avoid encounters with ants by swift evasive movements and (according to the authors) do not employ appeasement behavior (a special behavior in aleocharine myrmecophiles, which we describe later), but both species discharge their tergal gland contents when attacks by ants cannot be evaded. According to the authors, *P. cognata* produces only minute amounts of undecane, the alarm pheromone of *L. fuliginosus,* in its defense tergal gland, while producing large quantities of tridecenes, tridecane, and 7-pentadecene, compounds that do not elicit alarm behavior in the host ants. On the other hand, *L. funesta* produces larger amounts of undecane and a lesser amount of tridecane and sulcatone (6-methyl-5-hepten-2-one), which appears to be absent in *P. cognata.* Stoeffler et al. (2011) argue that *P. cognata,* when discharging tergal gland secretion, does not trigger an alarm response in host ant workers; nevertheless the secretion has a repellent effect on the ants due to the presence of quinones. Whereas *P. funesta,* when seriously harassed by the ants, will cause alarm in the ants, due to the undecane in their tergal defense gland, because of the simultaneously discharged sulcatone it is able to create an "ant free space" around the beetle. In contrast, *P. laticollis,* whose tergal gland contents are dominated by quinones and undecane, lives in refuse areas of ant nests, where it frequently encounters ants. Nevertheless, according to Stoeffler et al. (2011), *P. laticollis* rarely discharge tergal gland secretions. Instead, they employ an appeasement strategy that allows the beetles to avoid serious attacks by the host ants. Based on these results, Stoeffler and his colleagues postulate that these three *Pella* species evolved three different defense strategies.

We are not sure whether this conclusion is justified in its entirety. Indeed, as we pointed out above, undecane is contained in the Dufour's gland secretions of *L. fuliginosus* and functions, as in most other formicine ant species, as an alarm pheromone (Dumpert 1972; Hölldobler and Wilson 1990). The main function of the mandibular gland secretions of *L. fuliginosus* is most likely defense; in particular *Formica* and other *Lasius* species that live in close proximity to a *L. fuliginosus* nest are intensely repelled by the mandibular secretions of *L. fuliginosus* (Maschwitz 1964), whereas the alarm effect is comparatively weak. To repeat, the real alarm pheromone in *L. fuliginosus* is undecane, for which the ants possess specific olfactory sensilla on their antennae (Dumpert 1972). Neither Ulrich Maschwitz nor Hölldobler could observe a particular avoidance or repellent response in *L. fuliginosus* workers when presented with mandibular gland secretions of their own species. As these secretions also contain sulcatone, one would expect avoidance behavior according to the hypothesis proposed by Stoeffler et al. (2007).

In this context, it is important to note that Johannes Steidle and Konrad Dettner (1993) found undecane in the tergal gland secretions of several aleocharine beetles, even those that are not considered myrmecophilous, and it was suggested that undecane serves as a solvent for the quinones generally present in tergal glands of Aleocharinae. These authors also point out that, although it is intriguing to find sulcatone in some myrmecophilous aleocharine species and not in so-called free living species, sulcatone occurs in defensive secretions of many insect species.

We propose that the real function of sulcatone—together with the quinones dissolved in tridecane, undecane, and other hydrocarbons in the tergal defense glands of *Pella*—is to repel *Formica* and other ant genera, ant predators, and, when needed, the host species *Lasius fuliginosus*. We do not think it is justified in this context to speak of chemical mimicry.

Michael Stoeffler and his collaborators (Stoeffler et al. 2013) propose another case of chemical mimicry in the aleocharine beetles *Zyras collaris* and *Zyras haworthi*. The tergal gland secretions of these beetles are quite different from those of the *Pella* species discussed above. They mainly consist of α-pinene, β-pinene, myrcene, and limonene. Because these monoterpenes were also found in some aphid species from which ants collect honeydew, and because in bioassays *Lasius fuliginosus* workers exhibited increased antennation behavior toward filter paper balls treated with a mixture of these compounds, the authors hypothesize that these *Pella* beetles mimic chemical cues of aphids "and thereby achieve acceptance by their host ants." No further details concerning the behavioral interactions of these myrmecophiles with their host ants are known, therefore one cannot make a solid conclusion concerning the biological significance of this remarkable correlation. However, the hypothesis that this might represent a novel kind of chemical mimicry in myrmecophiles is certainly justified.

During the studies by Hölldobler et al. (1981) of the interactions of *Pella* beetles with their host ants *Lasius fuliginosus,* they were impressed by how rarely beetles employed the tergal gland defensive secretions. Usually, the fast-running beetles were very skilled in avoiding direct attacks by the ants, and in case host ants touched them, they used an appeasing defensive tactic; the repellent defense seemed to be applied only as a last resort.

In contrast, the appeasement behavior is particularly apparent in early spring, when most of the beetles can be found close to the nest entrance of the ants' nest or at the base of the carton structure, close to where the refuse area is located. Here the beetles displayed "death feigning" behavior when attacked by ants. They fell to the side, with legs and antennae folded tightly to the body and the body curved upward. The ants either ignored the motionless beetles or carried them around and finally discarded them on the "garbage dump." Only rarely did the ants injure the beetles.

Death-feigning behavior (tonic immobility) is a common passive defense in insects and has been observed in several myrmecophiles. Šípek et al. (2008) describe a buprestid beetle *Habroloma myrmecophila* that invades the leaf-tent nests of *Oecophylla smaragdina* in the western coast of India (Goa State) for oviposition. The larvae of this beetle species "mine in the leaves forming the nest wall and pupate within the mines." When touched or attacked by the ants, the adult beetles exhibit death-feigning behavior that appears to entice the *Oecophylla* workers to pick up the motionless beetle and discard it. Petr Šípek and colleagues suggest that these beetles imitate necrophoric chemical cues produced by dead ants (Howard and Tschinkel 1976) that trigger in ants the removal of dead nestmates. However, no experimental evidence for such chemical mimicry has been presented.

But let us continue considering the case of *Pella*. Later in the year, when the activity of ants and beetles was much higher, the beetles employed a different appeasement technique. As mentioned before, we only rarely noticed the discharge of tergal gland secretions by the beetles, although every time they encountered ants, they flexed their abdomens and pointed their abdominal tips toward their adversaries (Figure 5-16). Usually, the ants reacted by antennating the abdominal tip of the beetle and sometimes licking it. This ordinarily slowed down the ants' aggression, and the brief distraction, which usually did not last longer than one or two seconds, allowed the beetle to escape.

Histological investigations revealed that the posterior part of the beetle's abdomen is richly endowed with exocrine glands, in addition to the already mentioned tergal gland between the sixth and seventh abdominal tergites. In the four species of *Pella* investigated (*P. cognata, P. funesta, P. humeralis,* and *P. laticollis*), no major differences were noted (Hölldobler et al. 1981). Not all glandular structures could be anatomically characterized; however, we focus here on those structures that might be involved in the appeasement process. Besides the well-developed glandular hypodermal epithelia in the eighth, ninth, and tenth segments, there are paired clusters of secretory cells with duct cells that open dorsolaterally through the invaginated intersegmental membrane between the seventh and eighth or eighth and ninth segments. That is, it could not be ascertained whether a pair of small glandular cell clusters can be assigned to the seventh and eighth segments and the larger cluster to the eighth and ninth segments. Furthermore, paired clusters of secretory cells with duct cells were found between the ninth and tenth segments, which open into reservoirs composed of intersegmental invaginations, located laterally on each side of the hindgut. The reservoirs seem to open close to the anus. Hölldobler et al. (1981) called this structure the pygidial gland, following the first description of the defense gland in the staphylinid subfamily Steninae (Jenkins 1957), analogous to the pygidial defense gland in the beetle family Carabidae. However, this term can lead to confusion, because the pygidium is the dorsal tergite of the last external (visible) abdominal segment and can vary in different insect groups. Furthermore, this gland in the staphylinids does not open at the last external segment. We therefore tend to follow the nomenclature proposed by Konrad Dettner

5-16 Appeasement behavior in *Pella* species. The beetles "feign death" (upper picture), which causes the ants to either ignore them or pick up their motionless bodies and carry them to the "garbage dump." The lower picture depicts the typical chemical appeasement behavior of beetles who present the ant appeasing secretions on the abdominal tip. (Bert Hölldobler).

(1993), who called this kind of gland an "anal defensive gland." We should note that, in a more recent study of this gland in the staphylinid subfamily Steninae, Andreas Schierling and Konrad Dettner (2013) again called it the "pygidial gland." Whatever we call it, the pygidial gland in Carabidae and in the Staphylinidae are apparently not homologous, and the pygidial glands in beetles are very different from those found in certain ant subfamilies. In *Pella,* another

5-17   The *Pella* larvae are usually ignored by the host ants. When touched by the ants, they quickly raise the posterior tip and apparently offer an appeasement secretion from a gland located in this area (upper). The picture in the middle depicts a *Pella* larva feeding on a dead ant. The beetle larvae also exhibit cannibalistic behavior, shown in the lower picture. (Bert Hölldobler).

glandular structure appears to open sternally at the tip of the abdomen, but we could not identify exactly where. Because we cannot say which of these glands is involved in the chemical appeasement system, and because perhaps all of them are part of this process, we call these posterior glands the "appeasement gland complex" or appeasement glands. Of course, we do consider this appeasement process a defensive behavior, but it functions as a "gentle defense" and not one that triggers a repellent reaction in the ants. Contrary to Stoeffler et al. (2011), we observed appeasement behavior in all *Pella* species we studied. Whether appeasement is exhibited in one species more frequently than in others, or whether it is more commonly displayed in the refuse areas than on the trails, we cannot say, but we can certainly state that it occurs on the trails as well as in the refuse areas.

In the field and in the laboratory, *Pella* larvae were almost exclusively found near or in the refuse area of *L. fuliginosus* nests. Hölldobler, with colleagues, frequently observed the larvae feeding on dead ants. During feeding the larvae always attempted to stay "out of sight," either by remaining entirely beneath the booty and devouring it from below or by crawling inside the dead ant's body (Figure 5-17). Occasionally two to four larvae could be observed feeding on the same cadaver. But when they became too crowded, they frequently attacked each other, sometimes resulting in one larva eating the other. When ants encountered the larvae, they usually attacked them. Almost invariably, the beetle larvae reacted with a typical defense posture, by raising the abdominal tip toward the head of the ant. Often the ant responded by stopping the attack and palpating the larva's tip. In most cases this short interruption was enough to allow the larva to escape. Histological investigations revealed that *Pella* larvae have a complex glandular structure with a reservoir that opens dorsally near the tip of the second-to-last segment. Perhaps the secretions of this gland are involved in appeasement-defense behavior against the host ants (Hölldobler et al. 1981).

One other paper on the myrmecophilous genus *Pella* needs to be addressed. Toshiharu Akino (2002) published a brief study of the Japanese species *Pella comes* (it was previously called *Zyras comes;* see Maruyama 2006; Hlaváč and Jászay

2009). According to Maruyama, *P. comes* belongs to the "funesta group." Akino collected specimens on the trail of two *Lasius fuliginosus* colonies near Kyoto in Japan and placed seven individuals each in a glass dish together with five *L. fuliginosus* workers (most likely the ants were *Lasius spathepus* or *L. nipponensis,* because there is no record of *L. fuliginosus* occurring in Japan). Akino noted that

FORAGING PATHS AND REFUSE SITES

ant aggression toward the beetle did not occur when the beetles and ants came from the same colony, whereas the ants behaved hostilely when the beetles were housed with ants taken from a colony where no myrmecophiles had been found. Subsequently, the cuticular hydrocarbons of ants and beetles were compared: the hydrocarbon profiles matched closely. Whether this resemblance of hydrocarbon blends is the key mechanism that enables the myrmecophiles to live along the trails in close contact with the ants has not been answered, although it is suggestive.

However, Stoeffler et al. (2011) compared the cuticular hydrocarbon profiles of *Pella* species with those of their hosts (*L. fuliginosus*) and did not find any resemblance. Likewise, Thomas Parmentier and his coworkers (Parmentier et al. 2017a) have shown the cuticular hydrocarbon profiles of several myrmecophiles living in nests of *Formica rufa* do not at all, or only partly, resemble those of their host ants (see also Chapter 9). Presumably, *P. comes* feed on dead host ants or hunt ant workers, as several other *Pella* species do (Wasmann 1920; Donisthorpe 1927; Hölldobler et al. 1981; Maruyama 2006). In fact, Maruyama lists *P. comes* among the *Pella* species of which he observed predatory behavior. It is entirely conceivable that *Pella* beetles that feed on ant cadavers or prey on host ant workers ingest the hosts' hydrocarbons. An insect-rich diet has been shown to alter the cuticular hydrocarbon profiles of some ant species (Liang and Silverman 2000; Buczkowski et al. 2005). However, in a remarkable recent study, Adrian Smith (2019) demonstrated that the cuticular hydrocarbon profiles of *Formica archboldi* are a surprisingly close match to those of the native *Odontomachus* ant species, on which they prey. If the prey species is primarily *O. brunneus*, *F. archboldi* exhibits the *O. brunneus* cuticular hydrocarbon profile, and if the prey species is *O. relictus*, *F. archboldi* carries the *O. relictus* profile. Interestingly, *F. archboldi* retained their *Odontomachus*-like hydrocarbon signature after seven months in the laboratory without exposure *to Odontomachus* ants. Furthermore, *F. archboldi* that occurred in a field site with an abundance of both *O. brunneus* and *O. ruginodis* matched the hydrocarbon profile of only *O. brunneus*, not *O. ruginodis*, which is thought to be an exotic species in Florida. These two findings suggest that chemical matching is not always a by-product of having a significant portion of their diet consist of the *Odontomachus* species found in their habitat (Smith 2019). The mechanism of this kind of hydrocarbon matching is still unclear and under investigation.

To conclude, myrmecophilous *Pella* live as scavengers and predators, feeding on the ants' booty, or dead ants, or preying on live ants. There is no documented evidence that they can also solicit regurgitations from host ants. Attempts by Hölldobler et al. (1981), using radioactive tracer techniques, failed to register trophallaxis in *Pella,* and other marking techniques employed by Parmentier et al. (2016b) also failed to demonstrate transfers of regurgitated food from host ants to *Pella* beetles or beetle larvae. Nevertheless, Akino (2002) writes, "When the beetle begged from the workers, they often regurgitated fluid to it," and he presents a graph indicating nineteen observed events of regurgitation between ant and beetle. There is no mention at what time intervals these observations were made. From our experience with *Pella* beetles, we suggest the reported trophallaxis is perhaps a misinterpretation of some other behavioral interaction. No other scholar who has studied *Pella* reported trophallaxis between the beetles and their host ants. In fact, in a previous paper, Taniguchi et al. (2005) write that *Pella comes* is often attacked by the host ants *Lasius spathepus* (or *L. nipponensis*) on their trails.

On the Ants' Trails

One of the most spectacular experiences one of us (B. H.) had during a visit to the Costa Rican rain forest was witnessing the raids of the army ants of the genus *Eciton,* a very conspicuous group of ants in tropical forests from Mexico to Paraguay. These ants do not build nests like most other ants. They dwell in so-called bivouacs, which are partly sheltered cavities on the base of tree trunks or in the soil, where the queen, larvae, and pupae are covered and protected by the bodies of the workers. When the workers establish the bivouac, they link their legs and bodies together with strong hooked claws at the tips of their feet and thereby form layer upon interlocking layer of living bodies until finally a solid mass of about a meter across emerges (Kronauer 2020). As the pioneering studies by Theodore Schneirla have revealed, these army ant colonies alternate between a static period, in which the colony remains at the same bivouac site every day for about three weeks, and a migratory period. During the static period, the queen rapidly develops thousands of eggs in her ovaries. Her abdomen is grotesquely swollen, and in the following days she lays 100,000 to 300,000 eggs (Kronauer 2020). Toward the end of the static period, small larvae hatch from the eggs, and soon after, many new workers emerge from the pupal cocoons. They develop from the larvae produced in the previous static period. Apparently, the appearance of many new workers triggers the commencement of the migratory period, during which about half a million workers or more, with the queen and several hundred thousand larvae, move to a new bivouac site where they settle. During this migratory period, the workers conduct massive predatory raids, emerging from the bivouacs like gigantic pseudopodia catching small and large arthropods, sometimes even small vertebrates, or raiding nests of other ant species and pillaging their larvae, which all will serve as food for the rapidly growing army ant immatures. Although the army ants conduct prey raids in the static period as well, during the migratory period, with several hundred thousand growing larvae, the colony's nutritional needs are much higher. This is most likely the reason why they move from one bivouac site to another, to be able to exploit new hunting grounds.

A close look at the emigration and raiding columns of army ants reveals a multitude of other arthropods moving within or on the edges of the columns. Carl Rettenmeyer and his collaborators have recorded many species on the trails or around the periphery of the bivouacs and refuse deposits of the New World army ants (Rettenmeyer et al. 2011 and literature cited therein); David Kistner and coworkers described an equally rich group of species on the trails of New and Old World army ants (also called driver ants) (Kistner 1975, 1979, 1982, 1989, 1993,

1997, 2003; Jacobson and Kistner 1991, 1992; Kistner et al. 2003, 2008); and Ulrich Maschwitz, Volker Witte, Christoph von Beeren, and Munetoshi Maruyama and their collaborators discovered numerous new myrmecophiles in Southeast Asian army ants and other legionary ant species from the genus *Leptogenys* (Maruyama 2006; Maruyama et al. 2009, 2010a,b; von Beeren et al. 2011b, 2016). We will come back to some of these studies in later chapters. A very good selected account of the astounding diversity of army ant myrmecophiles and their natural history is provided by Daniel Kronauer in his recent excellent book, *Army Ants: Nature's Ultimate Social Hunters* (2020).

The army ant refuse area and emigration and raiding columns are special ecological niches occupied by a variety of myrmecophiles, and a typical instructive case of positional specialization is provided by the aleocharine staphylinid *Tetradonia marginalis*. Akre and Rettenmeyer (1966, 773) noted that this beetle species

> is the most frequently found staphylinid in the refuse deposits of *Eciton burchellii* and in the emigration. Sometimes more than 100 individuals can be found in a single large deposit of this host. Many of these myrmecophiles do not run along the emigration columns but apparently fly from one colony to another. *Tetradonia marginalis* was the species most frequently attacking uninjured, active workers as well as injured workers of both *E. burchellii* and *E. hamatum* along edges of the columns and near bivouacs. The staphylinids were not seen to attack workers running along the middle of the raid or emigration columns or those running quickly along the edges.

During field work in La Selva, Costa Rica, Edward O. Wilson and Hölldobler were able to confirm some of the observations of Akre and Rettenmeyer. It was in late March 1985, in the evening, when they encountered the colony emigration column of the army ant species *Eciton hamatum* with thousands of workers

5-18   Column of the army ant *Eciton hamatum* in Costa Rica. (Bert Hölldobler).

carrying larvae in the typical army ant fashion, the larva positioned underneath of the body of the carrier (Figure 5-18).

At the very end of the column, with no more than a hundred stragglers, some injured or moving slower than other nestmates and usually not carrying a larva, we spotted ten or more staphylinid beetles running swiftly along the trail, abdomens curled up. Most of these rove beetles, which turned out to belong to the genus *Tetradonia*, were attacking straggling *Eciton* workers (Figures 5-19, 5-20, 5-21). All the victims were slow and partly disabled, unable to walk swiftly. We speculated that they had fallen back because of injuries and therefore had been singled out by the beetles for attack. The rove beetles circled around, seizing a leg or jumping on the back of the ant, attempting to bite off the head or abdomen. Some dragged their victims swiftly along the trail. In one instance two beetles attacked an ant together, running along with it. On another occasion we observed these rove beetles near the refuse deposits of *Eciton hamatum*, eating the corpses of *Eciton* workers. Apparently, the rove beetles' major food source is injured or dead ants, and they move with the ants from one bivouac site to the other or, as Rettenmeyer and Akre suggested, may fly in from one colony to another (Hölldobler and Wilson 1990).

Akre and Rettenmeyer (1966) made a distinction between "generalized" species of myrmecophilous staphylinids, which look like most staphylinid species, and "specialized" species, which include so-called mimetic or myrmecoid forms—that is, the beetles' external morphology resembles that of the host

5-19 A *Tetradonia* beetle holding a handicapped worker by the leg at the end of the *Eciton hamatum* column. (Bert Hölldobler).

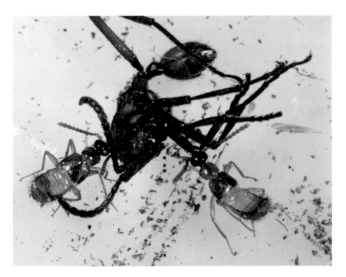

5-20 A *Tetradonia* beetle chasing a handicapped *Eciton hamatum* worker and subsequently jumping on its back (lower picture). The beetle moved up toward the head and bit the ant between head and alitrunk. (Bert Hölldobler).

5-21 Two staphylinid *Tetradonia* beetles pulling a dead *Eciton hamatum* worker away. (Bert Hölldobler).

5-22   The upper picture shows the rove beetle *Ecitophya* with the army ant *Eciton burchellii*. The lower picture depicts an unidentified rove beetle with the host ants *Aenictus* sp. In both cases the myrmecophiles splendidly mimic the appearance of their hosts. (Courtesy of Taku Shimada).

ants (Figures 5-22, 5-23). Akre and Rettenmeyer listed many behavioral differences between the two kinds of species. These data reveal a gradually increasing degree of behavioral integration with the host society, which may constitute the most obvious of the evolutionary pathways toward full integration with the host species. The two kinds of species can reasonably be interpreted as progressive stages of adaptation to the host ants and their way of life (Hölldobler and Wilson 1990).

Von Beeren et al. (2018), building on the Akre-Rettenmeyer hypothesis that host-symbiont associations range across the entire generalist-specialist continuum, compared the behavioral and chemical integrational mechanisms of the two kinds of species (generalists and specialists). Using bioassays, they

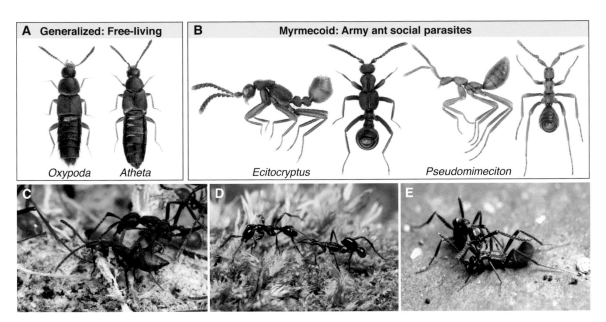

5-23   Myrmecoid syndrome in aleocharine rove beetles. (A) Examples of free-living Aleocharinae with generalized morphology, *Oxypoda* and *Atheta*. (B) Examples of army ant myrmecophilous Aleocharinae with myrmecoid morphology, *Ecitocryptus* (associated with *Nomamyrmex*) and *Pseudomimeciton* (associated with *Labidus*). (C–E) Living myrmecoids with host ants: *Ecitophya* with *Eciton* host (Peru), *Rosciszewskia* with *Aenictus* host (Malaysia), *Beyeria* with *Neivamyrmex* host (Ecuador). (Courtesy of Munetoshi Maruyama).

studied the interactions of the *Eciton* host species with the myrmecophilous beetles *Ecitophya simulans, Ecitophya gracillima, Ecitomorpha* cf. *nevermanni,* and *Ecitomorpha* cf. *breviceps.* These species resemble their *Eciton* host species in appearance, and, according to David Kistner (1982), such mimics are called "Wasmannian mimics." Each of these beetles was found to be specifically associated with a single host species. *Ecitophya gracillima* was exclusively collected from *Eciton hamatum* colonies. *Ecitophya simulans, Ecitomorpha* cf. *breviceps,* and *Ecitomorpha* cf. *nevermanni* were found only in *Eciton burchellii foreli* colonies. The authors did not find Wasmannian mimic species with any other *Eciton* species occurring in the La Selva study area; however, they point out that *Ecitophya rettenmeyeri* was previously collected from *Eciton lucanoides* colonies in La Selva (Kistner and Jacobson 1990).

The behavioral bioassays revealed that the Wasmannian mimic species frequently contacted their host ants. They were readily accepted by their hosts and even engaged in reciprocal grooming. In the test situation, which was a rather artificial setup, *Tetradonia* (generalist) beetles were also not attacked; though von Beeren et al. (2018) occasionally observed aggression in the field, they never saw aggression toward the specialist mimics. The comparative analyses of the cuticular hydrocarbon blends of beetles and host ants revealed close similarities between the Wasmannian mimics and hosts, whereas the generalist *Tetradonia* matched the hosts' cuticular hydrocarbons profile less well. These findings indicate that *Ecitophya* and *Ecitomorpha* species are more closely socially integrated with the *Eciton* hosts than are *Tetradonia*. However, the assumption that the similarity of the hydrocarbon profiles is the key mechanism enabling these Wasmannian mimics to deeply penetrate the *Eciton* society may not be warranted, based on the currently available information. It can be argued that the close match of the cuticular hydrocarbon profiles of guests and hosts is the consequence of the myrmecophiles' social integration. Most likely the beetles are frequently in contact with the ants and thereby become coded, "contaminated" with the ants' chemical colony signature. We assume there must be something else that allows the beetles to be adopted by the ants in the first place.

This brings us back to a brief discussion of Wasmannian mimicry, because it has previously been suggested that the remarkable morphological resemblance of beetles and host is one of the mechanisms by which the beetles fool the host ants. Quite a number of staphylinid myrmecophiles (and some diapriid wasps), marching with army ants in their emigration columns, exhibit such ant-like appearances. The resemblance evolved convergently through the modification of the abdomen and thorax, which resulted in the imitation of an ant-like petiole (Seevers 1965). In addition, the sculpturing of body surface and color of the host ants is often copied by some army ant myrmecophilous staphylinids. One might assume that these mimics appeal to the visual modality of their hosts, but the world of these ants is largely chemical and tactile. It was Erich Wasmann (1903) who first noticed and described these resemblances and proposed that it was tactile mimicry that deceives the host ants. Carl Rettenmeyer (1970) followed Wasmann's interpretation—he coined the term "Wasmannian mimicry"—and David Kistner (1966, 1979) argued that the antennation behavior during encounters of host ants with myrmecophilous beetles is "identical with that of an ant when it meets another ant." He claims that ants pay special attention to the petiole, and states, "Thus the morphological constriction permits the myrmecophile to function within the colony as though it were an ant." Wilson (1971) and Hölldobler (1971) found this interpretation unconvincing, because a number of myrmecophilous staphylinids that exhibit the highest degree of social integration, such as the genera *Lomechusa, Lomechusoides,* or *Xenodusa,* do not resemble ants at all, and in particular, they do not morphologically imitate a petiole (see Chapter 8). Wasmann (1903; and other references in Wasmann 1925) already indicated that perhaps the beetles' imitation of the host ant's coloration may have evolved to deceive birds and other predators that do not eat the stinging, poisonous ants (Batesian mimicry) (see Chapter 6). Obviously, this kind of protective mimicry would be much more effective if body shape and coloration resemble those of the ants. Though we do not want to dismiss Rettenmeyer's and Kistner's "Wasmannian mimicry" hypothesis—and most likely some kind of tactile mimicry exists in ant-myrmecophile interactions and in certain social parasitic ants (Fischer et al. 2020)—solid experimental evidence is still missing. It is noteworthy, however, that such striking cases of Wasmannian mimicry are most commonly found in army ant colonies, and in termites (Kistner 1969). In some of the latter cases, "deformations" of the staphylinid's abdomen are so grotesque, and resemble a termite's body so strikingly, that we are inclined to accept in these cases Kistner's interpretation of Wasmannian (tactile) mimicry.

In an excellent evolutionary phylogenetic study, Munetoshi Maruyama and Joseph Parker (2017) demonstrated that such morphological ant mimicry (myrmecoid body shape) by staphylinid myrmecophiles has evolved at least twelve times. "Each independent myrmecoid clade is restricted to one zoogeographic region and highly host specific on a single army ant genus. Dating estimates reveal that myrmecoid clades are separated by substantial phylogenetic distances—as much as 105 million years" (Figures 5-23, 5-24) (Maruyama and Parker 2017, 920).

Unfortunately, nothing is known about how the mimetic beetle species make their living in the army ant colonies. Presumably, they prey on the ant brood or feed on booty captured by the raiding ants. If they are just tolerated guests inside the colony and are not fed and tended by the host ants, maybe the acquisition of the host ants' cuticular hydrocarbon profiles and perhaps resembling the host ants' body shapes might be enough for profitable coexistence with the ants. This scenario would support the Wasmannian mimicry hypothesis. Kistner (1982) also lists some cases of Wasmannian mimicry of staphylinids living with non-army-ant species, although these mimicry cases are not as striking as those found with army ants. To reiterate and conclude, we do not oppose the Wasmannian mimicry concept, but believe more experimental evidence is needed. In fact, one experimental study has been published by Taniguchi et al. (2005) that supports the Batesian mimicry hypothesis. The authors realized and demonstrated experimentally that the Japanese tree frog (*Hyla japonicus*) preys on many ant species, except *Lasius spathepus*. Frogs that had eaten *L. spathepus* subsequently refused them as prey. The myrmecophilous staphylinid *Pella comes,* which lives along the foraging trails and refuse areas of *L. spathepus,* resembles its host ants in size, shiny black coloration, and way of locomotion. Frogs that had no recent experience with *L. spathepus* prey readily consumed *Pella* beetles; however, frogs that had previously tasted the host ants refused to prey on *Pella comes.* Apparently, *L. spathepus* is endowed with repellent substances that are distasteful to frogs. Previously, Maruyama et al. (2003) described a new species of the aleocharine genus *Drusilla* from Malaysia that lives on the trails and refuse sites of the myrmicine ant *Crematogaster inflata.* This ant is strikingly conspicuous, because its pronounced vault (bulla), which covers the storage chamber of the metapleural gland, has a bright yellow color. When disturbed, the ants discharge sticky repellent-defense secretions from this gland (Buschinger and Maschwitz 1984). Interestingly, the new species *Drusilla inflatae* features bright yellowish coloration at its third, fourth, and partly fifth abdominal tergites (Figure 5-25). At a quick

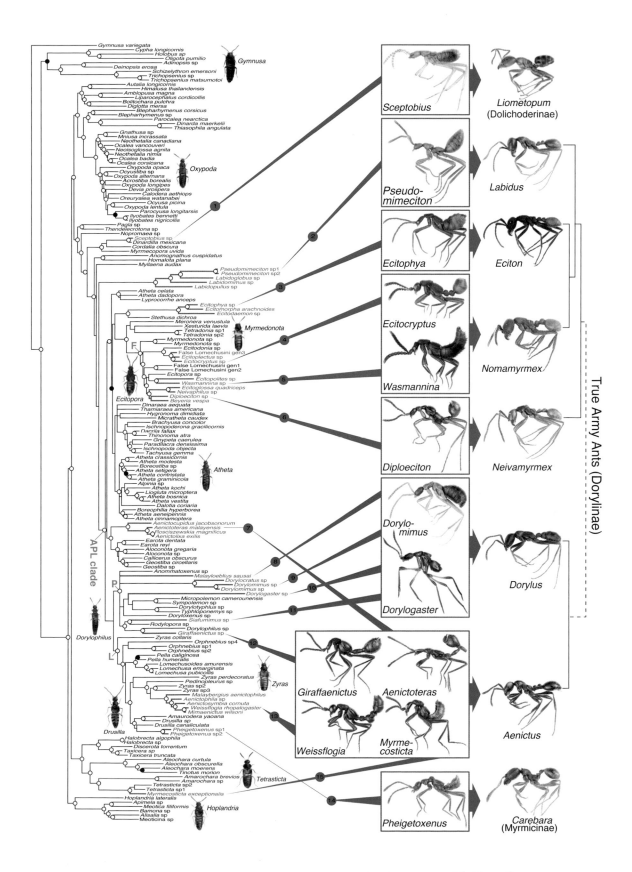

Gymnusa variegata
Cypha longicornis
Holobus sp
Oligota pumilio
Adinopsis sp
Deinopsis erosa
Schizelythron emersoni
Trichopsenius sp
Trichopsenius matsumotoi
Autalia longicornis
Himalusa thailandensis
Amblopusa magna
Liparocephalus cordicollis
Bolitochara pulchra
Diglotta mersa
Blepharhymenus corsicus
Blepharhymenus sp
Parocalea nearctica
Dinarda maerkelii
Thiasophila angulata
Gnathusa sp
Mniusa incrassata
Neothetalia canadiana
Ocalea vancouveri
Neoisoglossa agnita
Neothetalia nirnia
Ocalea badia
Ocalea corsicana
Oxypoda opaca
Ocyustiba sp
Oxypoda alternans
Acrostiba borealis
Oxypoda longipes
Devia prospera
Calodera aethiops
Oreuryalea watanabei
Ocyusa picina
Oxypoda lentula
Parocyusa longitarsis
Ilyobates bennetti
Ilyobates nigricollis
Pagla sp
Thendelecrotona sp
Nopromaea sp
Sceptobius sp
Dinardilia mexicana
Cordalia obscura
Myrmecopora uvida
Anomognathus cuspidatus
Homalota plana
Myllaena audax

Pseudomimeciton sp1
Pseudomimeciton sp2
Labidoglobus sp
Labidopullus sp
Atheta celata
Atheta dadopora
Lyprocorrhe anceps
Ecitophya sp
Ecitomorpha arachnoides
Ecitodaemon sp
Stethusa dichroa
Meronera venustula
Xesturida laevis
Tetradonia sp1
Tetradonia sp2
Myrmedonota sp
Myrmedonota sp
Ecitodonia sp
False Lomechusini gen3
Ecitoplectus sp
Ecitocryptus sp
False Lomechusini gen1
False Lomechusini gen2
Ecitopora sp
Ecitopolites sp
Wasmannina sp
Ecitoglossa quadriceps
Neivaphilus sp
Diploeciton sp
Beyeria vespa
Dinaraea aequata
Thamiaraea americana
Hygronoma dimidiata
Micratheta caudex
Brachyusa concolor
Ischnopoderona gracilicornis
Dacrila fallax
Thinonoma atra
Gnypeta caerulea
Paradilacra densissima
Ischnopoda objecta
Tachyusa gemma
Atheta crassicornis
Atheta modesta
Boreostiba sp
Atheta setigera
Atheta contristata
Atheta graminicola
Alpinia sp
Atheta kochi
Liogluta microptera
Atheta bosnica
Atheta vestita
Dalotia coriaria
Boreophilia hyperborea
Atheta aeneipennis
Atheta cinnamoptera
Aenictocupidus jacobsonorum
Aenictoteras malayensis
Rosciszewskia magnificus
Aenictolixa exilis
Earota dentata
Earota reyi
Aloconota gregaria
Aloconota sp
Calliceros obscurus
Geostiba circellaris
Geostiba sp
Anommatoxenus sp
Malayloeblius sausai
Dorylocratus sp
Dorylomimus sp
Dorylomimus sp
Dorylogaster sp
Micropolemon camerounensis
Sympolemon sp
Dorylotyphlus sp
Typhlopolemys sp
Doryloxenus sp
Siafumimus sp
Rodylopora sp
Dorylophilus sp
Giraffaenictus sp
Zyras collaris
Orphnebius sp4
Orphnebius sp1
Orphnebius sp2
Pella caliginosa
Pella humeralis
Lomechusoides amurensis
Lomechusa emarginata
Lomechusa pubicollis
Zyras perdecoratus
Pedinopleurus sp
Zyras sp2
Zyras sp3
Malaybergius aenictophilus
Aenictophila sp
Aenictosymbia cornuta
Weissflogia rhopalogaster
Mimaenictus wilsoni
Amaurodera yaoana
Drusilla sp
Drusilla canaliculata
Pheigetoxenus sp1
Pheigetoxenus sp2
Halobrecta algophila
Halobrecta sp
Discerota torrentum
Taxicera sp
Taxicera truncata
Aleochara curtula
Aleochara obscurella
Aleochara moerens
Tinotus morion
Amarochara brevios
Amarochara sp
Tetrasticta sp1
Tetrasticta sp1
Myrmecosticta exceptionalis
Hoplandria lateralis
Apimela sp
Meotica filiformis
Bamona sp
Alisalia sp
Meoticina sp

Gymnusa
Oxypoda
Myrmedonota
Ecitopora
Atheta
Dorylophilus
Zyras
Drusilla
Tetrasticta
Hoplandria

APL clade

F
P
L

1
2
3
4
5
6
7
8
9
10
11
12
13
14
15

Sceptobius
Pseudo-mimeciton
Ecitophya
Ecitocryptus
Wasmannina
Diploeciton
Dorylo-mimus
Dorylogaster
Giraffaenictus
Aenictoteras
Weissflogia
Myrme-costicta
Pheigetoxenus

Liometopum
(Dolichoderinae)
Labidus
Eciton
Nomamyrmex
Neivamyrmex
Dorylus
Aenictus
Carebara
(Myrmicinae)

True Army Ants (Dorylinae)

5-25   The rove beetle *Drusilla inflatae* (upper) that mimics the appearance of its host ant species *Crematogaster inflate* (lower). (Courtesy of Munetoshi Maruyama).

glance, the beetle can easily be mistaken for a host ant. It is a reasonable hypothesis that this is a case of Batesian mimicry, yet the experimental evidence has not yet been provided.

Many guests of army ants regularly accompany raiding columns, where they prey on captured booty of their hosts or on the host ants themselves. These myrmecophilous scavengers and predators, as exemplified by the above-mentioned *Tetradonia* beetles, often bring up the "rear guard" of the raiding columns, running along the newly vacated trails. It has been known for a long time that guests of army ants can orient by the odor trails without further cues provided by the hosts. Since army ants frequently change the location of their bivouacs, it is important for the myrmecophiles to be able to follow the emigration trails. Akre and Rettenmeyer (1968) investigated this phenomenon systematically by exposing a large variety of guests of *Eciton* army ants to army ant odor trails laid over the floor of the laboratory arena. They found that nearly all of the species tested, including samples of myrmecophilous Staphylinidae (rove beetles), Histeridae (clown beetles), Limulodidae (horseshoe crab beetles), Phoridae (hump-backed flies, scuttle flies), Thysanura (silverfish or common bristle tails), and Diplopoda (millipedes), were able to orient along the pheromone trail. Some myrmecophiles did better than others. For example, the staphylinids *Ecitomorpha, Ecitophya,* and *Vatesus,* which regularly accompany ant columns, followed experimental trails readily and accurately, whereas phorid flies deviated more frequently from the experimental trails, apparently seeking to contact live ants, which were absent on the experimental trails. Although the myrmecophiles exhibited a remarkable host-species specificity in choosing the trails—in fact, in some instances they were even repelled by trails drawn with pheromones of other army ant species—nevertheless, they did not distinguish the trails of their own host colony from that of another belonging to the same species. There are, however, some instances reported in which the myrmecophiles of *Eciton* army ants were less specific in their responses to trail pheromones, and this behavior was partly correlated with the myrmecophiles' acceptance of several host species in nature. David

Kistner, who studied Old World army ants for many years, observed a limu-lodid beetle *Mimocete* flying into a raiding column of *Dorylus,* which then me-andered along the trail for a distance of about ten centimeters, before it dis-appeared in the swarm of the progressing army ants (Kistner 1979). Many myrmecophiles are known to run along the trails of the *Aenictus* host ants (Maruyama et al. 2011) (see Figure 5-22). Volker Witte and his collaborators (1999) reported that the myrmecophilous spider *Sicariomorpha* (formerly *Gamasomorpha) maschwitzi* (Oonopidae), which lives in nests of the Southeast Asian ponerine legionary ant *Leptogenys distinguenda* (Figure 5-26), is able to

FORAGING PATHS AND REFUSE SITES

follow the host ants' trail for a distance of up to 20 cm (see Figure 6-17). In emigration columns the spider usually follows closely behind a host ant, but if lost, it can use the trail to catch up with the crowd. Apparently this myrmecophilous spider feeds on the prey captured by the host ants.

Myrmecophiles following their host ants' trails are known not only from species that live in colonies of army and legionary ants, but also from species associated with ants living in stationary nests. William Morton Wheeler (1910) described observations in the field of *Formica sanguinea* colonies migrating to a new nest site. The myrmecophilous staphylinid beetle *Dinarda dentata* was seen on the emigration trail in the company of its host ants. Hölldobler initiated nest emigrations of a *F. sanguinea* colony in the laboratory by making the ants move from one arena via a cardboard bridge to another nest arena. The *Dinarda* beetles were observed moving over the bridge during the latter section of the emigration process, suggesting that the beetles are also able to "read" the ants' trail pheromone.

Another interesting case of a myrmecophile following its host ants' trails is the cockroach genus *Attaphila,* of which nine species have been described (Nehring et al. 2016; Bohn et al. 2021). They all live in nests of the fungus-growing ants of the genera *Acromyrmex* and *Atta.* William Morton Wheeler (1900) described the first species, *Attaphila fungicola,* and he recognized the myrmecophilous nature of this cockroach, which lives in nests of *Atta texana.* Apparently these myrmecophiles feed on the cultured fungus of their hosts. The same *Attaphila* species has more recently been discovered in nests of a second host ant species, *Atta cephalotes,* in Colombia (Rodriguez et al. 2013). Shortly after Wheeler's report, four more *Attaphila* species from South America were described by I. Bolivar, who also observed that *Attaphila schuppi* follows trails of *Acromyrmex* in the field in Brazil (Bolivar 1905). About sixty years later, John Moser (1964) demonstrated that *Attaphila fungicola* follows the trail pheromone of its host ants, which he extracted from the ants' poison gland reservoirs and applied along artificially drawn routes in the foraging arena of the laboratory nest. Moser (1967) also

noted that individual cockroaches were riding on the backs of young *Atta texana* queens during the nuptial flights of the ants (Figure 5-27).

New attention was recently given to this fascinating myrmecophile by Zachary Phillips and collaborators, working in the laboratory of Ulrich Müller (Phillips et al. 2017), who studied the dispersal and interactions of *Attaphila fungicola* with its host ants *Atta texana* in greater detail, in both laboratory and field experiments. They confirmed that *A. fungicola* preferentially mount alate queens, rather than males, before they depart for the nuptial flight. They also demonstrated that their survival in founding nests of young queens or in brood chambers is very poor, whereas they do very well in the fungal gardens of established nests, and that even in foreign nests, they experience very little antagonistic behavior from *A. texana* workers. From these observations one can conclude, as Phillips et al. (2017, 2021) do, that the myrmecophilous cockroaches use winged *Atta* queens for hitch-hiking (vectors) but dismount them before or during nest construction by the young founding queens. During their dispersal to established colonies their ability to "read" their host ants' trail pheromone may serve as an important orienta-tion cue. However, Phillips et al. (2017) did not observe *Attaphila* enter established

FORAGING PATHS AND REFUSE SITES

A. texana colonies in nature, and even when they placed individual *A. fungicola* at the margins of *A. texana* nest mounds, the cockroaches did not enter the nests. Subsequently Zachary Phillips (2021) made the amazing discovery that *Attaphila*, after having dismounted the queen ant, travel to a foraging column of a mature *Atta* colony and leap onto the leaf of a forager, to ride into the nest.

5-27　The myrmecophilous cockroach *Attaphila fungicola* in the fungus garden of *Atta texana*. *Attaphila* migrate to new colonies by mounting alate queens and taking a ride with them when the winged queens leave the nest for mating flights. (Courtesy of Alex Wild / alexanderwild.com).

In this context the aforementioned study by Nehring et al. (2016) might be of special interest. They found that cockroaches share to a considerable degree the cuticular chemical substances of their specific host species, and in many ways their cuticular chemical profiles resemble those of their host colony. The myrmecophilous cockroaches are tolerated or ignored by workers of the host colony, yet when exposed to ants of a foreign, conspecific colony, *Attaphila* will be attacked. Although the authors could not investigate whether the myrmecophiles produce these chemical profiles on their own or acquire them from their host ants, the latter is most likely the case. Since *Attaphila* change host colonies, they have to ease their way into the new nest and acquire the new hosts' cuticular chemical profiles. They may do this already on the trails, and in fact, may lose some of their old cuticular profile when no longer exposed to the previous host ants. Although no exact observations exist, after having entered the nest as hitchhikers, they may first remain in the leaf deposit area before they venture into the fungus gardens, where they find plenty of food.

Although sometimes it has been claimed that myrmecophilous lycaenid caterpillars follow their host ants' trails, most experimental tests have failed to prove this (e.g., Fiedler 1990; Fiedler and Maschwitz 1989b). However, there seems to be one exception: Alain Dejean and Guy Beugnon (1996) demonstrated that the caterpillars of the myrmecophilous lycaenid *Euliphyra mirifica* follow the trails of *Oecophylla longinoda*.

Another intriguing example in which trail following would not be intuitively suspected was recently reported by Thomas Parmentier (2019), who studied the diversity of myrmecophiles in mound-building wood ants. Polygynous *Formica polyctena* (red wood ant) colonies are polydomous and frequently move from one nest site to another. In late spring, the time when *F. polyctena* colonies in Europe frequently move to new nest sites, Parmentier observed larvae of the chrysomelid beetle *Clytra quadripunctata* on the migration trail accompanied by *Formica* workers moving to their new nest site. *Formica* species mark their trails with contents from the rectal bladder, and the beetle larvae appear to be able to "read"

June and July these insects were seen running on the trails in the evening from dusk to midnight (see also Chapter 7). Thus, trail following by myrmecophiles is obviously much more common than previously assumed.

We conclude this chapter with a curious case of myrmecophilous specialization, which has been found to be a mutualistic association of the ponerine ant *Harpegnathos saltator* from India (see Figure 1-12) and a milichiid fly species. In order to understand this remarkable social relationship between the flies and the ants, we first have to describe the complex nest architecture of *H. saltator* (Peeters et al. 1994).

The nests of *H. saltator* are located close to the soil surface. A variable number of elegantly constructed chambers are stacked directly on top of one another. The chambers have flat floors, and their walls curve up to low, vaulted ceilings. A characteristic feature is the occurrence of a thick vaulted roof protecting the uppermost chamber. It is conspicuous, because a gap of six millimeters or more separates it from the surrounding soil. In bigger nests, the thick roof extends down the sides of the deeper chambers, and thus a flattened sphere comes into being. The intervening space, or "atrium," is then continuous around the outside of the shell, except that the latter abuts the soil in several places. Interestingly, the inside of the chambers is wall papered with brown papery material, consisting mainly of strips of empty cocoons enmeshed with pieces of cuticle, wings, and dry vegetable material. The entrance gallery, approximately eight to ten millimeters wide, leads downward and opens into the atrium. Small entrances to the nest chambers inside the sphere are located in the upper region of the sphere; some are encircled by neatly molded flanges, two to three millimeters thick and curling outward to form tire-shaped rings. Another tunnel extends from the lower reaches of the atrium to a refuse chamber approximately 50 to 250 mm deeper in the soil. It is filled with a moist, blackish-brown mass of prey remains (crickets, moths, spiders, other arthropods, but no *Harpegnathos* workers) together with numerous isopods and milichiid fly larvae. These fly larvae function as decomposers of the ants' garbage. Without them, this garbage chamber would soon be gridlocked, which would be a disaster, because this chamber, to-

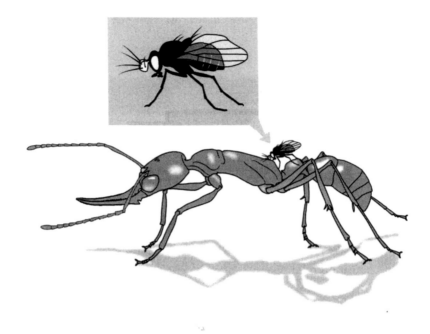

5-28 A milichiid fly rides on a *Harpegnathos* ant into the ant's nest. (Courtesy of Margaret Nelson).

gether with the atrium, also serves as a means of flood protection. In fact, this might be the main function of the chamber, which is always much deeper in the soil than the sphere that contains the main nest chambers. Both atrium cavity and garbage / flooding chamber are designed to drain water away from the nest sphere, and the use as garbage disposal might be secondary (Peeters et al. 1994).

How the two-millimeter-long adult milichiid flies locate their underground breeding site became clear after observing flies that were riding on the petiole of returning foragers (Figure 5-28). The flies hugged the ant's body and were not easily disturbed. Maschwitz (1981) made similar observations in Sri Lanka and identified the flies as belonging to the Milichiidae, and Musthak Ali et al. (1992) also reported seeing flies entering the *Harpegnathos* nests in this way. As we stated above, the fly larvae probably have a beneficial effect for the ants, because they eat the organic debris discarded by the ants, thereby keeping the refuse chambers from becoming clogged (Peeters et al. 1994).

Similar observations were published by Deborah Waller (1980), who observed the milichiid female flies *Pholeomyia texensis* hopping onto leaves carried by workers of the fungus-growing ant *Atta texana* and riding on the leaves into the ants' nest, where flies oviposit and the developing fly larvae apparently feed on the fungal debris. The mature flies mate outside on the nest mound, and after mating, the females wait next to the ants' trail for leaf-carriers to be used for hitchhiking into the nest.

The busy foraging trails of ant colonies provide ample opportunities for kleptoparasites and invading guests, many of which eavesdrop on the trail pheromone signals used for communication between the ants. Much work remains to be done to determine how myrmecophiles correctly decipher the "directional poles" of pheromones trails and whether, in two-way traffic, they pick out inbound foragers for a ride. In many cases, myrmecophiles found along ant trails or with nomadic army ant colonies have come to resemble the ants that they travel with. These "aboveground" associates likely avoid visually oriented hunters by mimicking the appearance of their well-defended ant models. In Chapter 6, we discuss some of the champions of morphological and behavioral mimicry of ants, the spiders.

6 Spiders and Other Mimics, Pretenders, and Predators

COMPARED TO THEIR free-living relatives, many myrmecophiles have shorter legs and thicker antennae. Major features like eyes and mouthparts are often reduced or absent, while strategic arrangements of setae, cuticular notches, and exocrine glands that aid in interactions with ant hosts are prominent (reviewed by Parker 2016; Parker and Owens 2018). In more extreme cases, disparate taxa have converged on suites of defensive characteristics that render them nearly indistinguishable. For instance, some silverfish and beetles share a slippery, teardrop-shaped (limuloid) body plan that shields the head and is difficult to grasp. Likewise, the slug- and limpet-like morphology of predatory *Liphyra brassolis* caterpillars (see Figure 4-22) and *Microdon* fly larvae (see Figure 3-8) allows the insects to cruise unimpeded across ant brood piles without being pried from the chamber floor (Dodd 1902; Hinton 1951; Seevers 1957; Duffield 1981; Yamamoto et al. 2016).

In contrast to myrmecophiles with defensive body plans, a second group of associates has come to mimic aspects of the coloration and anatomy of ants, with exquisite accuracy (Figure 6-1). Among them are a great variety of Aleocharinae staphylinid beetle species that arose in parallel from generalized ancestors to resemble their various army ant hosts (Maruyama and Parker 2017) (see Figures 5-23, 5-24). The false petioles (waists) of these myrmecoid beetles have been honed by evolution in at least seven different ways (Seevers 1965). This kind of mimicry is called myrmecomorphy, and in many cases the anatomical disguises of myrmecomorphs are further enhanced by persuasive behavioral imitations of their ant models.

6-1   Upper: A *Pseudomimeciton* rove beetle (Aleocharine) travels with a column of *Labidus* army ants. The beetle is shaped like an ant and lacks eyes and elytra (wing covers). Lower: In Peru, a *Colonides* clown beetle (Histeridae) blends into the emigration column of *Eciton burchellii* by mimicking the ants' coloration. (Courtesy of Taku Shimada).

Spiders are also among the premier myrmecomorphic and behavioral mimics of ants. In addition to mimicking detailed ant color patterns and arrangements of spines, many have extra constrictions of the opisthosoma and cephalothorax that divide their two-segmented bodies into three segments, like an ant (Figure 6-2). Some ant-mimicking spiders create the illusion of having antennae by waving their first pair of legs ahead of them, and even appear to walk with

6-2 Bodies of myrmecomorphic spiders like *Myrmecium latreillei* (upper), *Sphecotypus niger* (middle), and *Synemosyna formica* (lower) are divided into three segments, with constrictions of the cephalothorax and opisthosoma. Additional constrictions and shading can create the illusion of an ant's petiole, postpetiole, and metanotal suture. (Courtesy of Alex Wild / alexanderwild .com).

6-3 Left: The spider *Sphecotypus borneensis* (lower) mimics the defensive posture of ants in the genus *Polyrhachis* (upper). Right: *Myrmarachne* sp. (lower) mimics the defensive posture of a co-occurring ant, *Calomyrmex* sp. (upper). Both ant models produce formic acid and other pungent-smelling secretions. (Courtesy of Paul Bertner, Rainforests Photography).

an ant-like gait (Cushing 1997; Shamble et al. 2017). In this chapter we discuss some of the most extraordinary examples of myrmecomorphy and behavioral ant mimicry by spiders and other arthropods and consider the elements of their disguises that may be under selection by visually oriented predators and may have evolved to aid in their interactions with their ant models. Our discussion would not be complete without visiting the other relationships between spiders and ants, which are made more captivating by the ability of both groups to construct architectural masterpieces and form complex social groups.

Transformational and Compound Mimics

Poorly defended species that benefit from resembling a dangerous or unpalatable species found in the same habitat are known as Batesian mimics. Most myrmecomorphic spiders are Batesian mimics that evade ant-averse, visually oriented predators (Figure 6-3). In some cases, adult spiders belonging to a single species mimic multiple co-occurring ants, or even a variety of generalized ant-like forms and color patterns (Nelson 2010). The mating preferences of these distinct mimetic morphs, and the heritability of morphs, has been the focus of extensive detective work (Borges et al. 2007). Indeed, how mate preference is linked to personal appearance in polymorphic species is one of the central questions in the field of animal behavior.

Female spiders often advertise their readiness to mate by producing silk laden with sex pheromones (Baruffaldi et al. 2010), and males of the polymorphic ant-mimicking spider *Myrmaplata* (formerly *Myrmarachne*) *plataleoides* prefer the silk of female spiders that mimic the same ant species that they do. Although males do mate with alternative morphs at a lower frequency, preferences for morph-specific cues found on silk may lead to assortative mating and, ultimately, disruptive selection that reinforces mimicry of multiple ant models by a single spider species (Borges et al. 2007).

Spiders that accurately mimic the size and form of co-occurring ant species are less likely to be eaten by ant-averse predators with good vision. Given that

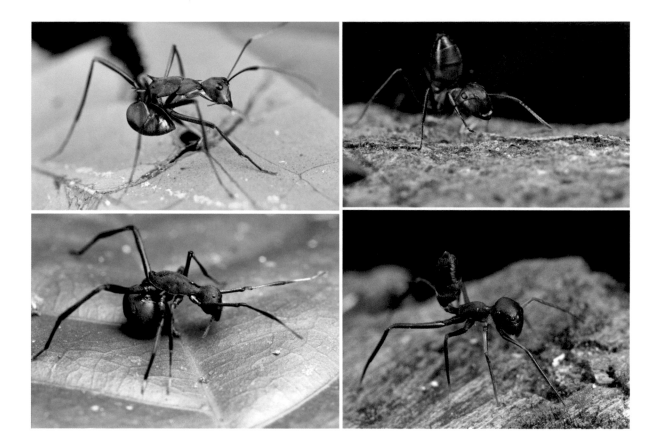

juvenile spiders are too small to match the proportions of ant species mimicked by adult spiders, an interesting question comes to mind: How do juvenile myrmecomorphic spiders avoid detection by predators as they grow from tiny hatchlings to much larger adults? The spiderlings of the ant-mimicking salticid *Toxeus magnus* seem to avoid detection by remaining cloistered inside their mother's nest during development (Figure 6-4). While they are safe inside a silk nursery, their mother provides them with a special fat and protein-rich "milk" that is deposited on the floor or imbibed directly from her epigastric furrow (Chen et al. 2018). This form of mammal-like provisioning has never been observed among nonmimetic spiders, and its existence may allude to the strength of the selective pressure imposed by predators on juveniles that have not yet attained the proportions of a larger, co-occurring ant species.

Rather than remaining hidden during development, the juveniles of other myrmecomorphic spider species undergo a remarkable and conspicuous developmental sequence. Across the successive stages of their development, these

SPIDERS AND OTHER MIMICS, PRETENDERS, AND PREDATORS

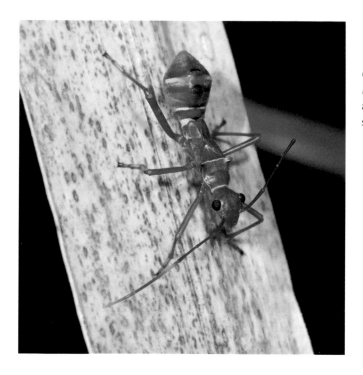

6-5 Nymphs of the pod-sucking bug *Riptortus serripes* (Alydidae) associate with and mimic the weaver ant *Oecophylla smaragdina*. (Courtesy of Lek Khauv).

M. plataleoides in containers with either *Oecophylla* worker ants or froghopper bugs. After two days, all the froghoppers were gone, and all the ants were still alive (Marson 1946). Despite nesting and hunting near their ant models, the spiders actively avoid contact with ants and will dart past workers to avoid them. *Myrmaplata plataleoides* does not show the same deference to juveniles of the ant-mimicking bug *Riptortus serripes* (Figure 6-5). Although the bug accurately mimics the ant *Oecophylla smaragdina, M. plataleoides* spiders easily recognize them as "non-ant" and do not move out of their way (Ceccarelli 2010). Evidence from several authors suggests that the cost of misidentifying an *Oecophylla* worker is high, as groups of *O. smaragdina* will capture and dismember *M. plataleoides* if given the opportunity (Figure 6-6) (Mathew 1954; Ramachandra and Hill 2018). Therefore, while *M. plataleoides* may avoid predators by resembling a weaver ant, it gains no food or protection from the ants themselves and is not recognized as a nestmate by its ant models.

In adulthood, male *M. plataleoides* face a conflict between mimicry and mating. Adult male *M. plataleoides* have enormously exaggerated chelicerae (mouthparts) that are used in male-male competitions (Figure 6-7) (Nelson and Jackson 2007). These chelicerae are so exaggerated that they extend the total body

6-6 The spider *Myrma-plata plataleoides* (Saltic-idae) is a Batesian mimic of the weaver ant *Oecophylla smaragdina*. It lives close to its ant model but actively avoids contact with the ants. If detected, *M. plataleoides* is quickly attacked and killed. (Courtesy of Pavan Ramachandra).

6-7 The spider *Myrmaplata plataleoides* mimics the weaver ant *Oecophylla smaragdina*. Males have enormously elongated chelicerae that are used in ritualized competitions with conspecific males. Note the eyespots on the marginal tips of the chelicerae. (Courtesy of Bharat Hegde).

length of their owners by 50%–70% and lack the normal venom delivery apparatus of female spiders (Pollard 1994). With such unusual chelicerae, it seems impossible that a male spider could accurately mimic a weaver ant. And yet, evolution has seamlessly incorporated the male's exaggerated mouthparts into a morphological bluff like no other. In some populations, the chelicerae of male *M. plataleoides* have come to resemble the shape and coloration of a second ant, complete with its own head and eye spots (Figure 6-8). Thus, not only is *M. plataleoides* a Batesian mimic and a transformational mimic, but the male is also a compound mimic, made up of two distinct ant components, one carrying the other (Nelson and Jackson 2006, 2012).

The "ant" formed by the male's chelicerae resembles an adult *Oecophylla* worker in its pupoid-carrying pose, common among nestmates that are transported between the numerous silk and leaf nests occupied by a single colony of *Oecophylla smaragdina* (Nelson and Jackson 2012). In this regard, the male spider's disguise echoes more than just the morphology of an ant; it also reflects the polydomous ecology of its ant model. Although the "carried ant" formed

6-8 Males of *Myrmaplata plataleoides* are compound (encumbered) mimics of *Oecophylla* ants. Their elongated chelicerae resemble an ant being carried by a nestmate. The upper picture shows a male *Myrmaplata plataleoides* morph with eyespots at the end of his chelicerae. (Courtesy of Melvyn Yeo). The lower picture depicts a male morph that features the imitation of the gaster of a carried ant at the end of the chelicerae. (Courtesy of Jeevan Jose).

by the male's chelicerae is situated backward by our assessment, a second *M. plataleoides* morph bears a "carried ant" that is in the correct position. The male of this *M. plataleoides* morph has a black opisthosoma that looks like the gaster of a local carpenter ant. In this case, the spoon-shaped tips of his chelicerae also bear the dark coloration of an ant's gaster (Figure 6-8) (Borges et al. 2007).

SPIDERS AND OTHER MIMICS, PRETENDERS, AND PREDATORS

6-9 The pronotum of the treehopper *Cyphonia clavata* (Membracidae) resembles a backward-facing ant, with antennae, propodeal spines or legs, and a shining head. (Courtesy of Iwan van Hoogmoed).

Although compound mimics like *M. plataleoides* succeed in evading predators that find ants unpalatable, there are some specialized predators that preferentially hunt ants. The myrmecophagous spider *Chalcotropis gulosus* (Salticidae) is a specialist hunter of ants carrying objects. Given the choice, it prefers "encumbered" ants over those whose dangerous mandibles are not occupied. Nelson and Jackson (2006) demonstrated that individuals of *C. gulosus,* in addition to consuming ants carrying prey and brood, readily prey on male *M. plataleoides,* which mimic worker ants carrying nestmates. The same predators avoid females of *M. plataleoides,* who lack exaggerated chelicerae and resemble unencumbered weaver ants. The predator's preferences are a testament to the accuracy of the male spider's disguise, though mimicry clearly makes *M. plataleoides* vulnerable to a new set of problems (Nelson and Jackson 2006).

Spiders are not the only myrmecomorphs to transform individual body parts into entire ants. The pronotum of the treehopper *Cyphonia clavata* (Membracidae) is sculpted into an ant seated in reverse, complete with its own leggy projections and propodeal spines (Maderspacher and Stensmyr 2011). The "ant" hangs like a strange ornament above the otherwise unassuming body of the bug (Figure 6-9). Likewise, the bodies of other membracid species seem to resemble individual ant gasters, making them difficult to single out among the ants that tend them for their honeydew (documented by Alex Wild, Figure 6-10).

6-10 Ants in the genus *Cephalotes* tend membracid bugs and gather the honeydew they produce. The bodies of some membracid bugs (possibly *Chelyoidea* sp.) blend in with the gasters of their well-defended ant attendants. (Courtesy of Alex Wild / alexanderwild.com).

6-11 *Myrmarachne melanotarsa* is a social spider and morphological mimic of *Cremato-gaster* sp. ants. The spider imbibes honeydew from coccids (scale insects) kept as livestock by the ants (upper), and gains access to the eggs of larger, ant-averse spiders. In turn (lower), the ants forage for discarded prey items in the spiders' communal webs. (Courtesy of Ximena Nelson).

Collective Mimicry

Ants are seldom alone, and ant-mimicking spiders likely benefit by joining their aggregations, either by reducing their own probability of detection or by avoiding predators that are deterred by the collective defenses of ants. One spider has capitalized on the collective appearance of its ant model by both mimicking their morphology and forming its own aggregations in interconnected silk hides that can house as many as fifty social spiders (Jackson et al. 2008). The communal dwellings of the salticid *Myrmarachne melanotarsa* are constructed near arboreal nests of highly defended *Crematogaster* sp., on the shores of Lake Victoria, in Kenya and Uganda (Jackson 1986; Wesołowska and Salm 2002). At a distance, the three-millimeter-long spiders are nearly indistinguishable from their *Crematogaster* ant models. During brief collisions along foraging routes, *M. melanotarsa* even "antennate" ants with their forelegs and raise their abdomens upward, so that that they are nearly perpendicular to the ground. These passing interactions between ant and ant mimic resemble the interactions between pairs of ant nestmates (Jackson et al. 2008).

Robert Jackson and colleagues (2008) observed another remarkable detail of the relationship between *M. melanotarsa* and its *Crematogaster* model. They found that the spiders commute up to a meter away from their nests to gather honeydew from the coccid livestock guarded by the ants (Figure 6-11). Although other salticids are known to consume nectar (Jackson et al. 2001) and even to place silk caps on extrafloral nectaries used by ants (Painting et al. 2017), this discovery represents a unique type of foraging among spiders. In addition to honeydew, *M. melanotarsa* consume the eggs of tree trunk spiders (Hersiliidae) and the eggs and juveniles of two other genera of social salticids (*Menemerus* and *Pseudicius*) that share their communal webs and likely gain protection from associating with *Crematogaster* (Jackson et al. 2008).

The story of *M. melanotarsa* and its *Crematogaster* model is made even more unusual by the observation that worker ants enter the silk nests of the spiders

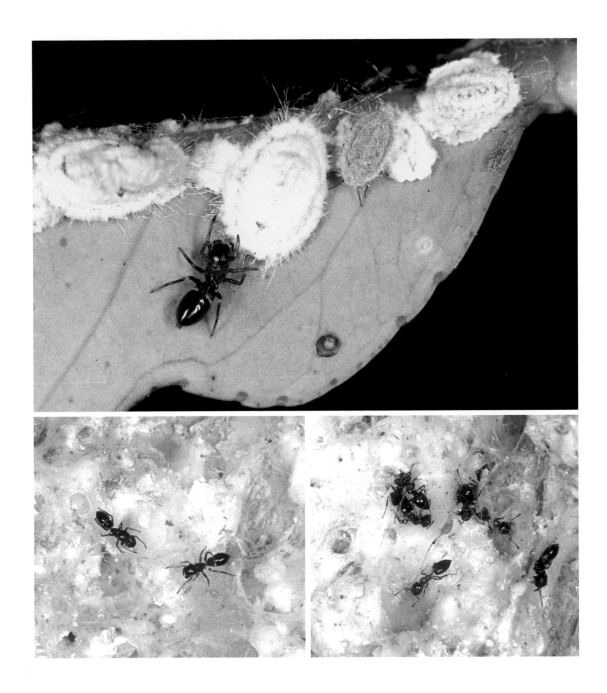

to forage on their discarded prey items, corpses, and other detritus (Jackson et al. 2008). The mimetic spiders are not aggressive toward their ant visitors, who have little difficulty walking within the public parts of their communal silk dens (Figure 6-11). Given that the spiders and ants provide services to one another, by some definitions, the relationship between the partners could be considered

SPIDERS AND OTHER MIMICS, PRETENDERS, AND PREDATORS

mutualistic. The main predators of spiders are often other spiders, particularly araneophagous salticids (Huang et al. 2011). The abundance of ant mimicry among salticid spiders may be driven in large part by an innate aversion toward chemically defended adult ants (reviewed by Cushing 1997). Ximena J. Nelson and Robert R. Jackson (2009b) found that larger, ant-averse salticid spiders were significantly more likely to abandon their egg sacs when approached by groups of ants or groups of mimetic *M. melanotarsa* spiders than when approached by lone ants or lone ant mimics. The araneophagous spiders did not flee from non-myrmecomorphic arthropods, and readily attacked gregarious salticids that do not mimic ants. The motion of potential prey species had little to do with the predators' aversion, as both living aggregations and immobile mounts of ants and ant mimics caused the predators to flee.

Salticid spiders are typically solitary hunters that form aggregations only for overwintering in temperate regions (Kaston 1948; Jennings 1972). While *M. melanotarsa* likely avoids predators by resembling a well-defended ant species (Batesian mimicry), groups of mimetic spiders also act as aggressive mimics and gain access to the eggs of larger, ant-averse salticids by resembling ants (Nelson

6-12   Several assassin bug (Reduviidae) species prey on ants and other small insects. The nymphs pile ant carcasses on their backs like a shell. The image shows a nymph of an unknown species (either *Acanthaspis petax* or *Inara flavopicta*) with its collection of ant corpses. (Courtesy of Nicky Bay).

and Jackson 2009a,b). The collective mimicry of *Crematogaster* by the myrmecophilous salticid *M. melanotarsa* is unusual, if not singular, among spiders.

Backpacks, Shields, and False Heads

A variety of true bugs (Hemiptera) prey on social insects, then gather their carcasses and pile them on their backs. These corpse backpacks have been shown to act as physical barriers, visual or chemical disguises for predators, and even as bait or camouflage directed at the bug's own prey (Brandt and Mahsberg 2002; Jackson and Pollard 2007). One such bug, *Salyavata variegata* (Reduviidae), loads the crumbs of termite carton nests on its back, and lures living termites out of the nest with the spent corpses of nestmates (McMahan 1983). The champions of ant-wearing are assassin bugs in the genus *Acanthaspis* (Reduviidae), who can amass up to 222 ant corpses in a "shell" that is glued to the body by sticky secretions (Figure 6-12). Bugs denuded of their shells suffer high rates of predation by nonvisual and visual predators alike (Brandt and Mahsberg 2002; Jackson and Pollard 2007). Miriam Brant and Dieter Mahsberg (2002) asked whether wearing ants might also increase the bugs' hunting success by allowing them to avoid detection by ants. They found that it is more likely the dust that accumulated on the bug's corpse collection than the corpses themselves that allow bugs to get close to living ants without being detected.

Spiders also make use of ant corpses in a more direct manner. The crab spider, *Aphantochilus rogersi* (Thomisidae), is a morphological mimic of the spiny and chemically defended ant *Cephalotes* (formerly *Zacryptocerus*) *pusillus*. Like its ant model, the spider bears a dull black cuticle, an oval-shaped abdomen, and a robust cephalothorax adorned with dorsal spines (Figure 6-13). It walks with an ant-like gait and wiggles its front legs ahead of its body like a pair of antennae. Yet, with only two body segments, *A. rogersi* is a truncated and imperfect replica of the posterior portion of an ant. This absence of an ant-like "head" is corrected by the spiders' unusual tendency to hold aloft the corpses of dead *C. pusillus* workers as they wander near the nests of *C. pusillus* (Piza 1937). Carrying an ant

SPIDERS AND OTHER MIMICS, PRETENDERS, AND PREDATORS

corpse not only creates a visual impression of an ant transporting a nestmate, but, as Paulo Oliveira and Ivan Sazima (1984) discovered, the corpse may also act like a chemical or tactile shield that allows the spider to go undetected among living ants.

Unlike other morphological mimics that are only spatially associated with ants, *A. rogersi* also feeds exclusively on *Cephalotes* and will reject a buffet of other ants and insect species when offered (Oliveira and Sazima 1984). During feeding, the spiders drain and hollow out the corpses of *C. pusillus* without damaging their exterior. The corpses are then carried and used as a shield for up to three days during interactions with living ants. In effect, *A. rogersi* is an aggressive mimic that employs shielding as a secondary disguise that appeals to the senses of the very species that it eats. If its ruse fails, the spider flees or resorts to suspending itself from a silk line until the ants end their search (Oliveira and Sazima 1984).

Couvreur (1990) observed another remarkable case of shielding in the spider *Zodarion rubidium* (Zodariidae). These spiders capture ants near the busy nest entrances of *Formica cunicularia* and *Tetramorium caespitum* in central Europe. If a spider meets another living ant while absconding with her prey, she extends her first pair of legs and drums on that ant's antennae. With the ant's full attention, the spider repositions the paralyzed victim in her chelicerae and presents it, like a shield (Couvreur 1990). Passing nestmates pause to inspect such decoys for an average of eleven seconds, and in a majority of laboratory trials, inspectors move on without further confrontation (Pekár and Král 2002).

Although *Z. rubidium* only loosely resembles its prey species in color and sheen, Stano Pekár and Jirí Král (2002) propose that the tactile signals of the spider are enhanced by more detailed morphological adaptions, including a lack of macrosetae on the distal part of the spider's front legs and the presence of flattened incised setae on its other limbs. Both patterns are suggested to correspond to the setosity of ant antennae and legs. Although the resemblance between the ant and spider appendages is not striking, in our opinion, such anatomical similarities could aid in tactile interactions with ants. Wasmann (1889b, 1903, 1925)

considered that, in some cases, ants might be the target of their guest's anatomical and tactile resemblance, though current evidence is largely anecdotal.

Two lines of evidence suggest that *Z. rubidium* do employ both olfactory and tactile signals associated with living ants while presenting corpse-shields to their hosts. First, paralyzed nestmates are met with alarm and intense inspection when

SPIDERS AND OTHER MIMICS, PRETENDERS, AND PREDATORS

6-15 When threatened, the spider *Pranburia mahannopi* extends its forelegs to unite a pair of semicircular brushes and waves its outstretched tibiae and metatarsi, like a pair of elbowed antennae (upper and middle: Courtesy of Paul Bertner). Together, the male's brushes resemble the head of the ant *Diacamma rugosum* (lower: Courtesy of Kwan Han).

6-16   The ant-mimicking spider *Myrmarachne formicaria* gives the illusion of walking on six legs, rather than eight, by raising its forelegs like antennae during 100-millisecond pauses, imperceptible to many visual predators. (Courtesy of Gil Wizen).

demonstrating that the mimetic spider *Myrmarachne formicaria* only gives the illusion of having an ant's six-legged gait (Figure 6-16). High-speed video revealed that the spider's walking trajectory approximates that of its ant model, but that the impression of an ant's gait and waving antennae is created as the spider lifts its first pair of legs above its body during 100-millisecond pauses, imperceptible to observers with slower visual systems. To watchful amphibians, reptiles, and other jumping spiders, the mimetic spider would appear to move as fluidly as an ant (Shamble et al. 2017).

While it seems that the ant-like appearance of *M. formicaria* is effective against araneophagous spiders, jumping spiders characteristically raise their front limbs to ward off other jumping spiders and during courtship displays. It is unclear whether araneophagous (spider eating) salticids are driven away by the impression of the ant-model or by the presentation of the mimic's raised legs. Whatever the reason, araneophagous salticids preferentially pounce on animated representations of nonmimetic spiders in motion at a rate 4.5 times higher than ant targets and three times higher than mimetic *M. formicaria* (Shamble et al. 2017). Likewise, spiders that do prefer to eat ants, like *Sandalodes bipenicillatus,* show a greater interest in myrmecomorphic *Myrmarachne* than in nonmimetic spiders (Nelson and Card 2015). These preferences can easily be assessed by showing predatory salticids the silhouettes of prospective prey species on video screens.

It is likely that the morphological and behavioral illusions of ant-mimicking spiders are under selection by more than one type of predator, and that certain elements are more effective against one predator than against another. For instance, the profile and dorsal view of a mimic may each be visible to a different size class of predator, or to a crawling or flying predator, respectively. In some areas, both ant-eating and spider-eating predators are common (Nelson and Card 2015), making the occurrence of ant-mimicking spiders an interesting puzzle of relative predator preference, sensory ability, appetite, and abundance.

Odor Cloaks

Pekár and Jiroš (2011) asked whether myrmecomorphic spiders, in addition to looking like ants, might also obtain the odor profiles of their ant models during their interactions. While these spiders do not live inside ant nests, they can be found in close proximity to their ant models aboveground and loosely resemble them in color, size, and movement. In general, the authors found that the cuticular hydrocarbons of the myrmecomorphic spiders and their models did not match. Specifically, the ants and spiders showed little similarity between di- and trimethyl-branched alkanes and alkenes, which are often used by ants to identify nestmates and nestmate status (Hefetz 2007; Martin and Drijfhout 2009b; Sprenger and Menzel 2020; see Chapter 3). Among the spiders sampled, only *Zodarion alacre* is a specialist predator of its ant model, *Formica subrufa*. Though *F. subrufa* were less aggressive toward freeze-killed mimics than toward unassociated spiders in behavioral trials, there was only a weak similarity between the odor profiles of the pairs (Pekár and Jiroš 2011). Like other *Zodarion* spiders, *Z. alacre* secretes a substance from a special femoral organ on its first pair of legs, which may play a more substantial role in its interactions with ants (Pekár and Sobotník 2007). Overall, the nonintegrated myrmecomorphs who are only loosely associated with ants, in space, seem to rely more on behavioral tactics than on odor mimicry to evade detection by their ant models.

A variety of arachnids do live as more integrated guests of ant colonies (Figure 6-17). The goblin spider, *Sicariomorpha* (formerly *Gamasomorpha*) *maschwitzi* (Oonopidae), is a common guest of the Southeast Asian army ant *Leptogenys distinguenda* (Wunderlich 1994; Ott et al. 2015). The little spiders are nearly blind and follow the frequent colony emigrations of their hosts by making physical contact with the ants in a sort of tandem run (Figure 6-18).

Volker Witte and colleagues (1999) also demonstrated that these spiders can follow freshly laid artificial trails drawn with the contents of their host's poison glands and pygidial glands. The spider remains in the bivouac rather than attending foraging raids, where it climbs over and rests on top of groups of workers and larvae, often mounting the ants while they are in motion. The ants are rarely aggressive toward their kleptoparasitic spider guests, which seem fully integrated with their hosts and are even groomed by them (Witte et al. 1999; von Beeren et al. 2012a). As we already discussed in Chapter 3, von Beeren and colleagues (2012a) discovered that *S. maschwitzi* obtains a significant proportion of its cuticular hydrocarbons from its host and loses them when placed in isolation. Although the ants use cuticular hydrocarbons as nestmate recognition cues, spiders that lose them during experimental isolation remain well integrated in host colonies. This could indicate that the spiders rely on "chemical insignificance" rather than mimicry, by normally maintaining a low abundance of cuticular hydrocarbons.

Another unusual arachnid guest can be found in the rainforests of Central America. The tailless whip scorpion *Phrynus barbadensis* (= *gervaisii*) (Amblypygi) is a facultative associate of the bullet ant, *Paraponera clavata* (LeClerc et al. 1987; Pérez et al. 1999; de Armas and Seiter 2013). Approximately half of all nests examined on Barro Colorado Island, Panama, house mature and immature whip

SPIDERS AND OTHER MIMICS, PRETENDERS, AND PREDATORS

6-17 Upper: The spider *Phruronellus formica* lives inside the nests of the ant *Crematogaster cerasi*. (Courtesy of Sean McCann). Lower: An unidentified spider travels with *Labidus* sp. army ants. (Courtesy of Taku Shimada).

6-18 The goblin spider, *Sicariomorpha maschwitzi* (Oonopidae), is an integrated guest of the Southeast Asian army ant *Leptogenys distinguenda*. (Courtesy of Christoph von Beeren).

scorpion tenants (LeClerc et al. 1987; Pérez et al. 1999; Peretti 2002). Bullet ants are so named because of their uncommonly painful sting and potent venom. Despite the moniker, the ants are not aggressive toward the large arachnids that occupy their nest chambers (Figure 6-19). Monica LeClerc and colleagues (1987) observed that "when disturbed ants erupted from the nest in defense, *P. gervaisii* [*barbadensis*] usually remained in the sides of the entrance unless ant activity became too intense, in which case they moved aside to accommodate the ants. Accommodation included lifting their long legs and body off the surface of the tunnel allowing ants to freely pass underneath. When *P. gervaisii* [*barbadensis*] was gently touched they typically retreated deeper into the entrance."

How the large-bodied tailless whip scorpions (12–22 mm) manage to go undetected is unknown (de Armas and Seiter 2013). Chapin and Hebets (2016) suggest that a second species, *Phrynus pseudoparvulus,* is only an opportunistic cavity dweller that has no special association with the ants it is found with. Although both species also shelter outside of ant nests, slow movements and odor (or lack thereof) almost certainly allow them to avoid eliciting the ant's typical aggressive response. Rolando Pérez and colleagues (1999) suggest the relationship between the *P. gervaisii* and its ant hosts may even be mutualistic or commensal in nature. In a long-term study of the biotic factors associated with colony survival, the presence of tailless whip scorpions was positively associated with colony survival: 61% of colonies that housed *P. barbadensis* during the first sample date of the study were still alive after two years, compared to only 17% of colonies that

6-19 The tailless whip scorpion, *Phrynus barbadensis* (Amblypygi), is found in the rainforests of Panama, where it is a facultative guest of the bullet ant *Paraponera clavata*. (Courtesy of Gil Wizen).

did not house the arachnids. Of course, it is possible that larger and more robust ant colonies might be more easily noticed by searching *P. barbadensis,* or provide more chamber space for lodging.

The first record of myrmecophily in a true scorpion, *Birulatus israelensis* (Scorpiones: Buthidae), was made in Israel's Jordan Valley in 2017. Yoram Zvik observed dozens of the one- to two-centimeter-long scorpions on the trails and around the nest entrances of the seed-harvesting ant *Messor ebeninus.* The scorpions dipped in and out of the ants' nests and wandered among the ants without alarming them, occasionally pausing to allow ants to inspect them. Workers deposited dead scorpions alongside dead nestmates in the organized corpse piles

6-20 The crab spider *Amyciaea* sp. resembles and preys on the Australian green tree ant, *Oecophylla smaragdina*. (Upper: Courtesy of Alex Wild / alexanderwild.com). The jumping spider *Cosmophasis bitaeniata* lives in and around the arboreal tent nests of *Oecophylla smaragdina*, where it feeds on ant brood. (Courtesy of Mark Moffett).

maintained by each colony. The fact that these scorpions were sorted in a manner distinct from food waste suggests that they may share their host colony's odor. In general, evidence that myrmecophiles receive an "ant's funeral" is scarce.

Several spiders form associations with ants in the genus *Oecophylla,* including the aforementioned transformational mimic *Myrmaplata plataleoides, Myrmarachne foenisex,* and the predatory crab spiders *Amyciaea albomaculata, A. lineatipes,* and *A. forticeps* (Thomisidae) (Allan et al. 2002). While each of these species bears a visual resemblance to *Oecophylla,* the salticid *Cosmophasis bitaeniata* does not (Figure 6-20).

SPIDERS AND OTHER MIMICS, PRETENDERS, AND PREDATORS

Cosmophasis bitaeniata lives in and around the arboreal tent nests of *Oecophylla smaragdina*. It is an abundant myrmecophile, and in 130 nests surveyed in Queensland, Australia, by Rachel A. Allan and Mark A. Elgar (2001), 36% contained as many as six spiders. Among the many leaf and silk nests occupied by a single host colony, spiders preferentially occupied older nests and nests with more of their favorite prey, ant larvae. *Cosmophasis bitaeniata* lives in close proximity to *O. smaragdina* and deposits its own egg sacs in the ants' tent nests, but it avoids close contact with the ants. Spiders approach minor workers only to steal larvae from their grasp. They do so by touching their forelegs to the antennae of the ant, which stimulates her to drop the larva held in her mandibles (Allan and Elgar 2001). In laboratory trials, spiders were more successful at stealing brood from minor workers belonging to their own nest than from minor workers belonging to foreign nests (Elgar and Allan 2006). Major workers seem to show little interest in *Cosmophasis bitaeniata*. While the ants react aggressively to discs marked with the cuticular odors of foreign workers (size-standardized) they show no aggression toward discs marked with the cuticular extracts of workers and spiders from their natal nest (Elgar and Allan 2006). Spiders do try to escape when confined with major workers from foreign nests, but it is not because they are harassed or noticed by the ants. Instead, the spiders reportedly use their own odor template to distinguish nestmate and non-nestmate ants (Elgar and Allan 2006).

To determine whether spiders might acquire their hosts' worker or larval recognition cues, Elgar and Allan (2004) took forty-five newly hatched *C. bitaeniata* from multiple mothers and randomly assigned them to three groups. Each group received larvae from one of three *O. smaragdina* colonies and was reared without any direct contact with ants. The cuticular hydrocarbon profiles of the isolated spiders were later compared to those of spiders reared alongside ants and larvae from each of the three *O. smaragdina* source colonies. The odor profiles of spiders raised in isolation resembled those of the colonies that provided them with larvae. They did not match the odor profiles of their own spider siblings, who were reared separately and fed on larvae from different ant colonies. The cuticular hydrocarbon profiles of spiders reared with and without workers were similar, though both groups of spiders shared only some of the cuticular hydro-carbons found on minor workers (Allan et al. 2002), and even fewer with major workers (Elgar and Allan 2006). Together these studies suggest that *C. bitaeniata* may pass unnoticed in its host nest by behaviorally avoiding ants and by sharing some of the recognition cues associated with larvae and minor workers.

Although we found these studies very interesting, the authors note, and we agree, that it is difficult to interpret which colony members spiders might be mimicking. From the gas chromatograms (Allan et al. 2002), it appears to us that the cuticular chemical profiles of ant larvae, minor workers, and major workers all share many of the same compounds in similar proportions. The spiders' cuticular hydrocarbon profiles do more closely resemble those of the ant larvae, because they also lack the shorter-chain compounds present on major and minor workers. However, Elgar and Allan (2006) report that there were additional, colony-specific components of the cuticular profiles of spiders and ant workers, and that these colony-specific components also differed between spiders and ants from the same nest. The spiders reportedly acquire their cuticular hydrocarbon blends from the ant larvae they eat (Elgar and Allan 2004), but it is still unclear whether the ant larvae's cuticular hydrocarbons show any colony specificity. If the spider odor does resemble that of an ant larva, then spiders should easily be accepted into colonies of the same species in the same way that larvae are universally accepted. Instead, the spiders' odors differ according to colony and region, and the spiders' own templates seemingly make them averse to joining foreign nests (Elgar and Allan 2006).

Phoretic Spiders Take Flight

As we have seen, a variety of spiders walk like ants or walk among ants, but there is at least one genus that prefers to ride them. The genus *Attacobius* (Corinnidae) includes more than a dozen spider species (Pereira-Filho et al. 2018), several of which are parasites of neotropical leafcutter ants in the genera

Atta and *Acromyrmex* (Mendonça et al. 2019). On the day of the ants' annual nuptial flights, winged *Atta bisphaerica* males and future queens emerge from their nests with up to three *Attacobius luederwaldti* spiders clinging to their backs (Ichinose et al. 2004). The pale-orange spiders are not harassed by the ants and do not disrupt the movement of the ants' wings as they ascend into the air and travel distances greater than 10 km (Jutsum and Quinlan 1978; Ichinose et al. 2004). Approximately 11% of unmated alates depart with spider passengers, but newly mated queens are never found with spiders still attached after mating flights, nor do they host them in their fledgling colonies. Instead, *A. luederwaldti* appear to use winged ants only to gain elevation and distance from their natal nests. It is presumed that they leap from their hosts midflight and balloon away on silk parachutes to seek out large, established colonies (Ichinose et al. 2004).

As we described in Chapter 5, Phillips (2021) observed a similar "two-part" mode of dispersal in the miniature wingless cockroach *Attaphila fungicola*. The roaches attach themselves to female *Atta texana* alates prior to mating flights but do not remain with the foundresses once they land to excavate new nests. Instead, the roaches "hitchhike" along the routes of established *A. texana* nests and leap onto leaves carried by inbound foragers. The suggested term for this type of dispersal is diplophoresy (Phillips 2021). The probability of a new leafcutter ant queen surviving is extremely low (Fowler 1992), and given this high chance of failure, there are clear benefits for spiders (and roaches) to pursue mating opportunities in already-established ant nests. This is likely why *A. luederwaldti* spiders board *both* female and male ants, the latter of which are all destined to die shortly after mating (Ichinose et al. 2004; Camargo et al. 2015). In contrast, the roach *Attaphila fungicola* prefers to ride female alates (Phillips et al. 2017), and rather than leaping off midflight like *Attacobius* spiders, the roaches remain with their alate until she has landed. In general, many female alates select suitable nesting habitat from the air, which may place the roaches closer to the habitat already occupied by other established colonies (King and Tschinkel 2016).

While spiders can move secondarily using silk and wind, *Attaphila fungicola* roaches are wingless and lack any long-range dispersal ability. A variety of phoretic mite species also prefer female alates over male alates for what may be entirely different reasons (Sokolov et al. 2003; Ebermann and Moser 2008; Uppstrom and Klompen 2011). Mites are much smaller and more prone to desiccation than *Attacobius* spiders and *Attaphila* roaches. Uppstrom and Klompen (2011) suggest that the mites may either gamble on the success of their chosen queen or wait with her corpse until it is transported into a new nest by cannibalistic ants that gather the bodies of failed queens.

Roberto da Silva Camargo and colleagues (2015) found that it is primarily female spiders that disperse on winged ants in a second *Attacobius* species, *Attacobius attarum* (Figure 6-21). We do not know the sex ratio of adult spiders within nests, but 96.9% of spiders dispersing on alates were found to be adult females. Rather than boarding the alates inside their nests, spiders select them on the nest mound and can be found on approximately 9% of alates just prior to departure. These are important observations, as there are very few reports of the sex-specific philopatry and dispersal of myrmecophiles, or even of their typical sex ratios. *Attacobius attarum* are regularly found clinging to the heads and gasters of similarly sized *Atta sexdens* host workers and may even enter large colonies this way following dispersal (Erthal and Tonhasca 2001).

When separated, the spiders scramble to return and continue their portage on the nearest available worker (Platnick and Baptista 1995). Spiders embedded with laboratory colonies readily consume ant larvae and pupa (Erthal and Tonhasca 2001), and this observation is consistent with that of related *Attacobius lavape* found in the nests of *Solenopsis saevissima* in Brazil. When locally abundant, *A. lavape* occur at an average of seven (and up to twenty-three) spiders per ant colony. Spiders from a single nest can consume an estimated thirty-five larvae and thirty-two pupae, along with numerous eggs per host colony per day (Mendonça et al. 2019). Together, spiders in the genus *Attacobius* were the first among all spiders to be described as parasitic (Mello-Leitão 1923). The fascinating story

6-21   The South American corinnid sac spider, *Attacobius attarum*, disperses on the winged sexuals of the leaf-cutting ant *Atta sexdens* during mating flights (phoresy). The upper pictures show the spider riding on unmated queens of *Atta sexdens* before taking off for the nuptial flight; the lower images depict the spiders riding on *A. sexdens* males. (Courtesy of Roberto da Silva Camargo).

of their discovery and classification can be found in Platnick and Baptista (1995) and Bonaldo and Brescovit (2005).

Spider Predators of *Veromessor pergandei*

Veromessor pergandei is a seed-harvesting ant found in the Sonoran and Mojave Deserts of the southwestern United States. Each morning, tens of thousands of shining black foragers depart their colony in a single column. Nestmates travel together in a ribbon for up to 40 m before fanning outward in search of seeds (Plowes et al. 2013). Our curiosity led us to study the foraging behavior of *V. pergandei*, and later to calculate the rate of worker production necessary to counterbalance daily worker mortality in these massive societies. Like so many serendipitous encounters with myrmecophiles, hours spent prostrate next to *V. pergandei* nests also revealed a distracting variety of spiders living alongside the ants.

6-22 Upper: A *Steatoda* sp. spider builds its web over a *Veromessor pergandei* foraging column. Left: Large-bodied ants dismantle the spider's web. They are attracted by the alarm pheromone released by a nestmate, suspended in the center of the web. Right: The ants cooperate to free their sister from the spider's silk. (Christina Kwapich).

Steatoda and *Asagena*

The large but ephemeral foraging columns of *Veromessor pergandei* are most noticeably exploited by false widow spiders, in the genera *Steatoda* and *Asagena* (Theridiidae). The spiders spend their nights beneath rocks next to *V. pergandei* nest entrances, where they also lay their egg sacs. At sunrise, the male and female spiders emerge and begin constructing small webs above or adjacent to their colony's newly established foraging column. Each spider anchors several messy tangle webs to the ground with sticky threads and secures additional lines between vegetation and stones. When a taut line of silk is touched by a passing ant, she is flipped upward and suspended into the spider's web (Figure 6-22). Captured ants are reeled in and wrapped in silk or left hanging as the spider attends to its bounty of up to seven additional captive nestmates, spread across multiple traps (Kwapich and Hölldobler 2019).

Although *V. pergandei* is found in one of the hottest habitats in the world, it is heat intolerant and prone to desiccation (Johnson 2000, 2021). Therefore, workers rush to complete multiple foraging trips within a narrow temperature interval

SPIDERS AND OTHER MIMICS, PRETENDERS, AND PREDATORS

usually lasting no more than two hours each morning (Bernstein 1974; Hunt 1977). Rather than constructing more permeant webs like many theridiids do, including widow spiders (*Latrodectus*), myrmecophagous *Steatoda* and *Asagena* construct temporary traps that persist for only a few hours on the desert floor. Constructing disposable webs allows the spiders to track the ants' ephemeral foraging columns, which depart in new directions each day.

Foragers in another dominant seed-harvesting ant genus, *Pogonomyrmex,* arrest their foraging activity or change their foraging patterns in response to predatory spiders (Hölldobler 1970a; MacKay 1982). In contrast, we found that *V. pergandei* cooperatively dismantle the webs built by *Steatoda* and *Asagena* and rescue sisters captured in the silk. A subset of large-bodied workers gather ensnared nestmates (living and dead) and groom away their silk bindings inside the nest (Figure 6-22) (Kwapich and Hölldobler 2019). Web-removers are even able to walk on the spider's webs and, despite the risk of capture, successfully detach the silk by gripping it in their mandibles and methodically backing away. Few other prey species seek out and destroy the traps designed to capture them. Indeed, web removal represents one of only four natural contexts for rescue behavior among ants. Similar silk removal and rescue behaviors have also been reported in *Oecophylla smaragdina* (Uy et al. 2019) and *Novomessor* sp. (Michael C. Clark personal communication, 2019).

Animals that perform rescue behavior typically form small groups with high-value individuals. In contrast, *V. pergandei* have enormous colonies that can replace a whopping 34,000 foragers during their peak foraging month and 230,000 in a single year (Kwapich et al. 2017). To determine why colonies might bother to rescue "disposable" workers, we calculated the costs and benefits of spiderweb removal at the colony level. By accounting for the length of a foraging career and number of foraging trips taken by an average forager population per day, we estimated that the loss of just five new foragers a day to spider predation could cost colonies 65,000 seeds per year. This is a high price to pay, because colonies already need to gather enough resources to replace more than 600 sisters each day. In addition to the direct costs of predation, seeds carried by foraging ants frequently become tangled in undetected spiderwebs, snarling traffic and further reducing the total number of foraging trips individuals can take per day. When their seeds become tangled in spiderwebs, the ants pull on them for the remainder of the foraging period, all without taking notice of the cause of the problem or removing the silk obstructing their progress (Kwapich and Hölldobler 2019).

Many ant species clear debris from their foraging routes, but *V. pergandei* foragers ignore novel objects and do not react aggressively to spider silk alone. Only when ensnared ants release a chemical alarm signal from the mandibular gland is a subset of large-bodied nestmates stimulated to remove surrounding webbing. Frozen "dummies" marked with the same alarm compound are also freed from their silk bindings (Kwapich and Hölldobler 2019). In essence, colonies benefit from the removal of webs only when workers are captured in them and subsequently signal their location. The tendency of *V. pergandei* to forage on a single route during a limited temperature window allows workers to encounter their alarmed sisters. This, coupled with the necessary scale of seed harvesting, likely led to the ant's unusual defenses against spiders.

Euryopis

Social prey species present a special problem for predators, who must account for additional aggression from vigilant group members. In the American west, *Euryopis* (Theridiidae) spiders are conspicuous hunters of harvester ants from the genera *Veromessor* and *Pogonomyrmex*. Each *Euryopis* species employs a unique, active hunting strategy that combines silk and venom. *Euryopis californica* subdue *Veromessor pergandei* workers by hog-tying their legs with silk and administering a bite to the gaster, before dragging them away on silk lines tethered to their spinnerets (Figure 6-23) (Hale et al. 2018).

Likewise, *Euryopis coki* ambushes its *Pogonomyrmex salinus* (= *owhyheei*) victims with a bite to the leg, then tethers them to the ground with sticky silk lines while

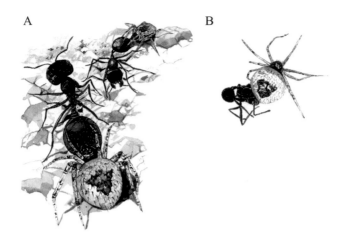

6-23   (A) *Euryopis californica* attack workers of the desert seed-harvesting ant, *Veromessor pergandei,* near the nest entrance. (B) Spiders transport their paralyzed prey on lines of silk attached to their spinnerets. (Courtesy of Amanda Hale).

the venom takes effect. Once the prey is paralyzed, the spider wraps it in a silk sling and carries it away (Porter and Eastmond 1982). A third *Euryopis* species observed by Kwapich in Arizona bites *P. barbatus* on the last pair of legs, waits for paralysis, and then feeds from the ant's head, just above the mandible. Finally, a fourth *Euryopis* species from Arizona bites its *V. pergandei* victims behind the head, then quickly binds the head and mandibles with silk, deactivating the ant's primary defenses and perhaps stifling its ability to signal nearby nestmates. While we are curious whether this behavior prevents the captured ant from releasing the chemical alarm signals from her mandibular glands, the observed variation in preferred bite location may simply represent the spider's preference for consuming one body part over another.

Stano Pekár and colleagues (2010) demonstrated that spiders offered either the foreparts of ants or the gasters of ants grew at different rates and had different mortality rates. Likewise, myrmecophagous *Euryopis episinoides* that were fed a diet of *Drosophila* rather than ants suffered reduced fecundity and reduced longevity (Líznarová and Pekár 2016). It appears that several *Euryopis* species are stenophagous, with preferences not only for a single prey species, but also for a particular body segment (Carico 1978). This may be a consequence of venom composition and its efficacy against the defenses of specific ant genera or subfamilies (Hale et al. 2018).

In addition to employing different silk and venom tactics, there is considerable variation in the daily rhythm of *Euryopis* hunting behavior. *Euryopis californica* (Theridiidae) ambushes *V. pergandei* workers after dusk once the ants' foraging has ceased, and the few workers scattered around the nest entrance move slowly in the cooling night air. As many as five spiders visit a nest at the same time (Hale et al. 2018). In southern Arizona, *Euryopis* sp. shelters in the upper chambers of *V. pergandei* nests during the heat of the day. Unlike their Californian counterparts, they actively hunt during the ants' peak morning foraging hours. Spiders wait at the periphery of *V. pergandei* nest mounds, where they position their back legs toward wandering ants, and release silk to snare their prey when their hind tarsus or metatarsus is grazed. Groups of paralyzed workers are then bundled together by the legs and hauled up plant stems, where they are consumed as the spiders hang from short lines of silk. Here, smaller males sometimes arrive to mate with females and share their meals (C. L. K., personal observations).

While the hunting forays of *E. coki* associated with *Pogonomyrmex salinus* are made by day, Kwapich observed that *Euryopis* sp. begin descending on *P. barbatus* nests an hour before sunset. Male and female spiders grab straggling foragers and guards on the surface or enter open nest entrances to drag out isolated workers. Up to nineteen tan or pink spiders assemble near *P. barbatus* colonies each evening, blending perfectly with the multicolored pebbles collected and arranged by the ants on their nest mounds. When female spiders capture ants, males rush over to mate with them and sometimes share their prey after copulation (C. L. K., personal observations). The tendency of some *Pogonomyrmex* colonies to close the nest entrances at the end of the workday also presents an opportunity for *Euryopis* spiders, who easily pick off the late-returning foragers forced to wait until the nest is unsealed in the morning.

New Spiders

Much remains to be discovered about spiders living inside ant nests. In 2019, Martín J. Ramírez and colleagues used molecular and morphological data to describe a new spider species, *Myrmecicultor chihuahuensis,* belonging to an entirely new spider family, Myrmecicultoridae. These spiders have a unique combination of morphological characters and are found on the surface and inside the nests of *Pogonomyrmex rugosus, Novomessor albisetosus,* and *Novomessor cockerelli* in Texas (USA). Over the past five years, our own work brought us into contact with another spider associate of harvester ants in the Southwest, tentatively identified as the corinnid *Septentrinna bicalcarata* (Bonaldo 2000). Kwapich found the bright red and yellow spider species dwelling inside the nests of *Veromessor pergandei, Pogonomyrmex rugosus, Pogonomyrmex barbatus,* and *Novomessor albisetosus* in Arizona (USA) and has studied the behavior and ecology of these spiders with growing interest.

The spiders can easily be found in the top chambers of host nests and frequently wander in the masses of ant foragers that gather on nest mounds just prior to the initiation of foraging. We excavated one *Veromessor pergandei* nest to a depth of 160 cm, until it entered the hard caliche layer, and found eleven adult and juvenile spiders in chambers full of ants across the subterranean breadth of the nest. Together with our colleague Ti Eriksson, we used the mitochondrial gene cytochrome c oxidase 1 (CO1) to obtain DNA barcodes of spiders from three host species across three field sites in Arizona and determined that the spiders are likely host generalists rather than cryptic species with a high degree of host specificity (Kwapich et al. in prep. a). Individuals collected from any *P. rugosus, V. pergandei,* or *N. albisetosus* nest within a site were more genetically similar than spiders collected with each host species across sites. In the laboratory, spiders were easily transferred between ant host species, though they were initially harassed by their new hosts and liberally autotomized (cast off) their own legs when pursued.

We used solid-phase microextraction fibers to sample the cuticular hydrocarbons of living *S. bicalcarata* and found that spiders lacked novel compounds in the range shared by their hosts and used by ants as nestmate recognition cues. Spiders did not develop matching odor profiles even three days after molting in isolation from their ant hosts. Instead, the spiders were like a "blank slate" and acquired only a portion of their host's profile in nests, over time. Three days after experimentally transferring adult spiders to the nests of new host species, their profiles included a mixture of both hosts' cuticular hydrocarbon profiles, but after two weeks they shared major cuticular hydrocarbon peaks only with the new host. Spiders likely acquire their odor from feeding directly on adult ants and ant brood, and although ant aggression decreases over time, it is still unclear whether sharing the ants' colony odor provides a major benefit to the spiders. Young spiders preferentially prey on ant pupae, and sometimes carry pupae for hours as they consume them (Figure 6-24). Adult spiders prefer to feed on adult ants while clinging to chamber ceilings. After feeding, the spiders groom themselves meticulously, sometimes for upward of thirty minutes.

We housed male and female spiders together in glass-topped observation nests with *Veromessor pergandei* colonies. On the glass lids of each nest, spiders laid thin, squiggly trails of silk, presumably for orientation within the nest. Adult male spiders were frequently found without one or both pedipalps after mating with females, suggesting that the pedipalps may act as a mating plug or that they are shed just as easily as legs during physical interactions. Females readily laid egg sacs on chamber ceilings, containing just four to ten eggs. The egg sacs were often placed in corners, where one or both spiders also sat during periods of inactivity. Whether or not this behavior represents true egg guarding is unclear. Male parental care is exceptionally rare in spiders (Moura et al. 2017), though living among ants may necessitate a higher investment in egg-guarding behavior. After all, nesting sites are spatially constrained by ant architecture, guests may have limited mating opportunities, and egg sacs are at risk from patrolling ants. The frequent loss of pedipalps by *S. bicalcarata* could also predispose males to parental care behavior, as males without pedipalps cannot deliver sperm to future mates.

Why myrmecophilous arachnids, and myrmecophiles in general, lay so few eggs is unclear. Could the spatial limitations of host nests or availability of resources lead to parent-offspring conflict? Are ants stimulated to abandon their nests if parasitic guests become too abundant? Must females invest in a few larger offspring that, as hatchlings, can successfully hunt large-bodied ants, instead of investing in many small offspring? The aforementioned army ant guest *Sicariomorpha maschwitzi,* for example, only produces one to five eggs per clutch (Witte et al. 1999). Small clutch sizes are best documented in myrmecomorphic spider

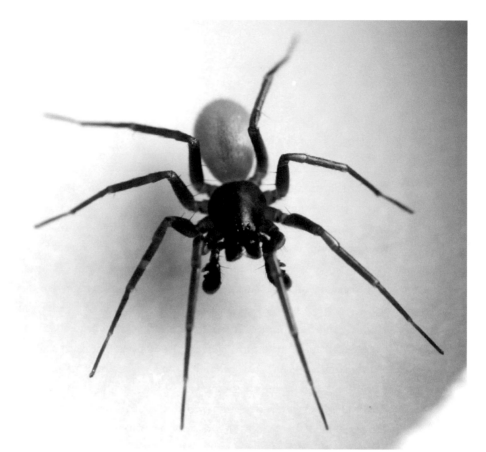

6-24   Upper: An unidentified spider species found throughout the nests of *Pogono-myrmex rugosus, Veromessor pergandei,* and *Novomessor albisetosus* in the desert Southwest of the United States. Lower: An adult male stalks its *V. pergandei* prey. The insert shows a newly hatched juvenile feeding on a *V. pergandei* larva. (Christina Kwapich).

species, where ant-like constrictions of the abdomen physically reduce the number of eggs females can carry (reviewed by Cushing 2012) and may even alter the size and arrangement of their silk-producing glands (Jessica Garb, personal communication). The physical trade-offs associated with myrmecomorphy are numerous. For example, when lunging at prey, jumping spiders must pump hemolymph into the cephalothorax to power a hydraulic catapult mechanism that extends their legs. The elongated bodies of ant-mimicking *Myrmarachne* seem to place constraints on the mechanics of this system, making very slender mimics among the poorest jumpers (Hashimoto et al. 2020).

The relationships between ants and spiders are more numerous and varied than we can express in this short chapter. For further reading, we recommend the comprehensive and compelling reviews of Paula Cushing (1997, 2012) and Jackson and Nelson (2012). In Chapter 7, we travel underground to consider the strange habits of ant-associated crickets. Although they are not Batesian mimics like many ant-associated spiders, it has been suggested that their smooth, rounded bodies aid in a chemo-tactile ruse employed in the darkness of subterranean nests. Many species also employ special tactile signals to pilfer food directly from the mouths of numerous host ant species.

7 The Mysteries of Myrmecophilous Crickets

MANY CRICKETS FROM the family Myrmecophilidae are found exclusively inside the nests of ants. These ant crickets measure no more than 1.47–8 mm in length, and are both the smallest and the only obligate parasites in the order Orthoptera (grasshoppers and crickets). They belong to one of the only ensiferan lineages without wings, without a song, and without a tympanal organ used for hearing (Song et al. 2020) (Figure 7-1).

Their round, wingless bodies are flanked by bulbous posterior femora and terminate in hairy cerci (Figure 7-2). Early in their taxonomic history ant crickets were misclassified as roaches (*Blatta*) (Panzer 1799), and more recently, a fossilized apterous roach was misidentified as a Cretaceous ant cricket (Martins Neto 1991; see Parker and Grimaldi 2014). The mistake would be easy to make, because ant crickets share many features with apterous roaches, including an incomplete metamorphosis, with juveniles that resemble tiny adults.

In 1825, ant crickets were again briefly misclassified, this time as bivalves. Toussaint de Charpentier intended to describe the insects' "corporis globositatem" with the new genus name *Sphaerium,* but was unaware that it already belonged to a group known colloquially as fingernail clams. The genus name was, therefore, an invalid taxonomic homonym. Today, six genera from two subfamilies are recognized. The subfamily Bothriophylacinae (*Eremogryllodes, Bothriophylax,* and *Microbothriophylax*) includes crickets that live in caves and the burrows of desert vertebrates rather than ant nests (Tahami et al. 2017), while the subfamily Myrmecophilinae includes the better-known ant cricket genera, *Myrmecophilus, Myrmophilellus,* and *Camponophilus.* The genus *Myrmecophilus* (formerly *Myrmecophila*)

7-1 Ant crickets (Myrmecophilidae) are wingless and lack tympanal organs, used for hearing. *Myrmecophilus pergandei* is found with numerous ant hosts across the eastern United States. An adult female is pictured. (Courtesy of Alex Wild / alexanderwild .com).

includes at least sixty-two valid species and is the focus of most behavioral investigations (Cigliano et al. 2020). Ant crickets can be found across Asia, Africa, Europe, and North America, but are conspicuously absent from Central and South America, except with ant host species introduced from other parts of the world (Saussure 1877).

In this chapter, we explore the enduring mysteries of ant crickets and the unusual consequences of their hemimetabolous development, a feature that sets them apart from the majority of insect myrmecophiles. Some ant cricket species can exploit numerous ant hosts from different subfamilies and size classes. The apparent influence of host identity on cricket body size has been a source of curiosity since the early twentieth century (Hebard 1920; Hölldobler 1947).

THE MYSTERIES OF MYRMECOPHILOUS CRICKETS

7-2 Ant crickets can instantly be recognized by their rounded bodies and robust posterior femora. Upper and middle: An adult female of an unknown ant cricket species from Malta. Her long ovipositor is visible. (Courtesy of Nikolai Vladimirov). Lower: A bristly adult *Myrmecophilus ochraceus* male from Sardinia, Italy. (Courtesy of Thomas Stalling).

Strigilators and Thieves

William Morton Wheeler (1910) described *Myrmecophilus nebrascensis* as strigilators, with mouthparts designed to lick and rasp waxes from the bodies of ants, like the blades of a strigil (used to scrape away bath oils in ancient Greece). In many ways, strigilation resembles the intensive grooming performed by partners in certain vertebrate mutualisms, and early myrmecologists considered that the relationship between cricket and ants might likewise be mutualistic (Savi 1819; Wasmann 1901, 1905). Later observations by Karl Hölldobler (1947) showed that *Myrmecophilus acervorum* also consume ant eggs, feed on insect prey collected by ants, and even solicit mouth-to-mouth trophallaxis with their hosts. He concluded that the relationship between ant crickets and their hosts was truly one-sided. Today, all ant-associated crickets are considered obligate kleptoparasites.

To solicit trophallaxis, the desert ant cricket, *Myrmecophilus manni*, duplicates the antennal drumming sequence used by ants during their exchanges with nestmates. Henderson and Akre (1986a) observed that, "when in trophallaxis, the cricket maintained a tense posture with its metathoracic legs outstretched. Trophallaxis usually lasted ca. five seconds before the ant became aggressive, and vigorous antennal contact by the cricket was required to maintain trophallaxis beyond this time. However, in three cases an ant remained in a position for trophallaxis for over a minute; the cricket left and returned several times within that period."

The authors made more than fifty such observations of trophallaxis between *M. manni* and some of its many hosts, including *Formica obscuripes, F. fusca, F. haemorrhoidalis, Camponotus vicinus, C. modoc, Tapinoma sessile,* and *Myrmica* sp. in Washington (USA). In the laboratory, crickets taken from nests of *F. obscuripes* also engaged in trophallaxis with unfamiliar ant species, like *F. fusca, F. haemorrhoidalis,* and *Camponotus vicinus,* demonstrating their lack of host specificity and their ability to exploit ant species never encountered during their lifetime (Henderson and Akre 1986a).

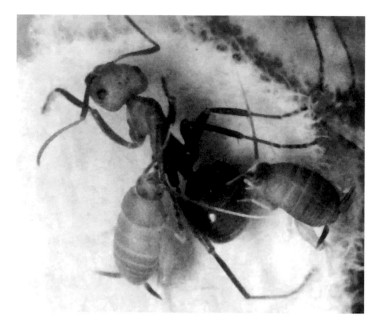

7-3 The desert ant cricket, *Myrmecophilus manni*, is found with numerous host species. A group of crickets grooms and strigilates a *Formica obscuripes* host worker. The crickets also consume host brood and initiate trophallaxis with them by mimicking antennal signals. (Courtesy of Gregg Henderson).

Although *M. manni* easily initiates trophallaxis with a variety of ants, it also exploits ant hosts that do not participate in liquid food exchange, including various *Pogonomyrmex, Veromessor, Aphaenogaster, Novomessor,* and *Pheidole* species in the southwestern United States (Hebard 1920; Kwapich et al. in prep. b). With these hosts, the crickets' diet is probably derived from a combination of strigilation and scavenging of food scraps, ant eggs, and dead workers (Henderson and Akre 1986a). *Myrmecophilus manni* seem to "take all they can get" from their various hosts, and groups of crickets will also strigilate hosts like *Formica obscuripes,* with whom they do exchange liquid food. During the procedure, workers shift to make parts of their bodies, including the head, available to the attending crickets, as though they are being groomed by ant nestmates (Figure 7-3).

For crickets that are host-generalists, the potential trade-offs associated with host choice are especially apparent when a subset of hosts does not participate in trophallaxis. As we will see later in this chapter, host diet may have strong effects on cricket phenotype and life history.

Specialized Mouthparts

The epipharynx of the host-generalist *Myrmecophilus manni* resembles that of some aquatic insects, with brushy projections that protrude past a recessed labrum (upper lip) and mandibles (Figure 7-4) (Henderson and Akre 1986b).

7-4 Upper: Scanning electron microscopic images of a ventral view of the mouthparts of the desert ant cricket, *Myrmecophilus manni* (bar: 100µm). Lower: The cricket's epipharynx has brushy projections that may break the surface tension of liquid food droplets and function like a scoop during strigilation (bar: 10µm). (Courtesy of Gregg Henderson).

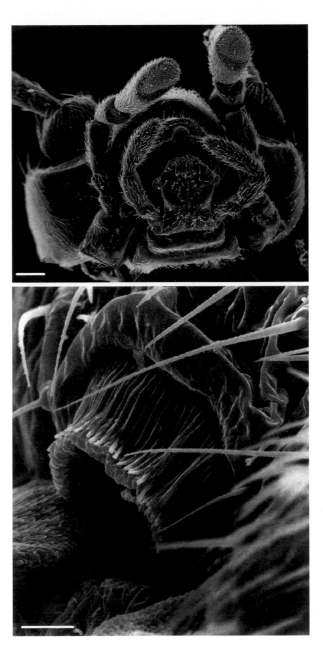

Because *M. manni* strigilates its hosts and participates in trophallaxis, Henderson and Akre (1986b) suggest that its unusual epipharynx may break the surface tension of liquid food droplets, while also acting as a kind of scoop during the crickets' near-constant scraping of ant bodies and nest surfaces.

Other ant cricket species also have specialized mandibles related to their feeding ecology. In a morphometric study of cricket mouthparts from four

THE MYSTERIES OF MYRMECOPHILOUS CRICKETS

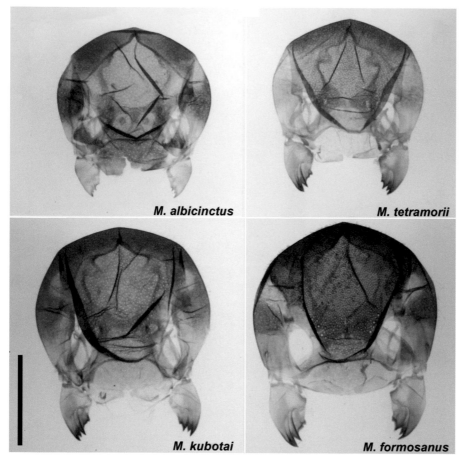

7-5   Slide-mounted heads of four *Myrmecophilus* species, with different diets and degrees of host specificity. *Myrmecophilus albicinctus* (upper left) has simple mandibles. It is a host specialist that relies exclusively on stealing liquid food from its hosts. *Myrmecophilus formosanus* is a host-generalist with robust mandibles. It feeds on host brood and insect prey gathered by ants (bar = 0.50 mm). (Courtesy of Takashi Komatsu).

Myrmecophilus species from Japan, Takashi Komatsu and collaborators (2018) discovered that the host-specialist *Myrmecophilus albicinctus* had less complex and largely nonfunctional mandibles compared to host-generalist species. *Myrmecophilus albicinctus* cannot feed itself and relies solely on stealing liquid food from its hosts, either by sneaking between ants that are engaged in trophallactic food exchange or by imitating the ants' food-begging signals to elicit regurgitation from them (Komatsu et al. 2009). In contrast, the most robust mandibles belong to the host-generalist *M. formosanus,* a cricket species that does not engage in trophallaxis with any of its ant hosts, but instead consumes ant brood and dead insects found in their nest chambers (Maruyama 2004). These predator-

7-6 *Myrmecophilus albicinctus* is a specialist, found only with the ant *Anoplolepis gracilipes*. It has spread across the globe with its host ant. (Courtesy of Taku Shimada).

scavenger ant crickets have comparatively large mandibles with well-developed teeth along the cutting edge (Figure 7-5) (Komatsu et al. 2018).

While it might appear that crickets with a single host species feed only by participating in trophallaxis, the full relationship between host specificity and diet is less clear. *Myrmecophilus tetramorii,* for instance, is a host-specialist found almost exclusively with the ant *Tetramorium tsushimae.* Yet it is not well integrated with hosts and does not participate in any close-contact behaviors, like trophallaxis or direct strigilation of host bodies (Komatsu et al. 2013). Despite a high degree of host specificity, *M. tetramorii* consumes only solid foods pilfered from its host's chambers, and its mouthparts are likewise adapted for the purpose of processing solids (Komatsu et al. 2018). Together, these studies suggest host specificity is not a key predictor of diet and feeding morphology in ant crickets.

Host-Specialists and Host-Generalists

As we have mentioned, some ant cricket species specialize on only one or a limited number of the ant species available in their environment. At least two such host-specialists have spread around the globe with "tramp ants" along human trade routes. Across an introduced range, which includes multiple continents, the ant cricket *Myrmecophilus americanus* is found only with the long-horned crazy ant, *Paratrechina longicornis.* Likewise, *Myrmecophilus albicinctus* is found only with the yellow crazy ant, *Anoplolepis gracilipes* (Figure 7-6) (Hugel and Blard 2005; Komatsu and Maruyama 2016; Hsu et al. 2020). In spite of its

7-7 *Myrmecophilus albicinctus* uses her legs and palps to drum on the mouthparts of an *Anoplolepis gracilipes* host worker. The cricket relies entirely on liquid food obtained by soliciting trophallaxis this way. (Courtesy of Taku Shimada).

name, the yellow crazy ant is a popular host that also harbors at least seven additional cricket species, all with varying degrees of fidelity (Hsu et al. 2020).

The host-specialist *Myrmecophilus albicinctus* elicits trophallaxis from *A. gracilipes* workers by beating on their mouthparts with its forelegs and maxillary palps (Figure 7-7) (Komatsu et al. 2009). Although it is fully integrated into the colony's food flow system, it still treads carefully around its host ants, and on rare occasions receives hostility from them. Komatsu et al. (2009) observed that when one cricket was pursued by an ant, "[it] stopped, became stiff (while staying on its feet), rounded its back, stretched itself, and tucked its antennae to the side of its body. The ant that caught up with the cricket stopped, sat on it, rounded its back a little, became stiff, and tucked its antennae also. The ant also showed other behaviors, such as rubbing the dorsal line of the cricket with its antennae and licking the cricket's cerci. After a few seconds, the cricket escaped by quickly weaving its way through the legs of the ant or jumping."

While *Myrmecophilus albicinctus* can dodge the hostilities of its normal host, *A. gracilipes,* it does not survive when experimentally transferred to the nests of other ants, even those that regularly host other crickets. For instance, *M. albicinctus* does not survive when placed with *Diacamma* sp., a ponerine favored by the host-generalist ant cricket *M. formosanus* (Komatsu et al. 2009). Indeed, *M. albicinctus* never invades the nests of alternative ant hosts, despite a variety of available host options deemed suitable by other endemic *Myrmecophilus* species. *Myrmecophilus albicinctus* relies so completely on trophallaxis with *A. gracilipes* that it dies when placed in isolation because it does not feed itself (Komatsu et al. 2009).

The mechanisms that reinforce cricket host fidelity are unknown, but host colony structure may control many aspects of myrmecophile life history and dispersal. It is possible that the specialist crickets found with *P. longicornis* and *A. gracilipes* seldom encounter alternative host species during normal dispersal events. Instead, *M. americanus* and *M. albicinctus* likely join ceding colony fragments of their natal colony when it undergoes a process called colony budding. Colony budding is a mechanism used by both *P. longicornis* and *A. gracilipes* to

establish new colonies from existing colonies, without the fuss of mating flights (Haines and Haines 1978; Pearcy et al. 2011).

In contrast to the specialist crickets we just discussed, other ant cricket species can and do exploit more than one ant species, and even bunk with hosts from different size classes and subfamilies. *Myrmecophilus oregonensis* is one such host-generalist, found with at least fourteen different ants in the far western United

The host-generalist *Myrmecophilus pergandei* is pictured with one of its many ant hosts, *Tapinoma sessile*. In addition to exploiting native ants from various subfamilies and size classes, it has also been found with a recently introduced fire ant species. (Courtesy of Alex Wild / alexanderwild.com).

States. It has even been found in the nests of the introduced Argentine ant *Linepithema humile,* a species with which it shares no coevolutionary or geographic history, and which does not normally host ant crickets in its native range (Hebard 1920; Elizabeth Cash, personal communication). Another host-generalist from North America, *Myrmecophilus pergandei,* was recently found in nests of a hybrid imported fire ant, *Solenopsis invicta* × *richteri,* whose parent species were introduced to the United States from South America less than a hundred years ago (Figure 7-8) (Hill 2009a). Such host-generalists are fascinating because, as Glasier et al. (2018) point out, parasitic myrmecophiles typically exploit fewer hosts than mutualistic myrmecophiles, and obligate myrmecophiles typically exploit fewer hosts than facultative species. Given these broad patterns, it is intriguing that all ant cricket species are obligate parasites, and that many can live with more than one ant species, including those they have never encountered in nature.

Wheeler (1900) found that another North American host-generalist, *Myrmecophilus nebrascensis,* could easily be transferred from the nests of *Formica neorufibarbis* into nests of the co-occurring harvester ant *Pogonomyrmex barbatus,* noting that the crickets' "adaptation to a new nest and to an ant of large size and belonging to an entirely different subfamily from their former host, was immediate and complete." Wheeler believed that the ability of some ant cricket species to exploit multiple hosts was due solely to their rapid, zigzagging escape movements, which make them hard for any ant species to capture.

Although they can exploit multiple hosts, not all host-generalists move between nests with ease. When Karl Hölldobler (1947) exchanged *M. acervorum* between host colonies, he observed that they were slow to invade, often taking four to six days before entering the entrance of their new host nest (Figure 7-9). Once inside, the crickets slowly adapted to the tempo of their new hosts and moved with rapid darting motions only when threatened. He reasoned that while ant crickets might use rapid movement as a last resort, they likely rely on tactile and odor mimicry as their primary defense against detection in the darkness of the nest. He speculated that the ants perceive both the feel of cricket body

7-9 *Myrmecophilus acervorum* lives with numerous host ants across Europe, including *Formica fusca*. This species lacks males and varies in size depending on host identity. (Courtesy of Pavel Krásenský).

characteristics and the chemicals on the surface of the crickets. This was perhaps the first suggestion that crickets could be chemo-tactical mimics of ants, and that the cricket's tempo and rounded body shape might also enhance their bluff (Hölldobler 1947).

Odor Mimicry and Trail Following

The head, thorax, and abdomen of ant crickets are adorned with long, serrated scales that give their bodies a shimmering quality (Figure 7-10) (Henderson and Akre 1986b; Maruyama 2004). In other insects, scales function as a disposable protective layer that can be shed in a predator's grasp or a spider's web (Nentwig

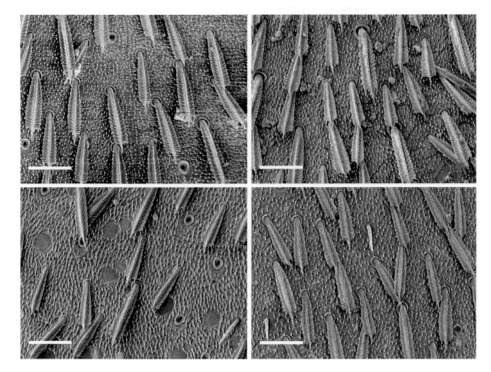

7-10 Crickets are covered in setae that resemble scales or feathers. The pronotal surface scales are pictured for *Myrmecophilus horii* (upper left); *M. kubotai* (upper right); *M. kinomurai* (lower left); and *M. ishikawai* (lower right) (bar = 20µm). (Courtesy of Munetoshi Maruyama).

1982). While scales might allow crickets to slip through ant mandibles, they may also enable crickets to hold the waxes gathered from hosts during strigilation and spread over their bodies during grooming.

Toshiharu Akino and colleagues (1996) demonstrated that crickets produce cuticular wax esters but do not produce their own cuticular hydrocarbons. Cuticular hydrocarbons function in waterproofing for most insects, but their odors have been co-opted as nestmate recognition cues by ants (see Chapter 3). As predicted by K. Hölldobler (1947), the presence of these odors decreases when crickets are isolated from their host ants (Akino et al. 1996). The implication being that crickets accumulate cuticular hydrocarbons almost entirely from interactions with ants, and that the accumulation of ant hydrocarbons might enhance their crypsis in ant nests.

Although ant crickets harvest ant cuticular hydrocarbons during strigilation, which could feasibly be used for both food and camouflage, most host-generalists are still treated as "persecuted guests" and escape the lunges and nips of their host ants only by rapidly darting away (Wheeler 1900) or autotomizing their enormous hindlegs (Henderson and Akre 1986a). Because cuticular hydrocarbons prevent desiccation, we suspect that long-range dispersal between nests in arid

habitats may be seriously limited by the risk of desiccation for some species, like the desert-dwelling ant cricket, *M. manni*. Though it is speculative, risk of desiccation could be one reason that *M. manni* is an extreme host-generalist, capable of diving into the next-nearest-nest of at least forty-one ant species in the arid western United States (Kwapich et al. in prep. b).

Ant crickets lack wings and, to our knowledge, do not disperse by riding on ant workers or clinging to host alates during nuptial flights. Instead, ant crickets must travel by foot when seeking a new ant nest. Some species, like *M. nebrascensis,* are reportedly "left behind" when colonies of *Formica neorufibarbis* relocate to new nest sites after floods (Wheeler 1900). However, Wasmann (1901, 1905)

7-11 *Myrmecophilus cyprius* can be found along the trails of *Messor structor,* in the Republic of Cyprus. (Courtesy of Thomas Stalling).

and other early myrmecologists reported that some *Myrmecophilus* guests do follow their host ants' emigration trails during nest relocations. More recently, Thomas Stalling (2017) observed *Myrmecophilus cyprius* following the trails of *Messor structor* in the Republic of Cyprus (Figure 7-11).

Henderson and Akre (1986a) likewise observed *Myrmecophilus manni* traveling alongside their host ants during colony emigrations, and suggested that they may disperse to new nests by veering off onto the trails of foreign ants that intersect with the emigration route. They found all life stages of *M. manni* along the chemically reinforced foraging trails of one of their hosts, *Formica obscuripes.* The crickets were most abundant on trails during the final portion of the ants' foraging period, between dusk and 23:00. Although *Myrmecophilus manni* can cover up to 40 cm in a single leap, leaping is not its normal mode of locomotion. Indeed, if crickets were to hijack ant trail pheromones to find new nests, they would need to keep their constantly vibrating antennae close to the ground.

To determine whether crickets can detect and follow the chemical trails of their hosts, Eva Junker (1997) designed a choice test for the European ant cricket, *Myrmecophilus acervorum.* She presented fourteen crickets with a T-shaped maze. One arm of the maze was marked with trail pheromone taken from the rectal bladder of a host ant, *Lasius niger.* The other arm was left blank, as a control. In fifty of sixty-four trials, the crickets elected to enter the arm marked with *L. niger* odors, regardless of source-colony identity. These results suggest that the trail pheromone of *L. niger* was attractive to the crickets, and that they might orient using it. However, because the alternative to the trail pheromone presented was an absence of an odor, an additional treatment is still necessary to determine whether crickets would also be more generally attracted to other unrelated odors.

Myrmecophilus acervorum live with a variety of ant host species. A potential problem associated with being a host-generalist is its need to identify and orient using the trails of dozens of hosts, from distantly related subfamilies, all using species-specific chemical signals. How these crickets break the ants' code during dispersal, and whether trail pheromones are the only cues used by invading crickets, remains to be studied. An experiment comparing cricket preferences

THE MYSTERIES OF MYRMECOPHILOUS CRICKETS

7-12   Left: *Camponophilus irmi* is the largest known ant cricket in the world, measuring 8 mm in length. It lives in Borneo with the largest ant species in the world, *Dinomyrmex gigas*. (Courtesy of Martin Pfeiffer). Right: For comparison, some adult *Myrmecophilus pergandei* grow to a maximum length of only 2 mm. (Courtesy of Brandon Woo).

5 mm

for the trail pheromones of familiar and unfamiliar ants, across a variety *M. acervorum* host species, would certainly be an interesting extension of Junker's (1997) work.

The Mystery of Cricket Body Size

There is a clear morphological relationship between crickets and their host species. For example, the largest ant cricket in the world, *Camponophilus irmi* (8 mm), is found with the largest ant in the world, *Dinomyrmex* (formerly *Camponotus*) *gigas* (Figure 7-12) (Ingrisch 1995). A graph showing the size of all known ant cricket species plotted against the size of their host species would reveal a strong positive correlation between the two variables.

In addition to varying in size across species, one of the most striking aspects of ant cricket biology is the apparent size variation within species. In his key to the North American *Myrmecophilus* (= *Myrmecophila*) species, Morgan Hebard (1920) bemoaned the morphological variation he observed within individual collecting sites, stating, "We have found little or nothing of value in the North American species, in size, form of segments, width of interocular space, size of eyes, length in proportion to width, form of caudal femora, form of external male genitalia or form of ovipositor." He carefully assessed body size with respect to geographic factors, finding that in all but one North American species, *Myrmecophilus oregonensis,* latitude and environment exerted no influence on cricket body size.

Instead, Hebard (1920) noted that large-bodied crickets were found with large hosts, and small-bodied crickets were found with small hosts. Crickets from different size classes were united as individual species based on the patterns of spinulae on their metatarsus and setae on the hind tibia. The smallest adult crickets he discovered measured a meager 1.47 mm in length and were found with colonies of *Tapinoma sessile.* He called these unfortunate specimens "depauperate" types because of their floppy and weakly sclerotized ovipositors and extreme reduction of typical adult features.

Correlations between myrmecophile size and host body size have been observed in at least one other myrmecophile species, the sexually dimorphic ant-associated isopod *Platyarthrus hoffmannseggii.* Terrestrial isopods also undergo a developmental sequence similar to that of ant crickets. In both cases the juveniles resemble adults and do not undergo a major rearrangement of features before adulthood. Like many ant crickets, *P. hoffmannseggii* is also a host-generalist capable of exploiting a variety of local ant species, as well as ant species recently introduced by human activity in its range (Dekoninck et al. 2007). Small male and female *P. hoffmannseggii* are found in nests of *Lasius flavus* (see Figure 9-9), while females up to 30% larger are found with *Formica polyctena,* an alternative large-bodied host (Parmentier et al. 2017b).

The plasticity in female *P. hoffmannseggii* size is suggested to exist because, while larger isopods can produce more offspring across ant species, myrmecophiles that are very small relative to their hosts tend to be ignored across a wide variety of ant species (Parmentier et al. 2016c, 2017b). *Platyarthrus hoffmannseggii* does not mimic the colony-recognition odors of ants and therefore, the authors hypothesize, should be as large as possible to reproduce and still avoid being noticed by each of its chosen host species. Similar relationships between small size and host tolerance have been observed among numerous phylogenetically distant myrmecophiles of *Eciton* army ants. In these cases, absolute size is a better predictor of host aggression than shared colony odor (von Beeren et al. 2021b).

This brings us to a simple question that is difficult to answer: Why are large ant crickets found with large hosts, while small ant crickets are found with small hosts? One possible answer is that ant crickets adhere to the same rule as *P. hoffmannseggii* and grow as large as possible while still maintaining relative anonymity with their hosts. In addition to tolerance, the benefits of living with a like-sized host may also involve appropriate scaling of the mechanical solicitations necessary for initiating trophallaxis with ants.

Wasmann (1905) observed that small-bodied *Tetramorium caespitum* tended to host juvenile *Myrmecophilus* sp., while comparatively large-bodied *Formica* sp. appeared to host the larger, adult crickets. He reasoned that if crickets must be an acceptable size to be tolerated in a colony, then juvenile crickets should be found with small-bodied hosts and later disperse to the nests of large-bodied hosts as they age and grow. This life history would require an adult female to leave the nest of her large-bodied host, invade the nest of a small-bodied host species, and stay long enough to deposit her eggs there. Alternatively, hatchling juvenile crickets would need to leave the nests of large-bodied ants in search of small-bodied host ants early in their development. The second mechanism was also suggested by Thomas Stalling and colleagues (2020), who collected eight juvenile and eleven adult *M. orientalis* living with separate host species. He reasoned that, although it is risky, cricket nymphs might move from their colonies of small-bodied hosts (*Crematogaster erectepilosa* and *Lepisiota frauenfeldi*) to large-bodied hosts, like *Camponotus samius,* in adulthood.

A third mechanism of host-size partitioning was suggested by Karl Hölldobler (1947) after he observed that small nymphs of *Myrmecophilus acervorum* were rarely seen in colonies of large-bodied ants but were found frequently with small-bodied ants, like *Lasius niger* and *Tetramorium* sp. (Figure 7-13). He proposed that large crickets might migrate twice during development, first from their natal nest (a large-bodied host) to a small-bodied host, then back to a large-bodied host as older nymphs and adults.

There are several alternatives that might explain why juvenile crickets are found in abundance with small-bodied ant hosts. First, it is possible that the nutrition crickets can obtain with small-bodied hosts is so poor that the rate at which nymphs progress through instars is altered, as is common among other hemi-metabolous insects, including numerous other orthopterans (Whitman 2008). If starved, crickets may molt into the "depauperate" adult form described by Hebard (1920) or stall in a juvenile instar until a food source becomes available. As a consequence, the relative frequency of juvenile crickets in the developmental distribution would be higher in the nests of small-bodied ant hosts. Another possibility is that large ants are more aggressive toward tiny cricket nymphs and, recognizing the extreme mismatch in size, cull more nymphs than small-bodied ant hosts.

In our studies of *M. manni,* we also detected a greater proportion of immature crickets living with small-bodied ants (<1.5 mm, Weber's length). However, immature crickets were collected with ant species of all sizes, including the large-bodied *Camponotus, Formica,* and *Novomessor* species. Juvenile crickets were also found with large- and small-bodied hosts across seasons, despite an overall lower abundance in nests belonging to large-bodied ants. Although host body size was a good predictor of cricket body size, worker number within a colony had no

effect on cricket size. For example, colonies with thousands of small-bodied workers still harbored small crickets, while colonies with only a few large-bodied workers still harbored large crickets (Kwapich et al. in prep. b).

An interesting scenario arises when ants that host crickets are themselves polymorphic. Within a single ant colony, there may be many differently sized and differently shaped workers. Polymorphism and monophasic allometry occur in a minority of ant species. In the southwestern United States, ant genera with polymorphic workers include *Pheidole, Formica, Camponotus, Liometopum, Veromessor, Myrmecocystus,* and *Acromyrmex.* The desert ant cricket, *Myrmecophilus manni,* is found with all these genera and with at least thirteen polymorphic host species out of thirty-three total hosts in our studies of Arizona and Northern Mexico alone (Kwapich et al. in prep. b).

While host species availability may explain the disproportionate occurrence of ant crickets with polymorphic hosts, there is also a tantalizing alternative: polymorphic ant hosts may have a broader template for the cues associated with nestmate size or hydrocarbon abundance. If ant vigilance toward parasitic myrmecophiles is relaxed because of natural variation of ant size within colonies, parasitic crickets may have a better chance of invading and surviving with polymorphic hosts. This leads us to another question: If ant size truly influences the scaling of ant crickets, then which worker size class will crickets resemble in a colony with polymorphic workers? Among the polymorphic hosts that we studied in detail, *M. manni* body size corresponded to the average size of workers in a nest, rather than the largest or smallest worker size class, or the most abundant worker size class, in a colony (Kwapich et al. in prep. b).

Phenotypic Plasticity and Cryptic Speciation

Let us now return to the original question posed at the beginning of this section: Why are large crickets found with large hosts, while small crickets are found with small hosts? As we have already discussed, it is less likely that crickets assort themselves between appropriately sized hosts as juveniles. An obvious but largely untested alternative hypothesis is that crickets identified as belonging to one species are actually several cryptic species that have adapted to ant hosts within a specific size class. Under the cryptic-species hypothesis, host preference should be strict and body size is expected to be relatively fixed and heritable.

Together with our colleagues Bob Johnson and Jeffrey Sosa-Calvo, we set out to investigate the mechanisms that underlie the enormous variation in body size and host preference displayed by the desert ant cricket, *M. manni*. Ant hosts included a full range of sizes, from tiny *Forelius pruinosus* and *Pheidole hyatti* to the comparatively enormous *Camponotus sansabeanus* and *Novomessor albisetosus,* which could hold a dozen smaller host species on their backs. We found that for every unit increase in host size (Weber's Length), there was a 0.32 increase in cricket width (second abdominal segment width), and corresponding volumetric increase. However, there were two interesting deviations from this predictable relationship. Ants from the genus *Pogonomyrmex* tended to host crickets that were smaller than expected, while ants from the genus *Trachymyrmex* tended to host crickets that were larger than expected. The former is a seed-harvesting ant that does not perform adult-to-adult trophallaxis, while the latter is a fungus gardener, whose nests contain a robust, domesticated nutritional resource. These clues led us to suspect that host identity and corresponding available nutrition might underlie the patterns in body size we observed (Kwapich et al. in prep. b).

Indeed, nutrition is often the determining factor affecting tolerance and abundance of particular myrmecophiles in host ant colonies. For example, the specific seed-milling behavior of the palearctic genus *Messor* appears to increase the prospects of a different group of myrmecophiles, namely various silverfish species. Molero-Baltanás and colleagues (2017) found that *Messor* colonies hosted more individuals and more silverfish species (especially host-specific silverfish), than co-occurring host genera like *Aphaenogaster* and *Camponotus*. This preponderance of silverfish (up to twenty species) could not be attributed to colony size, body size, or general aggression and vigilance by the host ants, but corresponded directly to the ants' granivorous diet. *Messor* cannot digest the tough husks of the seeds that they gather, but many Lepismatinae silverfish have the ability to digest cellulose and do so without the aid of endosymbiotic microbiota. As a result, silverfish seem to have diversified as commensals in the niche created by the copious husks discarded by the ants. Unlike ant crickets, the silverfish associated with *Messor* do not negatively impact the ants themselves. This too could decrease the amount of aggression ants show them and increase their abundance (Molero-Baltanás et al. 2017).

To test our hypotheses about available diet and body size in *Myrmecophilous manni,* we collected more than 300 crickets and extracted DNA from sixty-six individuals, representing eleven populations across Arizona and Northern Mexico. Collecting M. manni was an exciting venture, because new sites often yielded crickets paired with undocumented host species, as well as crickets of surprising size (so-called jumbo shrimp). We found *M. manni* living with eight different host species behind the Pinto Valley Mine, near Miami, Arizona. These included tiny *Pheidole hyatti* and comparatively massive *Novomessor albisetosus*. One fortuitous collecting event revealed crickets from two opposing size classes, living with ants from two separate subfamilies, beneath a single stone (*Camponotus sansabeanus* and *Pheidole hyatti*). Despite the variation in cricket size that we observed, spinulae and setae arrangement indicated that the crickets were all morphologically *M. manni,* according to Hebard's (1920) key.

To determine whether large crickets and small crickets represented separate, cryptic species, we built a Kimura two-parameter neighbor-joining tree using the mitochondrial DNA barcoding gene Cytochrome c oxidase I (CO1) (Moulton et al. 2010). We included a scaly cricket as the outgroup, as well as additional ant cricket species whose sequences were already available (Ortega-Morales et al. 2017). One of the challenges of working with orthopterans is the abundance of nonfunctional mitochondrial pseudogenes (numts), which can complicate traditional DNA barcoding methods used to identify cryptic species (Song et al. 2008; Moulton et al. 2010). Even so, our tree revealed that the geographic location of the *M. manni* was the best predictor of cricket grouping on the tree. In other words, most crickets from the same region were grouped into clades, independent of their body size or host preference. Within our Pinto Creek site, for example, a single clade included both large and small crickets found with multiple host species from several subfamilies. Between neighboring sites, there was a high degree of divergence between crickets, which might be a consequence of their poor dispersal ability. While these differences between collecting sites suggest that *M. manni* may be better described as a species complex, we can say with certainty that crickets show flexible host preference and express a range of adult body sizes within most populations tested (Kwapich et al. in prep. b).

To determine how host identity might influence cricket size more directly, Kwapich performed reciprocal rearing experiments between juvenile crickets captured with small hosts and juvenile crickets captured with large hosts. Each wild-caught juvenile cricket was placed into a laboratory colony with ants that were either larger or smaller than those from their natal nest, including *Crematogaster emeryana* (small) to *Formica occulta* (large); *C. emeryana* (small) to *Liometopum apiculatum* (large); *L. apiculatum* (large) to *Pheidole hyatti* (small); and *L. apiculatum* (large) to *C. emeryana* (small). As a control, juvenile crickets were also placed with new laboratory colonies belonging to the same species as those from their natal host.

When placed with large-bodied ants, crickets with small parents grew either larger than expected or to an intermediate size, between that of their parents and the expected size. Crickets taken as juveniles from large-bodied hosts spent an extended period of time as juveniles, and some never molted into adulthood during the course of a year. Those that did become adults fell into a range of size classes, indicating that external conditions could also affect their growth (Kwapich et al. in prep. b). While exchanging eggs between hosts of different sizes would be the ideal test for our question, we have not yet succeeded in doing so.

We next set out to describe the natural growth trajectories of *M. manni* found with a small-bodied host, *Crematogaster* sp. (nineteen adults, fifty-seven juveniles), and *M. manni* found with a large-bodied host, *Liometopum apiculatum* (thirteen adults, twenty-five juveniles). To do so, we collected crickets along a single forest road in the Chiricahua Mountains in Arizona (USA). Nests of the ant host species housed hatchling crickets of identical size, despite a considerable difference in the size of adults in the nest. Measures of cricket head capsule width at each instar revealed that head size in crickets from large ant hosts increased in size by an average of 11% per instar, while crickets found with small ant hosts grew only 7% between instars. The difference in growth rate was already apparent between the second and third instar, suggesting that body size is "lockedin" early in development.

Our first molecular findings suggest that crickets that can be morphologically assigned to *M. manni* exploit numerous hosts (forty-one and counting), while varying considerably in body size and host affiliation within populations. Even so, because a gene tree is not necessarily representative of a species tree, more work "below the cuticle" is still needed. It is likely that *Myrmecophilus manni* is a species complex, and that the crickets gathered in Arizona and Mexico might differ considerably from those studied by Henderson and Akre in Washington. In fact, Kwapich never saw *M. manni* engage in trophallaxis with its hosts, although Henderson and Akre (1986a) report that this behavior is common in *M. manni*. The extent to which body size is maintained across generations of these charismatic crickets requires further study. For *M. manni,* body size is a plastic trait that is affected by host identity and, most likely, by host diet and life history. Whether size plasticity can also be experimentally induced in host-specialist cricket species remains to be seen.

Gaster Mimicry and Egg Mimicry

The positive correlation between ant body size and adult cricket body size is clear, but any function of size-matching, if one exists, has remained elusive. Most casual observers will notice that the smooth, rounded bodies of adults resemble the posterior part of an ant's body, called the gaster (Figure 7-14). If tactile mimicry is important to the survival of guests inside ant nests, resembling a common morphological feature shared by ant nestmates may ring fewer alarm bells than

THE MYSTERIES OF MYRMECOPHILOUS CRICKETS

being an unfamiliar shape in motion. In our studies of *M. manni,* we found that host gaster size did frequently correspond to adult cricket size.

However, *M. manni* collected with very small host ant species were often larger than their host's gasters (Kwapich et al. in prep. b). The possibility that these crickets could instead resemble the gaster of a large queen ant, which would be an object of tactile familiarity in the colony, was suggested by Wetterer and Hugel (2008). They observed that *Myrmecophilus americanus* is the same size and shape as the gaster of a *Paratrechina longicornis* queen. Though a manipulative experiment is needed to determine whether the resemblance is important to the ants, a rough correlation between adult cricket size and worker gaster size does exist for at least a few ant cricket species measured.

7-14   In some cases, ant crickets are the same size and shape as a host's gaster (poste-rior section). Small adults and juvenile *Myrmecophilus acervorum* live with small-bodied ant species, like the *Lasius* sp. pictured here. Large adults are found with large-bodied host species. (Courtesy of Thomas Stalling).

The eggs of the desert ant cricket, *Myrmecophilus manni,* are the same size and shape as the eggs of at least one host ant, *Formica obscuripes.* Crickets deposit their eggs beneath ant brood chambers, where they are not destroyed by ants (Henderson and Akre 1986a). Given their similarity and location, we considered that cricket eggs might be tactile mimics of ant brood. If cricket eggs avoid detection by resembling the eggs or larvae of their hosts, then the positive cor-relation between adult cricket size and ant size might exist so that females could can lay appropriately sized eggs. To test this idea, Kwapich compared egg size between large-bodied *M. manni* found with large ant hosts and small-bodied *M. manni* found with small ant hosts.

We found no relationship between host and cricket egg size in *M. manni.* In fact, large-bodied crickets and small-bodied crickets laid eggs that consistently measured 1.2 mm in length, across a nearly twofold range in female body width and a 2.5-fold difference in dorsal length between the largest and smallest adult female crickets in our study. With larger hosts like *Liometopum apiculatum* and *Novomessor albisetosus,* *M. manni* eggs were closer in size to host ant eggs and early instar larvae. Yet the eggs of small-bodied crickets more closely matched the size of the late-instar larvae of small-bodied host ants, like *Crematogaster emeryana.*

In essence, the strong correlation between host body size and adult cricket body size is unlikely to be driven by a necessity to lay eggs that match the size of host ant eggs. It is more likely that egg size places a constraint on the lower size limit of *M. manni,* which might explain why the smallest crickets are still disproportionately large next to the tiniest ant hosts. Likewise, if egg-size mim-icry were driving adult female body size, one would not necessarily expect adult male body size to also be so tightly correlated with host ant size, since males can probably deliver spermatophores to females of any size (Henderson and Akre 1986c). Indeed, adult male *M. manni* are of identical size to adult females when found with the same ant host species. In this regard (and many others), *Myrmecophilus* crickets are unique. The absence of sexual size dimorphism is exceptionally rare in other crickets and grasshopper lineages, where females are almost always the larger sex (Hochkirch and Gröning 2008).

If cricket eggs are not size mimics of ant brood, then why do crickets deposit their eggs in the brood piles of ants? Depositing eggs in a brood chamber provides young myrmecophiles with all the benefits afforded to the ants themselves. After all, an ant nest is a nursery inside a fortress, with a carefully maintained hygienic environment and microclimate. A second benefit of being born into a brood pile was suggested by early observations of brood consumption by ant crickets (K. Hölldobler 1947). In 1986, Henderson and Akre photographed an adult *M. manni* that appeared to be engaging in trophallaxis with a *F. obscuripes* larva that it had hauled away from the brood pile. Closer inspection revealed that the cricket had lacerated the head region of the larva. Like Wheeler (Wheeler 1910), Henderson and Akre (1986a) also observed that ant eggs are frequently groomed and handled by crickets without apparent consumption. While we qualitatively observed decline in ant brood when ant crickets were placed with colonies in large numbers, we were not lucky enough to observe direct predatory behavior.

In soil laboratory nests of the ant *Aphaenogaster texana,* we found that *M. manni* did not oviposit in the absence of ant brood. When ant brood were returned to the colony fragments, the crickets commenced egg-laying in the soil below their host's brood pile. Eggs isolated from the ant's nest all failed to hatch, despite visible development and a well-armored chorion. In contrast, eggs kept with their hosts did hatch, suggesting that ants may clean the eggs or even assist with the process of hatching from the tough shell. When returned to *Aphaenogaster texana* colonies in the lab, isolated cricket eggs were gathered from the exterior of the nest by the ants and taken inside the nest (Kwapich et al. in prep. b). Unlike the soft and shining eggs and larvae of ants, the eggs of ant crickets harden rapidly after they are laid. They are so hard, in fact, that they are not easily deformed when pinched between the tips of a pair of watchmaker's forceps.

Any function of size-matching between ant crickets and the hosts remains a mystery. While body size in these unusual guests may simply be an epiphenomenon of colony nutrition or host aggression, a lower size limit may allow females to produce offspring large enough to engage in trophallaxis with their hosts after hatching, or to produce eggs that are tolerated in the ants' brood chambers.

Because egg size is invariant in the *Myrmecophilus manni* group, there are clear reproductive trade-offs associated with expressing the small-body phenotype induced by small-bodied host species. Why, then, are adult crickets still common in the nests of these "suboptimal" hosts? Perhaps harsh environmental conditions, poor dispersal ability, and/or a low abundance of waterproofing hydrocarbons make selective dispersal risky. The lack of host-specificity in M. *manni* may also correspond to other life history traits that differ between size classes, like age of sexual maturity. Careful experimental work is needed to determine how differently sized crickets interact with unfamiliar hosts, and whether host size-matching represents adaptive plasticity in the environment of an ant colony.

Island Endemics and Island Hoppers

Female crickets lay exceptionally large eggs for their size. The sight of a female cricket next to her enormous egg conjures the image of a kiwi bird of similar proportions. To pass their eggs, *Myrmecophilus* females employ an unusual ovipositor (egg-laying tube) that measures one-third of their body length and bears a point of articulation along the fused and elongated eighth and ninth terga (Schimmer 1909). The ovipositor is equipped with an extensible, membranous egg guide that is spiraled to allow the passage of one giant egg at a time (Henderson and Akre 1986b). *Myrmecophilus manni* females (ranging from 1.75 to 4 mm in length) produce eggs of identical size (1.2 mm). Therefore, small females must invest more per unit of body mass to produce an egg equivalent in size to that of a large female. There are likely some energetic trade-offs imposed by matching a host ant's body size and, more broadly, by host choice, as a consequence.

In *M. manni,* we found that each female carried a maximum of two eggs, one fully formed and the other still developing. The European *M. acervorum,* similarly, carries no more than four eggs at a time (K. Hölldobler 1947). Though the small clutch size and relatively large egg size of ant crickets is unusual for orthopterans, as we discussed in Chapter 6, reduced clutch sizes are common among a variety of myrmecophiles and myrmecomorphs across arthropod lineages.

For myrmecophiles with limited dispersal ability, being stuck in an ant colony of a particular size and quality may have a strong effect on clutch size. This topic has been studied at length in island-dwelling birds, where clutch size decreases and individual egg size increases on smaller islands (Higuchi 1976; Covas 2012). Among island endemics, female sex biases and parthenogenesis (asexual reproduction) are also more common than in mainland populations and species.

Like birds on islands, *Myrmecophilus* crickets in ant colonies frequently have female-biased sex ratios. The European ant cricket, *M. acervorum,* was presumed to be parthenogenetic (Wasmann 1901; K. Hölldobler 1947), until surveys in the central part of its range revealed a few populations containing both sexes (Iorgu et al. 2021). Wheeler (1900) also observed a strongly female sex bias in the *M. nebrascensis,* noting that across various hosts, adults were roughly 88% female and 12% male. Thelytokous parthenogenesis is rare in Orthoptera (crickets and grasshoppers), with the exception of just a few species, including isolated cave crickets and certain ant crickets (Hobbs and Lawyer 2003). While the often-skewed sex ratios could represent a true bias in the relative number of females and males born in a population, differences in biology or behavior could also lead to sex-specific survival rates. One possibility is that male crickets disperse from their natal nests, while females are philopatric. This sex-specific difference in behavior could expose male crickets to a greater risk of predation or desiccation aboveground. *Myrmecophilus manni* are also reported to compete in male-male duels and form dominance hierarchies within nests. These interactions, along with their interactions with females during courtship, can be injurious and may decrease the relative abundance of males (Henderson and Akre 1986c).

Across insect groups, infections by the intracellular bacteria *Wolbachia* also correspond with the production of female-biased offspring, either through the programmed death of male embryos, the feminization of genetic males, or by inducing thelytokous parthenogenesis, whereby infected females produce daughters from unfertilized eggs (Werren et al. 2008). Shu-Ping Tseng and colleagues (Tseng et al. 2020) discovered that *Wolbachia* are horizontally transferred between ants and at least three ant cricket species. In their study, crickets that participated in trophallaxis with their hosts, and were host-specific, had a higher *Wolbachia* prevalence and higher shared diversity than host-generalist crickets and host-specialist crickets with more diverse diets.

Myrmecophilus americanus, a strict specialist of *Paratrechina longicornis,* had the highest *Wolbachia* prevalence and diversity, such that 43% of crickets were infected by three strains, and 55% with four strains. Only one of these strains was shared between the crickets and their host ants, suggesting that *M. americanus* may be more naturally susceptible to *Wolbachia* infection. Indeed, host specificity and trophallaxis are not the only predictors of transmission of *Wolbachia* strains between crickets and hosts. For example, the host-generalist *Myrmophilellus pilipes* shares a strain of *Wolbachia* with the ant *P. longicornis,* but no *Wolbachia* are shared between the host-specialist *Myrmecophilus albicinctus* and its single ant host species, *Anoplolepis gracilipes* (Tseng et al. 2020).

Although Tseng et al. (2020) do not report sex ratio for the crickets in their study, the occurrence of intraorder *Wolbachia* infections between ants and their myrmecophiles is a metric of great interest. In addition to potentially affecting sex ratio, *Wolbachia* infections are known to induce cytoplasmic incompatibilities between male and female insects that harbor different strains (Werren et al. 2008). Such incompatibilities could be an unseen mechanism of speciation among myrmecophiles that participate in trophallaxis with their hosts. Unexpectedly, Iorgu and colleagues (2021) found that *Wolbachia* was present in sexually

8-1 The staphylinid beetle *Pella humeralis.* (Courtesy of Pavel Krásenský).

with 40%, but Hemiptera (true bugs, aphids, scale insects, cicada), Diptera (flies), Hymenoptera (ants, bees, wasps), Acari (mites), and Araneae (spiders) were also commonly found. Parmentier et al. (2014) assert that the astounding diversity of myrmecophiles can best be explained by the nest structure of the host ants, because their big nest mounds provide stable and long-lasting habitats with various temperature and moisture zones and constant availability of food.

Within *F. rufa* nests, highly integrated myrmecophiles usually live close to the ants' brood chambers, and many of them prey on the immatures of the hosts, freeload on booty brought in by forager ants, or partake in the social food flow of the colony. Striking examples of such fully integrated myrmecophiles include the rove beetle genus *Lomechusa* (formerly called *Atemeles)* and the larvae of the syrphid genus *Microdon* (see Chapter 3). Examples of the less-integrated inqui-lines are species of the rove beetle *Dinarda,* which appear less frequently inside the brood chambers but instead roam the peripheral nest departments and re-fuse chambers. Among the Diptera, Parmentier et al. (2014) list the milichiid midge *Phyllomyza formicae* and the ceratopogonid midge *Forcipomyia myrmecophila,* but very little is known about their association with their *Formica* hosts. And then there are the co-inhabitants that are not directly involved with the ants but that live close to the ants' nests and prey on Hemiptera, which are tended by the ants because they produce valuable honeydew, an important food source for the ants.

It is not our intention to report the entire faunistic account provided by Parmentier et al. (2014); instead we wish to focus on some selected studies that have analyzed the behavioral mechanisms that enable the well-integrated myrmecophiles to exploit the "social acquisitions" of their host ants. To hypothetically consider the possible transitions in behavior and morphology that allowed certain myrmecophiles to become well integrated with their hosts, we will first compare four extant aleocharine beetle genera that represent different degrees of integration with their host ants.

Pella humeralis: The Predator and Scavenger

We begin at the periphery of the *Formica rufa* nest, with a beetle that acts as a scavenger and predator but is also capable of exuding appeasement and defensive compounds. We have already discussed examples of myrmecophilous aleocharine beetles of the genus *Pella,* whose main niches are the foraging tracks of one of its host ants, *Lasius fuliginosus* (see Chapter 5). In addition to its association with *L. fuliginosus, Pella humeralis* (which has also been called *Zyras humeralis* or *Myrmedonia humeralis*) (Figure 8-1) is also found with ants of the *Formica rufa* group.

According to Wasmann (1912, cited in Wasmann 1920), *P. humeralis* has two host ant species: the primary host is *L. fuliginosus,* with which it commonly lives in late spring, summer, and early fall, whereas during the winter and early spring it lives mainly with species of the *Formica rufa* group. Donisthorpe (1922, cited in Donisthorpe 1927) made similar observations. He corroborated Wasmann's reports concerning *P. humeralis* being a predator of *Formica* and *L. fuliginosus* workers, although Kolbe (1971) could not confirm these observations. We think, however, that Wasmann and Donisthorpe were correct. In fact, Donisthorpe's observations of predatory behavior of *P. humeralis* are quite detailed and match those of Hölldobler et al. (1981) for other *Pella* species preying on *Lasius fuliginosus* (see Chapter 5). According to Donisthorpe, *P. humeralis* is a predator and scavenger in the colonies of both host species. As reported for *Pella* spp., the beetles present appeasement secretions from the appeasement gland complex located in their abdominal tip when contacted by host ants. Only on the rare occasions of serious attacks by the ants do *P. humeralis* discharge strong-smelling defense-repellent secretions from the tergal gland. The larvae of *P. humeralis* seem to develop in the refuse sites of *L. fuliginosus* nests.

Dinarda Beetles: The Sneaky Thieves

We next describe the behavior of beetles in the genus *Dinarda,* which also live at the nest periphery of *Formica* hosts but are not found on their foraging tracks. These beetles are endowed with appeasement and defensive glands and, in addition to scavenging and occasional predation, can exploit the social food flow of their hosts by snatching liquid food during trophallaxis events between ants. The aleocharine genus *Dinarda* belongs to the tribe Oxypodini, whose known species are all obligate myrmecophiles associated with different ant species (Wasmann 1889b, 1920). *Dinarda dentata* lives primarily in nests of *Formica sanguinea; D. maerkelii* is associated with *Formica rufa; D. hagensii* lives with *Formica exsecta; D. pygmaea* lives with *Formica rufibarbis;* and *D. lompei* has been found with *Formica gagates.* Although *Dinarda* appears to be quite host-ant-specific, there are exceptions. For example, *D. dentata,* which usually live in *F. sanguinea* nests, can also be found in nests of other *Formica* species. In fact, in addition to *F. sanguinea,* Päivinen et al. (2003) list *F. fusca, F. rufibarbis, F. exsecta, F. cinerea,* and *F. aquilonia* as hosts. Wasmann, who also noticed occasional differences in host identity, called these "abnormal" hosts, with *F. sanguinea* being the "normal" host. During our own research we found *D. dentata* (Figure 8-2) only in *F. sanguinea*

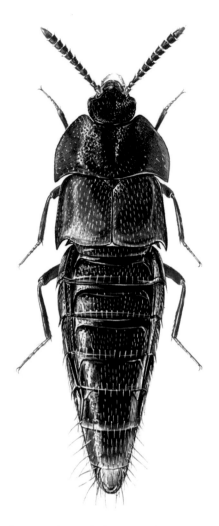

nests (Hölldobler and Kwapich 2019). In general, we can state that species of *Dinarda* are not strictly host-specific, but certain host species appear to be preferred by the various *Dinarda* species.

Wasmann and subsequently Donisthorpe (1927) were the first to provide observations of the myrmecophilous behavior of *Dinarda dentata* and other *Dinarda* species. Wasmann postulated that this genus represents an intermediate evolutionary state between the myrmecophilous predators and scavengers, such as the genus *Pella,* and the brood nest myrmecophilous parasites, such as *Lomechusa* (formerly called *Atemeles*) and *Lomechusoides* (formerly called *Lomechusa*) (we discuss these two genera in the next sections). Wasmann did not consider these cases as various degrees of adaptation to particular ecological niches in ant nests and colonies but argued that *Dinarda* was still in the evolutionary process of reaching the highest level of myrmecophily. Of course, we know nowadays that evolution does not proceed along a ladder, directed toward one form or the other, but we can envision the evolutionary transition from a free-living predatory lifestyle to an integrated social-parasitic lifestyle. But before we delve into that topic, let us first consider in greater detail the way of life of *Dinarda dentata,* which can be conveniently studied with intact host ant colonies in formicaria. This approach would be considerably more difficult with mound-building *Formica* host species. *Dinarda dentata* (Figure 8-3) dwells mainly in the more accessible, peripheral nest chambers and nest refuse areas of *Formica sanguinea* colonies.

The level of integration of, and niche occupied by, *Dinarda* can best be understood by considering the beetles' diet and location, as well as its response to aggression received from host ants. The beetles have never been seen hunting ants. Instead, Wasmann (1889b, 1920) reports that the beetles eat debris discarded by the host ants and are occasionally seen inserting themselves between two food-exchanging ants, intercepting a food droplet that was about to be passed from one ant to the other. He also reported one incidence of having observed a *Dinarda* beetle with an egg between its mandibles. In a more recent paper on *Dinarda maerkelii,* Parmentier et al. (2016b) observed occasional egg and larvae predation and presented evidence for trophallaxis with the host ants *Formica polyctena* and *F. rufa.* They used dyed liquid food fed to the host ants, which were housed together afterward with *D. maerkelii* beetles. After forty-eight hours, the guts of the beetles were dissected and the dyed liquid in the beetles' guts indicated that they had received food from the ant workers.

In the late 1960s, Hölldobler did similar experiments with *Dinarda dentata* and its host species *Formica sanguinea.* However, he employed honey-sucrose water labeled with the radioisotope ^{32}P, which was added as orthophosphate. This made it possible to quantitatively measure food transfer from the ants to the beetles without the need to dissect the beetles' guts. Ant foragers that enter the nest with a full crop seek to deliver the collected liquid to the nestmates, and occasionally one can see the large regurgitated droplet between the gaping mandibles. *Dinarda*

8-4 *Dinarda dentata* beetles snatching regurgitated food from the host ants *Formica sanguinea*. Upper: An ant forager returns with a full crop and offers food to nestmates. A large, regurgitated food droplet appears between the gaping mandibles of the donor ant. Middle: Beetles seek out food-exchanging ants. Lower: They insert themselves between food-exchanging ants and attempt to snatch some of the regurgitated liquid. (Bert Hölldobler; Turid Hölldobler-Forsyth; ©Bert Hölldobler).

beetles tend to sneak between the ants engaged in trophallaxis and snatch a share of the regurgitated food (Hölldobler and Kwapich 2019) (Figure 8-4). Similar behavior was previously reported by Wasmann for *D. hagensii,* and as stated above, Parmentier and his coworkers provided evidence for trophallaxis in *D. maerkelii*. In addition, we occasionally observed the beetle surreptitiously approaching a food-laden forager and, by touching the ant's labium, inducing the regurgitation of a small droplet (Figure 8-5).

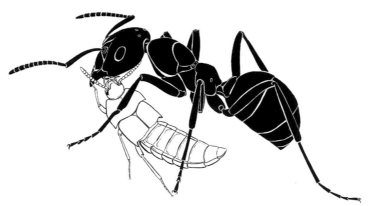

8-5 *Dinarda dentata* beetle snatching food from host ants. The beetle sneaks underneath a food-laden ant (upper) and stimulates the lower lip (labium) of the ant (lower). This sometimes elicits regurgitation of crop contents by the ant. (Bert Hölldobler; Turid Hölldobler-Forsyth; ©Bert Hölldobler).

By the application of the radioactive tracer technique, we were able to demonstrate that the likelihood of the beetles receiving food from host ants is significantly higher when food-laden foragers attempt to deliver their crop content to nestmates in the nest than when beetles are kept together with a group of well-fed ants in which trophallactic interactions are much less frequent. Although the beetles are much smaller than their host ants, they still received as much as 32% of the food share normally delivered to host workers. This is likely because

8-8 *Lomechusoides strumosus* larva "gently" handled by a *Formica sanguinea* worker (upper) and a *Lomechusa* larva carried by a *Formica* worker to another destination (lower). (Bert Hölldobler).

of the radioactive food was taken up by the ants and transferred to the larvae. These experiments showed that in a mixed population of beetle and ant larvae, the beetles obtained a disproportionately large share of the food. The presence of beetle larvae reduced the normal flow of food to the ant larvae; on the other hand, the presence of ant larvae did not affect the food flow to the beetle larvae. The measurements were taken with last-instar beetle and ant larvae. In this stage, the body mass of both beetle species and ants is not significantly different.

8-9 The myrmecophilous beetle larvae frequently rear and wave their heads up and down or sideways. This behavior is especially obvious when the larva is touched by a host ant worker, indicating that the larva is seeking to contact the ant's head. These images show *Lomechusoides* larvae and their hosts *Formica sanguinea* and *Formica fusca*. (Bert Hölldobler).

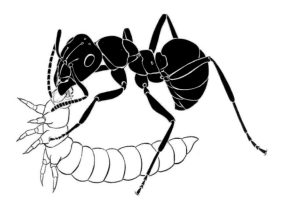

8-10 *Lomechusa pubicollis* larva in mouth-to-mouth contact with the ant, receiving regurgitated liquid from the host. (Turid Hölldobler-Forsyth; ©Bert Hölldobler).

8-11 The larvae of *Lomechusa* and *Lomechusoides* also prey on the host ants' brood. The upper picture shows *Lomechusoides strumosus* larvae in the brood nest feeding unresisted on the ant larvae; the lower picture depicts a larva of *Lomechusa pubicollis*. (Bert Hölldobler).

In addition, *Lomechusa* and *Lomechusoides* larvae prey on the ant larvae without intervention by the nurse ants (Figure 8-11). This observation raises a question: How does the ant colony manage to survive the beetle larvae's competition for food and their intense predation on the ant larvae in the brood chambers? A possible explanation might be the cannibalistic behavior of the beetle larvae. Apparently, they are unable to distinguish their fellow beetle larvae from ant larvae. They therefore cut down their own population, while the ant larvae do not. Whereas ant larvae are clustered inside the brood chambers, the beetle larvae usually do not reside in dense neighborhoods with their conspecifics.

8-12   A freeze-killed *Lomechusoides* larva is fully attractive to the host ants *Formica sanguinea* and readily retrieved into the ants' brood nests, where it will be tended for several days. (Bert Hölldobler).

How is it possible for the beetle larvae to be treated by their host ants like ant larvae? Surely, their mimicry of larval food-begging behavior plays a role, but this cannot be all, because freeze-killed beetle larvae, when presented in the foraging arena, are invariably picked up by ant workers and carried to the brood chambers. Several subsequent investigations suggest that chemical communication is involved. When freeze-killed beetle larvae were washed with diethyl ether or dimethyl ketone (acetone) and, after drying, placed together with a freeze-killed but untreated beetle larva, the latter were intensively groomed (Figure 8-12) and readily carried to the brood nest, whereas the former were either ignored or eventually deposited in the refuse area.

Inside the brood chambers, live beetle larvae were frequently groomed by the ants, which usually led to the typical food-begging behavior described above. Hölldobler was able to demonstrate that approximately two to four days after the larvae have been fed by the nurse ants with radioactively labeled food, the grooming ants pick up small amounts of radioactivity from the beetle larvae. Whether they obtain it from the surface, or from the anal area, or during

GRADES OF MYRMECOPHILOUS ADAPTATIONS

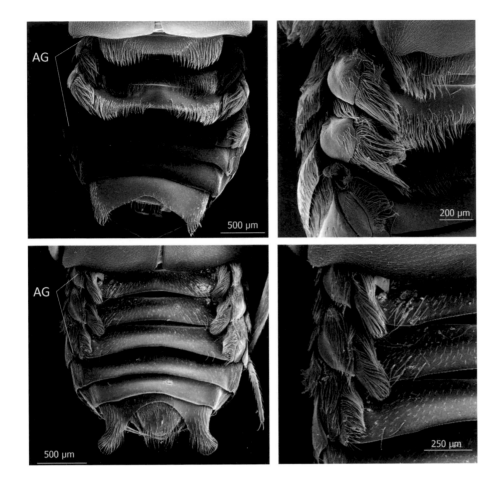

the abdomen of *L. pubicollis* and *L.emarginata* with large intracellular reservoirs, but we could not determine the exits of the duct cells. In addition, the posterior segments are endowed with glandular hypodermal epithelia. Finally, we noticed in the hindgut tissue an unusually thick glandular epithelium, the cells of which have large nuclei. We do not know from which of these exocrine glands the appeasement secretion originates. Perhaps one or all these glandular structures are involved in this gentle defense process; therefore, we call this remarkable assemblage of exocrine glands close to the abdominal tip, like in the previously discussed myrmecophilous staphylinids, the "appeasement gland complex."

The *Lomechusa* beetle reaches backward and keeps antennating the ant that licks its abdominal tip, apparently "verifying" that it is the correct host ant species. Simultaneously, the beetle often exhibits slight tremble movements and

8-18   Scanning electron microscopic images of the abdomen of myrmecophilous staphylinid beetles *Lomechusoides strumosus* (upper) and *Lomechusa pubicollis* (lower). AG indicates the lobes of the paratergites with the dense endowment with trichome hairs. This area is richly endowed with exocrine glands, which are called "adoption glands." (Bert Hölldobler).

gentle drumming with its legs. Finally, the beetle lowers its abdomen so that the ant can access the abdominal margins, where the so-called adoption glands are located. The margins of the tergites II, III, IV, and V are endowed with striking tufts of golden hairs, called trichomes (Figure 8-18).

These trichomes are especially dense on the lobes of the paratergites. In contrast to Erich Wasmann (1903, cited in Wasmann 1915), Karl H. C. Jordan (1913) recognized that these golden bristles are closely associated with exocrine glands, which open through cuticle pores adjacent to the golden bristles on the elevated edges of the paratergites and pleurites, and he recognized that the bristles are innervated. Jordan described the glands as flask-shaped hypodermal gland cells. This characterization of the glandular cells is not correct. Instead, the secretory cells are combined with duct cells that open through ducts in close vicinity to the bases of the trichome setae (Pasteels 1968; Hölldobler 1970b; Hölldobler et al. 2018; Figure 8-19). Similar morphological and glandular structures exist in the genus *Lomechusoides* (Figures 8-18, 8-20), and most likely also in the North American genus *Xenodusa,* which exhibits identical trichome structures.

During the adoption process, the ant is attracted to these gland trichomes and grasps some of the bristles, lifting the beetle off the ground. The beetle assumes a pupal position, with legs and antennae tightly folded to the body, and is carried by the ant into the host's brood chambers (Figure 8-17). Apparently, the secretions of these trichome glands are essential for the adoption process. In a series of experiments in which Hölldobler covered the glands with colophonium wax, the adoption process usually failed. He therefore called these glands "adoption glands" and hypothesized that they may mimic, perhaps in a "supernormal" way, the brood pheromones of the host ants. Inside the brood chambers, the beetles prey unimpeded on the ant brood and successfully solicit regurgitation from their *Myrmica* hosts (see Figure 8-14), as they did previously with their *Formica* hosts.

Other *Lomechusa* species such as *L. emerginata* or *L. sinuate* (Figure 8-21) exhibit the same behavioral patterns although they use different host species.

GRADES OF MYRMECOPHILOUS ADAPTATIONS

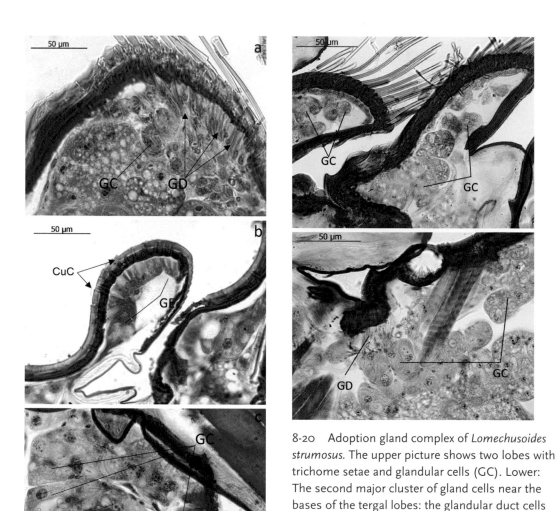

8-20 Adoption gland complex of *Lomechusoides strumosus*. The upper picture shows two lobes with trichome setae and glandular cells (GC). Lower: The second major cluster of gland cells near the bases of the tergal lobes: the glandular duct cells (GD) open through the cuticle near a tracheal tract. These glands are most likely part of the adoption gland complex. (Bert Hölldobler).

8-19 Parts of the adoption gland complex of *Lomechusa pubicollis*. Longitudinal section through the abdominal paratergal lobes. (a) A lobe with trichomes and many glandular cells (GC), the duct cells (GD) of which open through cuticle channels between the trichome setae. (b) Some areas of the lobes that have no trichomes have glandular epithelia (GE), the cells of which open through cuticular channels (CuC). (c) The second cluster of gland cells on the base of the trichome lobes, the duct cells of which open through the cuticle near a major tracheal tract. (Bert Hölldobler).

8-21   *Lomechusa emarginata* (upper, courtesy of Pavel Krásenský) and *Lomechusa sinuate* with *Myrmica* host queen (lower, courtesy of Taku Shimada).

8-22 *Lomechusa sinuate* beetle solicits food from the *Myrmica* sp. host. It first stimulates the mouthparts of the host ant with its forelegs (upper). This triggers the regurgitation reflex in the ant (lower). (Courtesy of Taku Shimada).

We are delighted to be able to show some of the photographic documentation of *L. sinuate* provided by Taku Shimada. Here too, the larvae are raised by *Formica* species and the winter hosts are *Myrmica* species (Figures 8-22, 8-23, 8-24, 8-25).

Hölldobler verified the food transmission from the *Myrmica* host to *Lomechusa* guests (*L. pubicollis and L. emarginata*) by employing radioactively labeled food. Indeed, *Lomechusa* beetles very successfully participate in the social food flow of their host colonies, yet we have no evidence that during trophallaxis food also flows from the beetles to the ants. In contrast to *Formica* species, *Myrmica* colonies also contain brood during the winter period, so the beetle has plenty of food for completing sexual maturity. In spring, after hibernation, *Lomechusa* migrates again to *Formica* colonies, just at the time period when *Formica* raises their first brood, and the social food flow is plentiful. Mating takes place in the *Formica* nests, where females also deposit their eggs, and, as we already discussed, the

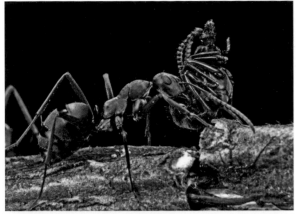

8-23   *Lomechusa sinuate* migrate from the *Myrmica* "winter host" to the *Formica* "summer host." *Formica japonica* carry the beetle into their nest (upper). Inside the nest the beetle solicits food from *Formica* workers in the same way it did with its *Myrmica* hosts (middle and lower). (Courtesy of Taku Shimada).

8-24   The larvae of *Lomechusa sinuate* develop inside the brood nest of the *Formica hayashi* host, where they prey unimpeded on the ant larvae. (Courtesy of Taku Shimada).

8-25   *Lomechusa sinuate* larvae are groomed and fed by the *Formica* host workers, and when brood chambers are relocated, the *Formica* workers carry the parasitic myrmecophiles to the new location. (Courtesy of Taku Shimada).

beetle larvae are raised by the ants, even though these larvae also prey on the ants' brood.

Lomechusa are not the only staphylinid myrmecophiles capable of making themselves at home with more than one ant species. William Morton Wheeler (1910) reports that the staphylinid beetles of the genus *Xenodusa* change their domicile with the seasons. The larvae live in *Formica* nests through the summer, and the adults overwinter in nests of the carpenter ants of the genus *Camponotus*. It is interesting to note that the carpenter ants also maintain larvae throughout the winter. It may well be that the evolutionary history of *Xenodusa* beetles parallels that of *Lomechusa* in selecting and adapting to a winter home. Six species of the Nearctic *Xenodusa* are known to science, and for all of them, *Formica* species are listed as summer hosts and *Camponotus* species as winter hosts (Hlaváč et al. 2011).

Wasmann (1920) considered the strategy of using different host species in the winter and summer a derived state that evolved from ancestor species that parasitized only one host species. Species of the phylogenetically closely related genus *Lomechusoides* do not change their *Formica* host species, although they also migrate in the fall and spring to different *Formica* colonies of the same species. Fifteen *Lomechusoides* species are known to science, and all use *Formica* species as hosts. The best-studied species is *Lomechusoides strumosus*. No one has published more papers or has observed *L. strumosus* longer than Erich Wasmann. He wrote more than 200 papers on myrmecophiles, many of them on *L. strumosus*. Much of this work is summarized in an overview published in 1915. Over a period of three decades, Wasmann discovered many phenomena regarding the natural history of *L. strumosus*. He also attempted to understand the evolution of this intricate parasitic-symbiotic relationship between this myrmecophile and its host ants, *Formica sanguinea*. He proposed that the ants evolved a symphilic instinct, an aberration of the brood care instinct. He also argued that the host ants actively select the most desirable beetle individuals for breeding, because the ants became "addicted" to the glandular secretions exuding from the "trichome glands." He termed this specific selection behavior "amical selection." Although Wasmann's hypothesis was attacked by several contemporary entomologists, such as Jordan (1913), Escherich (1898a), and Wheeler (1910), Wasmann responded by listing many behavioral facts he assembled, which he thought solidly supported his theory. Finally, in 1948, Karl Hölldobler published a detailed analysis of the facts and theoretical arguments Wasmann listed, concluding that, although most of the facts are not in question, Wasmann's evolutionary interpretation lacked a logical foundation. Nevertheless, even he proposed that the ant hosts might eventually develop a kind of "addiction" to the trichome exudate, and therefore tolerate the beetles, even though they can become detrimental to the colony. Indeed, Wasmann found a strong correlation of aberrant worker morphology (so-called pseudogynes) and striking decline of the production of alate reproductives in colonies that housed *L. strumosus.* Although the exact physiological cause of the aberrant development of workers and decline of production of sexuals (reproductives) is not known, given that the beetles divert food from the colony, it is likely this phenomenon is caused by starvation of the larvae. It should also be pointed out that the correlation between the occurrence of *Lomechusoides* larvae and pseudogynes in *Formica sanguinea* colonies could not be confirmed by Horace Donisthorpe (1927). However, Donisthorpe himself states that his data base is by far not as extensive as that of Wasmann. In summary, K. Hölldobler recognized that the myrmecophiles exploit stimuli the ants employ in intraspecific interactions within the colony and called the myrmecophilous beetles "psychoparasites."

Our comparative studies of *Lomechusa* and *Lomechusoides* revealed that both genera have almost identical glandular equipment and both myrmecophiles employ the glands in similar ways (Hölldobler et al. 2018). In addition, *Lomechusoides* has well-developed trichome glands in the legs, particularly in the femurs, which seem also to be involved in the adoption process (Figures 8-26, 8-27).

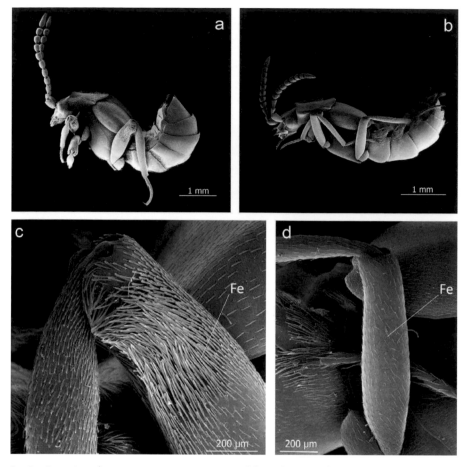

8-26　Scanning electron microscopic images of the side view of *Lomechusoides strumosus* (a) and *Lomechusa pubicollis* (b). Close-up of the femur (Fe) of *L. strumosus* (c), the trichome setae clearly visible; (d) close-up of the femur of *L. pubicollis*, trichome setae are absent. (Bert Hölldobler).

When we placed a beetle collected from a field colony in the arena of a laboratory colony, we noticed that the beetle, when contacted by the ant, positioned its legs outward so that the relatively large femur reached out in an almost horizontal position. Usually the ants licked the legs, especially the femurs (Figure 8-28). The beetle bent backward or sideways with head and thorax, apparently attempting to contact the ant with its antennae. The myrmecophile often rolled up its abdomen and pointed the tip of it toward the ant (Figure 8-29).

Our more detailed observations of the encounter phase showed that the ants contact the posterior of the beetle's abdomen most frequently, followed by the

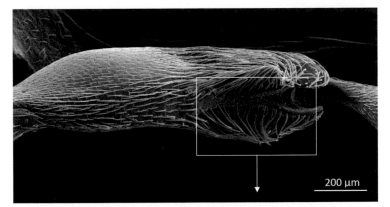

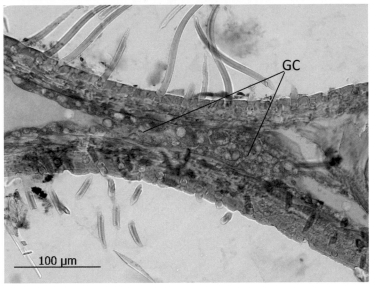

8-27 Trichome glands in the femur of *Lomechusoides strumosus*. Scanning electron microscopic image of the ventral side of the femur. Box indicates the location of the histological preparation shown in the picture below, with glandular cells (GC) inside the femur. (Bert Hölldobler).

200 μm

GC

100 μm

legs. The ants often exhibited slight aggressive behavior, yet *L. strumosus* did not employ the repellent secretion from the defense tergal gland. Instead, the beetle presented its abdominal tip, which the initially aggressive ant licked whilst the beetle continued to reach backward with its antennal tips, the last segments of which are packed with chemo-sensilla. Occasionally, a white, opaque droplet appeared at the spots where we suppose the "anal gland" (or pleural gland) opens. We postulate that this gland, and other exocrine glands in the abdominal tip, together with the rectum, may all be involved in the appeasement process that has the effect of muting the ant's aggression and changing it to docile licking. Only after this initial phase, which usually lasts a few minutes or less but may sometimes take up to twenty minutes, do the beetles allow the ants to have full

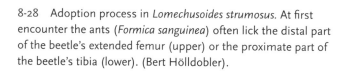

8-28 Adoption process in *Lomechusoides strumosus*. At first encounter the ants (*Formica sanguinea*) often lick the distal part of the beetle's extended femur (upper) or the proximate part of the beetle's tibia (lower). (Bert Hölldobler).

8-29 In the next step in the adoption process of *Lomechusoides strumosus*, the beetle presents the abdominal tip to the *Formica sanguinea* host (upper). The ant licks the posterior parts of the beetle's abdomen, and during this process the beetle attempts to touch the ant with its antennal tip (lower). (Bert Hölldobler).

access to the tufts at their abdominal margins where the adoption glands open (Figure 8-30). Finally, the beetle is carried by the ant into the brood nest of the host colony, and the beetle's behavior during this transport is identical to that of *Lomechusa* beetles.

The participation of *Lomechusoides* beetles in the food flow inside the host ant colony has been known since Wasmann's extensive studies. One can frequently observe mouth-to-mouth contact between the beetles and ants, and, by marking the food provided to the ants with a dye and subsequent dissections, Wasmann demonstrated that food was transferred from ant to beetle (Figure 8-31). The

THE GUESTS OF ANTS

8-31 Inside the nest, the *Lomechusoides* beetles solicit regurgitated food from the host ants and occasionally engage in trophallaxis with one of the *Formica sanguinea* host ants, while simultaneously appeasing another host ant that licks the beetle's abdominal tip. (Bert Hölldobler).

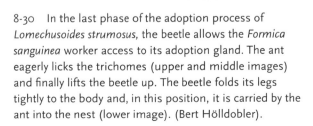

8-30 In the last phase of the adoption process of *Lomechusoides strumosus,* the beetle allows the *Formica sanguinea* worker access to its adoption gland. The ant eagerly licks the trichomes (upper and middle images) and finally lifts the beetle up. The beetle folds its legs tightly to the body and, in this position, it is carried by the ant into the nest (lower image). (Bert Hölldobler).

8-33 Two clavigerite myrmeco-philes *Claviger testaceus* (upper) and *Claviger longicornis* (lower). (Courtesy of Pavel Krásenský).

8-34   The clavigerite genus *Adranes* sp. feeding on the brood of its *Lasius* host ants. (Courtesy of Alex Wild / alexanderwild.com).

Lasius umbratus, but sometimes also with *L. flavus, L. niger,* and occasionally with *Myrmica* species. Donisthorpe reports one finding, in a *Lasius fuliginosus* nest, but he assumes this was a relatively young colony that was founded parasitically in a *L. umbratus* nest, and the *C. longicornis* was a "left-over" of the former *L. umbratus* colony. According to Donisthorpe, *Claviger* beetles prey on the ants' brood, the eggs, larvae, and even pupa, although the last has not been confirmed by Roger Cammaerts, who conducted the most thorough recent studies of these ant guests. The beetles were observed feeding on booty brought into the brood nest by their hosts, and they were frequently groomed by ants. In particular, the beetle's trichome setae at the basal abdominal margins are attractive to the host ants. The exocrine glands associated with these trichomes have been described by Cammaerts (1974). To our knowledge, nothing is known about the larvae and life history of *Claviger* beetles, although mating of beetles inside the nest has been observed by several authors (Donisthorpe 1927); yet recently, photo images were published on the internet by Taku Shimada that appear to show the larvae of

GRADES OF MYRMECOPHILOUS ADAPTATIONS

8-35   The developmental stages of the clavigerite beetle *Diartiger fossulatus:* (top left) last-instar larva; (top right) silk pupal cradle; (bottom left) the cradle has been opened to expose the early pupal stage; (bottom right) adult *D. fossulatus* beetle with host ant larvae. (Courtesy of Taku Shimada).

the clavigerine genus *Diartiger fossulatus,* but no further details have been reported (Figure 8-35).

The *Claviger* beetles seem to be unable to fly. In fact, Donisthorpe (1927) reports that the first specimen of *C. testaceus* collected in Britain was attached to a winged female ant of *Lasius flavus,* on the underside of the body, and he further notes that Hetschko and Janet have found *C. testaceus* attached to both males and winged females of their hosts. This suggests that this myrmecophile achieves its propagation by riding as a hitchhiker on the alates of their host colony. Should the beetle land, during such "odysseys," in the nest of another ant species, this apparently does not endanger its life. In fact, when transplanted into laboratory nests of other ant species, the beetles were accepted by a diversity of host genera with which they have never been found in nature (Donisthorpe 1927; K. Hölldobler 1948). Karl Hölldobler was able to keep *Claviger testaceus* for many months in a nest of the tiny thief ant *Solenopsis fugax*.

For a long time, scientists were puzzled by how the *Claviger* beetles are able to solicit regurgitation in the host ant workers, because, contrary to previous reports, the beetles do not employ their antennae for enticing trophallaxis in the ants, and neither are the beetles' mouthparts especially equipped for stimulating the ants' labium. Thus, previous claims by Erich Wasmann (1891, 1898 cited in Wasmann 1920) that *Claviger testaceus* solicits regurgitation by stimulating the host ants with its antennae and licking the ants' mouthparts appear not to be entirely correct. According to Cammaerts's observations, the beetle pushes its mouthparts against the labium (lower lip) of its host only when the ant, arriving with a full crop, "spontaneously" regurgitates a droplet, "or when remains of regurgitated food are still present on the host's labium." In a series of clever experiments, Roger Cammaerts (1995) found the solution to this puzzle. He discovered that regurgitation in the ants is chemically elicited by secretions originating in the beetle's head (Cammaerts 1996). After a careful morphological analysis of the opening of the duct cells of the exocrine glands in the head, Cammaerts covered certain sections of the beetle's mouthparts with varnish, and he observed under the stereomicroscope whether ants regurgitated food to the treated beetles. In this way he was able to zoom in on the effective parts. For example, when only the upper lip (labrum) of the beetle was not treated, but all the other mouthparts were, there was no difference in comparison to untreated beetles in frequencies of ant regurgitations. However, when only the lower lip (labium) remained uncovered, and the upper lip (labrum) was covered, workers did not regurgitate to the beetles. Cammaerts hypothesized that glandular secretions that originate from cephalic glands on or near the labrum serve as chemical

8-36   Workers of *Lasius* sp. unload a piece of prey next to the myrmecophilous beetle *Claviger testaceus*. (Courtesy of Pavel Krásenský).

signals (allomones) that release regurgitation in the host ant workers. Previously, Cammaerts (1974) showed that most of the duct cells of so-called labral glands and some of the duct cells of the mandibular glands open on the surface of the labrum, and, respectively, some labral gland duct cells and most mandibular gland duct cells open on the surface of the mandibles. Further exclusion experiments strongly suggest that the labral gland secretions are responsible for releasing regurgitation in the ants. In addition, and surprisingly, the host ants also regurgitated onto the beetle when they licked the trichomes on the base of which glandular duct cells open. However, the regurgitation rate was lower than that onto the beetle's mouth parts. As is the case in the head-to-head trophallaxis, the ant's regurgitation onto the trichomes is preceded by the worker's intense licking of the beetle's head or abdomen trichomes, respectively.

Considering how small the *C. testaceus* beetles are (about two to three millimeters in length), the selective covering of the glandular openings on the labrum (upper lip) and labium (lower lip), or mandibles, is quite an achievement by Roger Cammaerts. At first glance these results appear puzzling and odd. Why should workers of *Lasius flavus* regurgitate onto the beetle's abdomen? How and why do the trichome gland secretions elicit such a regurgitation reflex?

In a detailed subsequent study of the regurgitation behavior of *L. flavus,* Cammaerts (1996) discovered that ant workers, which are about to present pieces of insect prey to their larvae, regurgitate a droplet onto this cadaver, and he noted that this particular regurgitation behavior is similar to that exhibited when workers deliver regurgitated liquid to *Claviger* beetles. He suggests that this regurgitation behavior is different from behavior that workers employ when feeding larvae, nestmate workers, and queens, and he proposes that the *Claviger* beetles are treated by the host ants like a piece of prey to be provided for larval consumption. Cammaerts states that, as the ants "regurgitate in response to particular secretions of the *Claviger,* it may be inferred that the regurgitation allomone emitted by the beetle mimics the actions of a substance produced by decaying insect corpses." Cammaerts hypothesizes "that this behavior is

THE GUESTS OF ANTS

instrumental in ensuring the extraoral enzymatic digestion of solid meaty food given to larvae or in providing them with more balanced diet." Pavel Krásenský provided us an interesting image of a *Lasius* host ant that seems intent on unloading a piece of prey near the *Claviger* beetle (Figure 8-36).

This is a very intriguing hypothesis and, if confirmed, would be a totally new mechanism of behavioral adaptation of myrmecophiles making their living in the brood chambers of their host ant colonies, the ultimate evolutionary grade of myrmecophily. The slight caveat we propose is the question, Why is the *Claviger* beetle not treated as prey by the ants themselves? There are no reports suggesting that the ants attempt to feed on the beetle or tear it to pieces, as they do with prey objects. No one has reported any aggressive behavior by the ants toward the beetle; even when the beetle moves around the nest, it is mostly ignored or groomed by the ants. However, the ants often pick up beetles and carry them to and place them among the larvae. According to Hölldobler's observations (Hölldobler and Wilson 1990), *Claviger* actively seeks mouth-to-mouth contact with

GRADES OF MYRMECOPHILOUS ADAPTATIONS

is known about the behavior of paussine myrmecophiles. Stefanie Geiselhardt, Klaus Peschke, and Peter Nagel (2007) presented a comprehensive review concerning the morphology, systematics, phylogeny, distribution, and myrmecophily of the Paussinae. We focus here only on some studies that describe and analyze myrmecophilous behavioral interactions with host ants.

Paussines, also called ant nest beetles, are mostly associated with ant species of the subfamilies Myrmicinae and Formicinae (Nagel 1987), but host ant specificity appears not to be a general trait of these myrmecophiles (Nagel 1987; Di Giulio and Taglianti 2001). For example, *Paussus megacephala* was reported as guest of *Messor barbarus, Camponotus lateralis,* and *Ponera* sp. (Kistner 1982; Wasmann 1894, cited in Geiselhardt et al. 2007). On the other hand, Geiselhardt and her colleagues state that a single ant species may host several different paussine species, and "on the genus level, species of the myrmicine genus *Pheidole* have been reported most frequently as hosts of ant nest beetles." One of the best-studied genera is *Paussus,* especially *P. favieri;* the following deliberations focus mainly on this species.

George Le Masne (1961a,b,c) studied the feeding and predatory behavior in the *Paussus favieri* beetles that live inside the nests of the myrmicine *Pheidole pallidula* (Figure 8-37). The beetles feed on the host colony's eggs, larvae, and on adult ants (Escherich 1899a,b, 1907; Le Masne 1961a,c; Nagel 1979). With their pointed mandibles, they puncture the integument of their prey and extract hemolymph and soft tissue. The beetle's mouthparts appear to be very well suited for this kind of "Dracula-like" preying. M. E. G. Evans and T. G. Forsythe (1985) describe the morphology of the head structures, which appear to be especially adapted for this feeding process. They write, "Species of *Paussus* have small, sharply pointed mandibles, and a well-developed prementum, flanked by (internal) mental pillars supporting a suspensorium; there is also an enlarged cibarium-pharynx. This is all consistent with a sucking pump for a fluid feeding capability" (Evans and Forsythe 1985, 116). Although this functional morphological explanation is hypothetical, it is nevertheless very suggestive. Interestingly,

8-37 The paussine beetle *Paussus favieri* with its host ants *Pheidole pallidula*. This *Paussus* species served as model for many behavioral studies within the genus *Paussus*. (Courtesy of Pavel Krásenský).

the ants that were attacked by paussine beetles did not react aggressively or attempt to rid themselves of the attacker, and after the beetle stopped feeding, the ants stayed close to the beetle and finally died within a few days (Geiselhardt et al. 2007). *Paussus* beetles were never observed feeding on the ants' prey or scavenging in the midden of the host ants' nest.

Why are the *Paussus* beetles accepted into the host ants' nests and tolerated there? The beetles are richly endowed with exocrine glands, with and without trichome hairs, in the antennae, head, thorax, elytra, legs, and posterior abdomen, first thoroughly investigated and beautifully illustrated by Y. C. Mou (1938); additional studies concerning these exocrine glandular structures in paussine myrmecophiles were provided by Reichensperger (1948), Nagel (1979, 1987), Di Giulio et al. (2009), and Maurizi et al. (2012). Most of these glands are probably so-called myrmecophilous organs, which are licked by the ants and may be responsible for the integration of the beetles within the host ants' colony, though no experimental evidence has been published to date. Especially during the

GRADES OF MYRMECOPHILOUS ADAPTATIONS

adoption process of the adult beetles attempting to enter a new host colony, extensive licking and pulling of the beetle has been described by Karl Escherich (1898a). The secretions of the glands located in the antennae appear to be especially attractive to the ants.

The acts of social adoption and integration proceed differently in various *Paussus* species. *Paussus arabicus* beetles are initially treated aggressively by their *Pheidole* hosts. The ants attempt to remove the beetles from the brood chambers or remove the brood from the beetle, but eventually the ants ignore the guest and do not prevent it from preying on the larvae (Escherich 1898a, 1907; reviewed in Geiselhardt et al. 2007). Comparing with other *Paussus* species, Stefanie Geiselhardt and her colleagues write that, "in contrast, after short initial aggression, *Paussus turcicus* is intensely groomed by its *Pheidole* host ants, which seem especially attracted to the antennal cavities and always surround or even cover the slowly moving beetle (Escherich 1898a, 1899a,b). Nevertheless, *P. turcicus* is dragged along the tunnels of the nest like *P. arabicus*" (Geiselhardt et al. 2007, 883). The myrmecophilous relationship of *Paussus favieri* with its host ant, *Pheidole pallidula,* is strikingly different from the abovementioned cases. Geiselhardt et al. (2007, 883) noted, "The beetles are usually ignored by the ants, moving quickly and undisturbed within the tunnels and touching ants and objects with the antennae. Although *Paussus favieri* may also be faintly attacked and licked during adoption, the ants only very rarely touch, quickly groom or drag the beetle later on" (Le Masne 1961b). Geiselhardt and colleagues hypothesize that,

> from an evolutionary point of view, the integration type of *Paussus arabicus* can be considered as the most basal, as attacks occur, but cease during the contact with the ants. A possible explanation would be that the beetle adopts the colony odor during the interaction with the ants. In contrast, *P. turcicus* provides a reward in the form of a myrmecophilous secretion, which may or may not entail a nutritive benefit for the ants. Nevertheless, the production certainly is costly for the beetle, as the secretion is not derived from waste material, like aphid honeydew, but a product of specialized glands.

This is why we consider the integration type of *P. favieri* as the most derived. These beetles are allowed to move freely among the ants without apparent costs, and one may assume advanced chemical mimicry as the mediating mechanism in this association. (Geiselhardt et al. 2007, 884)

These are interesting speculations, but since nothing is known about the nature of the glandular secretions of paussine myrmecophiles, the assumption that these secretions serve as "rewards" for the ants is questionable. Early myrmecologists have even contemplated that a mutualistic, or reciprocal, relationship between highly adapted paussine myrmecophiles and host ants may exist, a kind of "give-and-take" relationship, assuming the beetle's glandular secretions compensate for the loss caused by the beetle's predation. But these speculations were later almost unanimously rejected. Nevertheless, the term "reward secretions" remained in the literature (e.g., Maurizi et al. 2012). To us, the more appropriate characterization of these secretions would be appeasement secretions, or, if this could be experimentally demonstrated, adoption secretions that might mimic brood pheromones.

Emanuela Maurizi and colleagues (2012) provided the first attempt of a behavioral repertoire (ethogram) of *Paussus favieri* inside the ant nest. They recorded the frequency and duration of five behavioral categories: rewarding, antennal shaking, antennation, escape, and "no contact." The term "rewarding" stands for the ants licking the beetles on the trichome glands. As we stated above, the term can be misleading. Interestingly, this is the most frequently occurring behavior, and this underlines the significance of these appeasement or adoption substances for the social integration in the host ant society. A striking observation is the repeated close contact of *P. favieri* beetles with the ant queen. The authors write that the beetles occasionally "remained in the queen's chamber for some days, antennating and rubbing against the queen's body without any aggressive reaction from the queen or the workers." This behavior is apparently

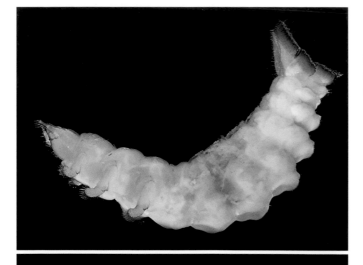

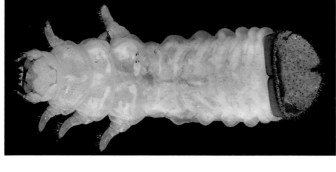

8-38 Third-instar larva of *Paussus siamensis;* upper, lateral view; lower, dorsal view. Note the terminal disc at the rear end of the larva. (Courtesy of Munetoshi Maruyama).

fully tolerated by the workers; it appears the beetles do not hurt the queen in any way. Perhaps this behavior enables the beetles to acquire some of the specific cuticular hydrocarbons that characterize the queen. Andrea Di Giulio and his colleagues (2011) report that female beetles lay their eggs inside the host ants' nest, and the almost immobile beetle larvae develop in the brood chambers of the ants. They are morphologically highly modified: "The components of their terminal disk are appressed and the discal surface is covered with modified structures that presumably help spread substances attractive to the ants" (see Figure 8-38) (Di Giulio 2008; Di Giulio et al. 2017; Moore and Di Giulio 2019). As we pointed out above, the *Paussus* larvae have sharp, pointed mandibles, and during feeding they puncture the integument of the ant larvae and suck the hemolymph out of their victims. They also solicit regurgitation of food from the host ants, perhaps by imitating the begging behavior of the ant larvae. The myrmecophilous behavior of the *P. favieri* larvae strikingly resembles that of *Lomechusa* and *Lomechusoides,* and it is tempting to speculate the same behavioral mechanisms convergently evolved in these phylogenetically distant myrmecophi-

8-39 Adult beetle of *Paussus siamensis* (upper). The lower picture shows the third-instar larva of *P. siamensis* exuding a large droplet of a clear liquid that is imbibed by a worker of the host ant species *Pheidole plagiaria*. (Courtesy of Takashi Komatsu).

lous beetle families. In both cases, the larvae seem to mimic the host ant's larval communication signals. Eventually the larvae will dig burrows in the soil of the ant nest, where they pupate (Di Giulio et al. 2017; Moore and Di Giulio 2019).

A most remarkable discovery was reported by Andrea Di Giulio and his colleagues (2017) in *Paussus siamensis* third-instar larvae, which exude large droplets of transparent liquid from the thorax (Figures 8-38, 8-39). The host ants (*Pheidole plagiaria*) readily imbibe this liquid. No glandular structures appear to be associated with the release of this secretion. It seems to be a kind of autohemorrhaging (reflex bleeding) employed by some insects in defense against predators. In the case of the *P. siamensis* larvae, it might be a "gentle defense" or appeasement behavior. This very remarkable and peculiar behavior has never been seen in other *Paussus* larvae and deserves to be further investigated.

It is well known that many carabid beetle species are endowed with stridulatory cuticle structures and that they produce stridulation sounds when harassed. These acoustic or vibrational stimuli are often combined with chemical defense (Alexander et al. 1963; Freitag and Lee 1972; Masters 1979; Eisner 2005). The adult *Paussus favieri* also possess elaborate stridulation organs. Andrea Di Giulio and his colleagues (2014) described three types of stridulatory organs in three lineages of Paussini, and they hypothesize that acoustical communication has played an important role in the evolution of myrmecophily in these ant nest beetles. They state, "While the roles of stridulation in this group remain speculative, we verified that all three types of stridulatory organs are present in both sexes and are similar to stridulatory organs known in their host ants which also use stridulation as a method of communication" (Di Giulio et al. 2014, 692). To our knowledge, communication by stridulatory vibrations in *Pheidole* has not yet been experimentally demonstrated, yet we do not deny that vibrational signals in *P. pallidula* might also modulate the response threshold to chemical signals, as has been shown in other myrmicine species (e.g., Markl and Hölldobler 1978; Baroni Urbani et al. 1988).

In a subsequent paper, Di Giulio et al. (2015) seemed to have demonstrated this, especially with respect to the myrmecophilous interactions of *Paussus favieri* and its host ants. Di Giulio and colleagues recorded the airborne sounds produced by minor and major workers (soldiers) and queens and found no differences in pulse length and frequency of sounds produced by major workers and queens, whereas workers' pulse trains were longer. However, the sounds of each of the two worker subcastes and the queen differ in intensity. In *Paussus favieri* beetles, which are endowed with three different stridulation structures, stridulation sounds consist of three different kinds of pulses. The sequences of pulses lasted approximately three seconds, on average, and both sexes emitted all three types of signals. A comparison of the stridulation package containing the three pulse types of *P. favieri* beetles with *Pheidole pallidula* sounds revealed that it features the same pulse lengths as those emitted by major and minor workers, while another pulse type emitted by the beetle had the same intensity and pulse length as that of the queens.

To test the impact of these stridulation sounds on the behavior of *Pheidole* workers, the authors employed a very similar behavioral bioassay to that which Barbero et al. (2009) used with the larvae and pupae of the cuckoo lycaenid *Phengaris 'rebeli,'* which putatively produce sounds that mimic those of the *Myrmica* queens (see Chapter 4). In playback experiments, only two workers were placed into the test arena, measuring 7 cm in diameter. In the center of the arena a miniature speaker was installed through a hole in the floor, with the surrounding edges sealed and the speaker covered with a thin layer of sand. Controls consisted of either a silent speaker or white noise. In randomized sequences employing a double-blind observation protocol, sounds taken from five individual *P. favieri* males and four females, and of five specimens each of minor and major *Pheidole* workers, and five *Pheidole* queens were tested. During playback experiments, the following behavioral acts were recorded: "(1) walking (the worker was attracted to the speaker and walked over it); (2) antennating (the worker antennated the speaker for at least three seconds); (3) guarding (the worker showed an alert on-guard poise especially directed toward the queen or other 'valuables' for the colony) on the speaker for at least three seconds; (4) digging (the worker dug into the soil over the speaker); (5) staying (workers stayed on the speaker without performing any movement or assuming a particular pose)" (Di Giulio et al. 2015, 16).

Reporting the results, the authors write,

In playback experiments, we did not observe any antagonistic or alarmed ant behavior, but always non-aggressive responses involving attraction (walking, antennating and staying) and interaction with other ant castes (guarding, digging). . . . Behaviors such as guarding, digging, and staying were not produced by controls. *Pheidole pallidula* ants were attracted to and induced to "walk" on the speaker by all the sound stimuli, showing no differences in the frequency of responses to the beetle's or ant castes' stridulation. Interestingly, playback of *P. favieri*'s stridulation produced a number of antennations similar to that elicited by sounds emitted by *Pheidole pallidula* queens. Soldiers' stridulation elicited a smaller amount of antennations, but not statistically different from those elicited by *Paussus favieri* single pulses or worker stridulation. Results for antennation are remarkable because this behavior is known to be linked to nest-mate recognition, recruitment, or to facilitate trophallaxis or pheromone emission. Guarding was only induced by sounds of *Paussus favieri* and queens. Workers responded to these stimuli by assuming a posture similar to that adopted when they attended queens or objects of great value to their society. The queen's sounds produced the highest occurrences of guarding, which is consistent with the high status and protection afforded to queens in the colony's hierarchy. (Di Giulio et al. 2015, 10–11)

The results are summarized by the statement "Our data suggest that, by mimicking the stridulation of the queen, *Paussus* is able to dupe the workers of its host and to be treated as royalty" (Di Giulio et al. 2015, 1).

These results sound extremely interesting and exciting, but are the conclusions justified, based on the behavioral investigations performed? We think the exact morphological description of the stridulatory organs and the comparative characterization of airborne stridulation sounds in *Paussus* beetles and host ants are valuable contributions. However, we consider the behavioral studies less convincing. To study social behavioral responses in two ant specimens, completely isolated from the colony or larger group of nestmates is, in our view, questionable. Such studies should always be combined with observations in the intact colony. The ethogram provided by Maurizi et al. (2012) briefly mentions *Paussus* females moving their hindlegs as "a behavior possibly connected to the emission of stridulation during precopulatory behavior" of *P. favieri* beetles. But no such possible stridulation behavior in interactions with ants has been mentioned as part of the ethogram, nor have the authors attempted to record stridulation sounds in the intact colony. Stridulation movements in workers and queens are visible, even in much smaller ant species such as *Temnothorax*. But they are not mentioned in the ethogram study, nor has "the royal guarding behavior" been noted, which the *Pheidole* workers supposedly perform around their queen and the myrmecophilous beetles. The work on stridulation signals is also compromised by the fact that ants do not perceive airborne sounds, but rather react very sensitively to substrate-borne vibrations (see discussion in Chapter 4). The authors placed the miniature speakers in direct contact with the floor so that some of the vibrations were transmitted through the substrate; however, this is not the ideal way to test the specificity of substrate-borne vibrational signals. We also wonder how the stridulation sounds were recorded. The authors state, "During recording sessions, the specimens were placed on the microphone." We wonder how the experimenter made the ants and beetles stay on the microphone and how stridulation was elicited? Were the specimens affixed or tethered onto the microphone? Usually, myrmicine ants stridulate when prevented from moving freely, and this is most likely also the case in *Paussus* beetles. Ants are generally very sensitive to substrate vibration and react by either startling response, antennation, or increased locomotion. An interesting—in fact, proper—control would have been to present the substrate vibration caused by stridulation of another species, such as *Messor barbarus,* the workers of which are about the size of *P. pallidula* queens.

We propose a more parsimonious explanation for the specific responses of *Pheidole* workers to the stridulation vibration caused by *Pheidole* queens and *Paussus* beetles. Because of the larger body mass of queens and beetles, each probably produces a stronger substrate vibration within a certain frequency range that elicits the typical "attention behavior" in *Pheidole* workers, known from a number of ant species. The ants freeze with head and thorax slightly raised, the antennae waving and the mandibles somewhat gaping. Such behavior usually indicates a lower response threshold in the ant to react to chemical signals such as alarm pheromones. Stimuli that cause such reactions are usually called modulating signals, or modulatory communication signals (Markl 1985).

All experimental evidence to date suggests that the attractiveness of the queens in ant societies is due to chemical cues and signals, which might be mimicked or acquired by some myrmecophilous beetles, such as *Paussus favieri*. Vibrational stimuli might be involved in modulating or enhancing the chemical communication, and some of the above-described results could be interpreted in this way, yet this has still to be demonstrated.

In conclusion, we think Andrea Di Giulio and his collaborators have made many very valuable contributions to our understanding of numerous aspects of myrmecophily in the Paussini, and perhaps vibrational communication within the host ant society plays a significant role; however, we think this has not yet been demonstrated.

In this chapter we explored hypothetical pathways of aleocharine rove beetles, from generalist predators in the leaf litter to more specialized predators of ants, roaming along the trunk routes and debris sites outside the nest. Then we considered the scavengers and predators that live mostly in the kitchen middens of ant nests. The next adaptive grade is found in the more specialized myrmecophiles that mainly live in the peripheral nest chambers, from where they occasionally sneak into the brood nests to prey on ant eggs and small larvae, and steal regurgitated food from returning foragers. Finally, we explore the most advanced grade of brood-nest specialists, which are adopted and fed by the ants, prey unimpeded on the ants' brood, and have their larvae raised by the ants. These brood-chamber parasites have conquered the heart of the host ants' nest, presumably by mimicking brood signals in a supernormal fashion. We also point out that other myrmecophilous beetles, in the course of evolution, succeeded in invading and being cared for by the ants in the brood nest. These beetles may trick their host using other means—for instance, by imitating a piece of booty.

In Chapter 9 we return to the ecosystem of an ant nest and explore how diverse myrmecophiles that occupy different nest niches are interconnected, and how the social organization of an ant colony can affect infestation by myrmecophilous social parasites.

9 Myrmecophiles in the Ecosystem of Ant Nests

"THE 'NICHE' OF an animal means its place in the biotic environment, its relations to food and enemies." These words by Charles Sutherland Elton (1927) defined and classified an ecological niche according to the foraging activity of a population of a species. According to Elton, the niche is defined by the species' response to, and impact on, the environment. Later ecologists (in particular G. Evelyn Hutchinson 1959) extended this definition, stating that an ecological niche is the role and position a species has in its environment: how it meets the species' needs for food, shelter, survival, and reproduction. "A species' niche includes all of its (spatio-temporal) interactions with the biotic and abiotic factors of its environment." Our interpretations are contextualized within Elton's simplified spatial niche concept, because the application of the more comprehensive niche concept would have been impossible in these empirical studies that investigated only some niche parameters, such as position in the host ants' nest and food flow. Importantly, ants construct niches by modifying the physical environment across a range of species-specific nest architectures, and through the social and physical organization of individuals and resources therein.

Hölldobler and Wilson (1990) suggested that the greatest diversity of myrmecophiles is to be found in ant species for which mature colonies are exceptionally large and contain a greater breadth of niches to be invaded. For example, the massive colonies of the Neotropical army ant species of the genus *Eciton* or the African *Dorylus* species are host to an exceptional variety and abundance of myrmecophilous species. Considerably fewer myrmecophilous species are expected with ants that form comparatively small colonies (for example, species of *Strumigenys* and *Orectognathus;* of *Leptothorax* and *Temnothorax;* certain amblyoponine, heteroponerine, and ectatomine species; and many others).

The rationale for these assumptions was that large colonies with complex social organizations and nest structures provide a greater diversity of niches for myrmecophiles, which in turn leads to a higher diversity of these symbiotic species that each specialize on particular niches. The abundance of myrmecophiles of one species is usually controlled by the spatial dimension of the particular niches and intra- and interspecific predation among the myrmecophiles or the host ants. Similar suggestions, though with a different emphasis, were proposed by Hughes et al. (2008), who hypothesized "that the unique conditions in nests of large insect societies are expected to produce highly diverse communities of relatively avirulent pathogens (and myrmecophilous parasites) and moderately benign mutualists." To our knowledge almost no data exist with respect to degree of virulence of particular myrmecophilous parasites in large colonies. One of us (B. H.) compared the social food flow directed to *Lomechusa pubicollis* larvae and host ant larvae and found that parasites receive significantly more food from the nurse ants than the ant larvae, and similar results were obtained for *Lomechusoides strumosus* (Hölldobler 1967). In addition, as we already pointed out in Chapter 8, Wasmann (1920) described an aberrant worker morphology which is likely caused by the presence of many *Lomechusoides* beetles and their larvae. There is circumstantial evidence that due to the intensive food consumption by the parasitic myrmecophiles, the host ants are unable to raise viable reproductive females. These findings speak against the hypothesis by Hughes et al. (2008) that highly adapted myrmecophiles in large complex colonies exert only very low or no virulence. Their hypothesis is intriguing and mainly based on theoretical contemplations, but still lacks empirical support.

The number of myrmecophiles is limited in the nests of some ant genera and subfamilies whose species form small colonies. However, recent surveys revealed a surprisingly large diversity in fractions of the myrmecophilous community found with the Neotropical species *Ectatomma tuberculatum* (Pérez-Lachaud et al. 2011) and across the entire myrmecophilous fauna of the Neotropical *Neoponera villosa,* which build nests in the bromeliad plant *Aechmea bracteata* (Rocha et al. 2020). These species form relatively small colonies, with monomorphic workers and poorly developed division of labor, yet house a remarkably rich community of ant guests. For example, the colonies of *N. villosa* consist of one to three queens, and about one hundred workers (large colonies can have approximately 300 workers), and although the numbers of individual myrmecophiles are relatively small, the diversity is astounding. Although not all listed species were found in every nest, the accumulated records of species retrieved from the eighty-two colonies or colony groups is amazing. The authors report to have collected "organisms of six classes distributed in at least 43 different taxa belonging to 16 orders and 24 families." Myrmecophiles found in direct physical association with the ant brood or workers in the central part of the nest included Hymenoptera (Diapriidae, Eucharitidae), Lepidoptera (Riodinidae), Diptera (Syrphidae), Coleoptera (Staphylinidae, Tenebrionidae), several mites, pseudoscorpions, and fungi (Ophiocordycipitaceae).

It is likely that similar thorough surveys in the future will change our view and theoretical contemplations concerning the proneness of ant species to infestations by myrmecophilous parasites, parasitoids, predators, and commensals. The following studies make detailed accounts of the biodiversity within the spatial and social landscape of nests from several ant species.

The Nest Ecosystem of the Harvester Ant *Pogonomyrmex badius*

Pogonomyrmex badius is a seed-harvesting ant, common in the longleaf pine forests of the Southeastern Coastal Plain of the United States. Its spiraling nests can extend more than two meters below the surface and are home to a small but charismatic group of commensals, predators, and parasites. Generalists and endemics found with no other ant species make their living across the vertical

9-1 An emigration trunk trail connects the old and new nests of a Florida harvester ant, *Pogonomyrmex badius* colony (left). (Photo: Courtesy of Walter Tschinkel). The picture on the right shows *P. badius* workers closing their nest entrance at the end of the day. The ants collect sand pebbles and charcoal pieces (residues from natural burns) for plugging the nest entrance hole. The copper-wire rings on the ants' petiole were used for marking the ants. (Christina Kwapich).

strata of *P. badius* nests and foraging trails. To appreciate the niches these myrmecophiles occupy, we begin at the top of the nest.

Each evening the ants close the nest entrance with debris and sand kicked backward, before diving through the loosely filled entrance hole (Figure 9-1). By morning, the assassin bug *Apiomerus crassipes* (Hemiptera: Reduviidae) (Figure 9-2) can already be found waiting for the nest to reopen (C. L. K. personal observation). As the ants gather on their nest mound, the bugs begin picking them off. The bugs grab them with their front legs, pierce the ants behind the head (Figure 9-3), and discard their drained bodies like a pile of crushed soda cans (Morill 1975). The bugs can hunt this way for hours, easily dodging living ants by shuffling from side to side. It is unknown whether the bugs use chemical cues to locate unopened nests, or visual cues associated with the striking black discs of charcoal collected by the ants.

The bugs are not the only hunters that frequent *P. badius* nest mounds. The false widow spider *Asagena* (= *Steatoda*) *fulva* (Theridiidae) also builds its webs

9-2 The assassin bug *Apiomerus crassipes* (Reduviidae) is a predator of the Florida harvester ant. (Courtesy of Roy Cohutta).

9-3 The assassin bug *Apiomerus crassipes* feeds on a *Pogonomyrmex badius* worker, by piercing her behind the head. (Christina Kwapich).

9-4 A male of the spider *Asagena fulva* in front of the entrance of an ant nest. (Courtesy Lynette Elliott). The lower picture shows a female of *Asagena fulva*. (Courtesy of Leonard Vincent).

on the nest mounds of *P. badius* in Florida (USA) (Hölldobler 1970a) (Figure 9-4). In the afternoon, females balancing huge red and white abdomens can be seen moving toward *P. badius* nests. One of us (B. H.) observed that the spiders capitalize on a daily lull in the ants' foraging activity to build their webs. After the midday heat has subsided, ants that reemerge to forage during the afternoon shift are quickly captured in the spider's web. Attempts to attend alarmed nestmates are thwarted by the spider's sticky silk and rapacious appetite. In response to the presence of a spider, colonies may close their nest entrances for up to three days before reopening a new one a full meter away. In this way, the spider succeeds in garnering only a single feeding bout from each colony it visits (Hölldobler 1970a).

When not interrupted by *Asagena fulva*, *P. badius* foragers depart the nest on chemically reinforced trunk routes and return with seeds and other items,

9-5 The spider *Micaria delicatula* can be found along the trails of *Pogonomyrmex badius*. It resembles an ant from above and in profile view. The photos depict a male of the spider. (Christina Kwapich).

including insects and charcoal from burned pine needles (Hölldobler and Wilson 1970; Hölldobler 1971). Foragers can also be tempted to visit strategically placed piles of cookie crumbs and bird seed, which allows us to mark them and estimate their numbers and longevity (Kwapich and Tschinkel 2016). On many occasions, one of us has plucked an unusual-looking forager from a pile of cookie crumbs that was in fact a small gnaphosid spider, *Micaria delicatula* (C. L. K. personal observation, species determined by Nadine Dupérré). This interloper is russet-colored like its ant model and walks on stilted legs, occasionally flagging or tucking its abdomen in a manner characteristic of the genus (Figure 9-5). Males have a true constriction of the abdomen that resembles the narrow waist of their presumed model, while females give the illusion of a constriction with two angled stripes on the dorsal surface of the abdomen, the precise color of the sand below them. This dimorphism may allow females to accommodate eggs while still maintaining an effective disguise. Though the spider shows no interest in eating the ants themselves, it slowly loops in and out of the trails of *P. badius,* revealing itself as a spider only when startled into running. In this regard, the

MYRMECOPHILES IN THE ECOSYSTEM OF ANT NESTS

spider is likely a Batesian mimic that benefits from resembling a well-defended model species.

Incoming *P. badius* foragers deposit seeds and scavenge insects in shallow entrance chambers, no deeper than 12 cm below the surface (Kwapich and Tschinkel 2013). These items are later shuttled downward to deeper storage chambers by a different group of workers. Their stepwise vertical progress toward seed chambers can be tracked by presenting colonies with dyed seeds at various time intervals and then excavating the nest. The ants also carry excavated material and trash to the surface in a series of steps. This was determined by placing colonies on top of layer cakes of differently colored sand, then monitoring the boluses of colored sand that they excavated. The high degree of color-mixing in each bolus revealed that ants temporarily cache sand below the surface before it is reformed and brought to the surface by a different group of ants (Rink et al. 2013; Tschinkel et al. 2015). Thus, there is a confluence of incoming food and outgoing debris that resides temporarily in the uppermost chambers of each *P. badius* nest. Colorful neuropteran larvae and wireworms (Coleoptera: Elateridae) can frequently be found just behind the walls of these chambers (C. L. K. personal observation) (Figure 9-6). Further inspection reveals that the floors of the entrance chambers are also perforated with dozens of like-sized holes. The holes lead to small vertical tunnels, each occupied by a single larva or pupa of a new, not-yet-described species of beetle in the genus *Hymenorus* (Tenebrionidae: Alleculinae) (Kwapich and Johnson, in prep.).

To determine how this *Hymenorus* beetle makes its living in near-surface chambers, colonies were offered a buffet of whole and partially crushed seeds (harvested from other colonies), termites, and tabanid flies (captured feeding on the researcher, C. L. K.). Equal portions of each item were submerged in either rhodamine-B or methylene blue dye, then washed clean with water and dried on a paper towel on the hood of a burning hot car. Twenty-four hours after the items were gathered by the ants, the colonies were hand-excavated, and the guts of beetle larvae and ant larvae were searched for the presence of dye. Beetles never consumed crushed seeds or whole seeds. In contrast, dyed seed pulp was found

9-6 Colorful neuropteran larvae and other insect immatures can be found just behind the walls of the upper chambers in *Pogonomyrmex badius* nests (upper picture). Seed-storage chambers are found deeper below the surface and contain thousands of large seeds and their associated myrmecophiles (middle picture). Farther below, the brood chambers provide another niche for guests. (Christina Kwapich).

in the guts of many ant larvae (65%) and some callow workers, as revealed by fluorescence under UV light. An average of 87% of beetle larvae fed on the insect protein collected by ant foragers before that prey ever made it below the top 5 cm of the nest. Beetles likely consume most of the insect prey before the ant larvae ever have a chance to do so. In one nest, twenty-two of twenty-three beetle larvae consumed insect protein, whereas only three of 180 ant larvae had the chance to do so. In a neighboring nest without the *Hymenorus* beetle, fifty-eight of sixty-three ant larvae contained dyed insect protein in their guts. Though *Hymenorus* sp. occurs in colonies of all sizes, the colonies used in our experiment were immature (<700 workers) and perhaps more susceptible to the costs of hosting beetle larvae (Kwapich and Johnson, in prep.).

Given these results, we can conclude that *Hymenorus* sp. larvae are kleptoparasites that exploit the tendency of *P. badius* to shuttle items up and down the vertical strata of the nest in multiple steps. Beetle larvae are never found wandering in chambers, and both larvae and pupae are quickly attacked and consumed by the ants if exposed during excavation. It is likely that the beetle larvae reach out of their bore holes to grab insect prey that has been set down by the ants, like puppets in a miniature whack-a-mole game. The beetle larvae are found in nests in all seasons, including winter, and newly eclosed adults can sometimes be found sealed in their pupal chambers, just below the upper chambers of nests. Their strong odor often precedes them during nest excavations. Fully sclerotized adults are rarely found in nests and presumably exit their pupal chambers in a cloud of quinones to pursue an alternative diet on the surface.

Deeper in *P. badius* nests, thousands of large seeds are stored in seed chambers. Farther below them, eggs, larvae, and pupae are kept in separate brood chambers (Figure 9-6). The ants cannot open most of their stored seeds themselves and instead wait for the seeds to germinate to access their nutrients (Tschinkel and Kwapich 2016). Seed pulp and remaining bits of insects (not scavenged by *Hymenorus* sp.) are carried to the bottom of the nest and placed on the upturned bellies of waiting larvae. Here, Sanford Porter (1985) discovered that at least four mite species from four different families make their living (Belbidae, Uropodidae, Laelapidae, and Rhodacaridae). As we discussed in Chapter 2, incoming insect protein delivered to larvae is also the likely source of mermithid worm infections in *P. badius*. Callow workers infected with worms are ejected from the colony and wander on the surface until the worms emerge from their hollowed gasters (Kwapich in prep. a).

Small white springtails (Collembola) can be found darting around the brood chambers and seed storage chambers of *P. badius*. These include the entomobryid *Pseudosinella rolfsi* and an isotomid, both commensals that graze on waste and fungus on the seeds and chamber floors (Porter 1985). Although the ants take no notice of the springtails, a two-millimeter-long spider, *Masoncus pogonophilus*, makes its living by feeding on them (Porter 1985; Cushing 1995a,b). The spider is found only in the nests of *P. badius*, where all of its life stages reside, year-round. Both sexes build miniature prey-capture webs and produce sticky silk into adulthood, which is uncommon for male spiders. Females deposit egg sacs containing one to six eggs in small, silk-covered depressions on the ceilings of *P. badius* chambers, where they hatch after just three weeks (Porter 1985; Cushing 1995b).

Pogonomyrmex badius relocate their nests an average of once per year (up to four times) and carry their entire collection of seeds and charcoal (which they collect and place on top of their nests) to each new nest location (Gentry and Stiritz 1972; Tschinkel 2014). Collembolans and the spider *M. pogonophilus* follow the emigration trails of their hosts and are found in abundance in freshly moved colonies, suggesting that both may be attuned to the trail pheromone of *P. badius* (Porter 1985), though direct tests have been inconclusive (Cushing 1995b). When spiders were experimentally introduced to foreign *P. badius* colonies, they were immediately attacked and killed, implying that they may also share the colony-odor of their hosts (Porter 1985).

Paula Cushing (1998) predicted that the myrmecophilous spiders would live in isolated populations within their host nests. After all, *P. badius* colonies occur at an average distance of 16 m from one another (Tschinkel 2017), and the small desiccation-prone spiders could gain access to new nests only when the entrances are open during the heat of the day. Instead, molecular data revealed high similarity between neighboring nests and high intra-nest genetic diversity, indicating that spiders move frequently enough between neighboring colonies in each generation to mitigate any signals of genetic drift (Cushing 1998). The passage of spiders between nests most likely occurs during colony emigration events, as spiders have the opportunity to depart and seek the trails of neighboring host colonies. The spider's extreme female-biased sex ratio (7.5 females per 1 male [Cushing 1995a]) suggests that it may be males who make this journey, and more frequently become lost along the way.

In summary, we can conclude that the myrmecophiles of *P. badius* depend on the ant's daily foraging schedule, frequent nest relocations, diet, and vertically stratified division of labor. This constellation of colony features differs from that of the dozens of other ant species occupying the same forest, each of which holds its own wellspring of diversity.

The Role of Myrmecophiles in the Ecosystem of Mound-Building Wood Ants

Let us now return to the ecosystem studies of myrmecophiles in nests of the mound-building red wood ants of the *Formica rufa* group. In 2013, N. A. Robinson and E. J. H. Robinson published the results of a survey of invertebrates living in nests of *Formica rufa,* which they conducted over a period of three years. They found twenty-two obligate myrmecophiles and more than seventy invertebrates of unknown associations with red wood ants; among these were the xenobiont ant species "*Formicoxenus nitidulus* [which can be found with almost all species of the *F. rufa* group], wasps, beetles, flies, springtails, woodlice, millipedes, centipedes, spiders, pseudo-scorpions, mites and worms." The *Formica* nests surveyed belonged to an isolated population of *F. rufa* "at the northern limit of its British range." This might explain why the number of obligate myrmecophiles, although high, is considerably lower than that reported later by Parmentier et al. for the same host species (2014). For a comprehensive review of cohabitants in *Formica* wood ant nests and close surroundings, see Robinson et al. (2016).

As we recounted in Chapter 8, Thomas Parmentier and his coworkers conducted a literature review of all arthropods found in red wood ant nests of the *Formica rufa* group and listed a total of 125 obligate myrmecophiles, which occupy a variety of niches constructed by the particular nest architecture and social organization of the host ants (Parmentier et al. 2014). In order to achieve some sort of categorization of these myrmecophiles for subsequent studies, the Parmentier group adopted the nomenclature of "integrated" and "non-integrated" species, first introduced by David Kistner (1979). However, Parmentier used different parameters for the characterization of these categories. The main parameter was the location within the nest where the myrmecophiles reside, rather than the symbionts' behavior or the behavior of the host ants toward the symbionts. According to their definition, an integrated species is "able to penetrate into the dense brood chambers, whereas non-integrated species occur in sparsely populated nest chambers without brood at the periphery of the nest." They investigated whether niche specialization develops by different degrees of tolerance of the hosts toward symbionts and explored the question of whether symbionts that cause "lower potential costs" to the host are better integrated in the host colony and are treated with less aggression by the ants. Indeed, in the Asian army ant species *Leptogenys distinguenda,* this appears to be the case (Witte et al. 2008; von Beeren et al. 2011). We have to keep in mind that almost all myrmecophiles investigated in Parmentier's study of the *F. rufa* group are "non-specialized"; for example, they do not specialize in making their living exclusively in the brood chambers of the ants, as do *Lomechusa* beetles and their larvae.

Parmentier et al. (2016a) investigated the trophic interactions among a relatively large group of myrmecophiles by employing direct feeding tests and stable isotope analyses of carbon and nitrogen. One of the main revelations of this study was that numerous trophic interactions among the myrmecophiles exist, and that "the host ants can indirectly benefit from these interactions because brood predators are also preyed upon by other myrmecophiles." We already discussed such an effect caused by the cannibalistic behavior of larvae of *Lomechusa* and *Lomechusoides* beetles. In fact, ant nests are not always a "safe haven" (or enemy-free space) for the symbionts, as has occasionally been proposed (Kronauer and Pierce 2011). Though this might be the case for some symbionts, others are exposed to intra- and interspecific predation.

Based on behavioral and trophic interaction studies, Parmentier et al. (2015a,b; 2016a,b,c) argue that niche specialization among myrmecophilous parasites in the host ants' nest can emerge from competition among different species or can be "structured directly by the host when its defense strategy depends on the parasite's potential impact." If the latter is the case, one should expect that parasitic species that exhibit no or only minor brood predation will be treated less aggressively by host ants than will parasites that are specialized on brood predation. Thus, we should expect that the former group of symbionts is better integrated in the host ants' social system than the brood predators. However, as we emphasized before, we have to keep in mind that the authors' investigation focused exclusively on so-called unspecialized ant symbionts, most of them staphylinid species, of which most belong to the subfamily Aleocharinae (*Dinarda maerkelii, Thiasophila angulata, Notothecta flavipes, Lyprocorrhe anceps, Amidobia talpa*), two are species of the Staphylininae (*Leptacinus formicetorum, Quedius brevis*), and one is of the Steninae (*Stenus aterrimus*) (Figures 9-7, 9-8). In addition, they compared five non-staphylinid beetle species, two spiders, and one springtail. From all the listed staphylinid species, *D. maerkelii* appears to be the most specialized myrmecophile, yet by far not to the degree of *Lomechusa pubicollis,* which we discussed in Chapter 8.

It has been known for some time that the ant brood (mainly the eggs and larvae) is a much-desired source of high-quality food, but the extent to which this valuable resource is exploited by a variety of myrmecophilous arthropods

9-7 The rove beetles *Notothecta flavipes* in the left picture and, in the right, *Thiasophila angulata.* (Courtesy of Pavel Krásenský).

9-8 The rove beetles *Quedius brevis* (upper) and *Stenus aterrimus* (lower). (Courtesy of Pavel Krásenský).

9-9 Two isopods: the wood louse *Platyarthrus hoffmannseggii* (courtesy Christophe Quentin) and the rough wood louse *Porcellio scaber* (lower) (Jymm / Wikimedia Commons / CC BY 4.0).

has not previously been documented. A good example is the isopod *Platyarthrus hoffmannseggii,* a so-called wood louse, which is an obligate guest in nests of *Lasius* and *Myrmica* species but also occurs in *Formica* nests (Figure 9-9). It was generally considered to be a commensal in the ant nests, feeding mainly on decaying nest material and other debris. However, Parmentier et al. (2016a) demonstrated that this isopod also feeds on ant eggs. In contrast, *Porcellio scaber* (the so-called rough wood louse) is a facultative myrmecophilous isopod that can frequently be found outside of ant nests but also occurs in *Formica* mounds,

9-10 The springtail *Cyphoderus albinus.* (Courtesy Pavel Krásenský). The lower picture is a close-up of *C. albinus.* (Andy Murray / Wikimedia Commons / CC BY 2.0).

where it appears to feed on rotting debris or prey on small commensals in and around the nest.

It is, indeed, surprising how widespread brood predation by myrmecophiles in *Formica* nests is. "With the exception of the rove beetle *Stenus aterrimus* and the springtail *Cyphoderus albinus,* all (tested) myrmecophiles were found to prey on the host ant eggs" (Parmentier et al. 2016a) (Figure 9-10). Most of the tested symbionts also fed on ant larvae, except for the rove beetle *Lyprocorrhe anceps,* the wood louse *Platyarthrus hoffmannseggii, Stenus aterrimus,* and *Cyphoderus*

albinus. The latter two species do not prey on ant brood at all. In addition, this study demonstrates that all tested myrmecophiles, although to different degrees, fed on prey brought into the nest by ant foragers, and a large proportion of them were "unspecialized" in the choice of the locations where they spent most of their time in the ant nest.

Only the rove beetle *Dinarda maerkelii* was shown to solicit regurgitated food from its host ants. This species appears to be more dependent on this social re-

9-11   The spider *Thyreosthenius biovatus* (courtesy of Pavel Krásenský) and *Mastigusa arietina* (© Hannu Määttänen).

source than previously suspected, yet it also feeds on prey retrieved by the ants and on other insect cadavers. In fact, the beetles and their larvae (as previously described for the rove beetle genus *Pella*), by feeding on insect cadavers in the refuse area, may indirectly render a service to the host ants by contributing to "early corpse decomposition," and thus "controlling fungi infestation" (Parmentier et al. 2016a).

Finally, what role do the two spider species (*Thyreosthenius biovatus, Mastigusa arietina*) (Figure 9-11) play in these trophic interactions? Parmentier and his coworkers discovered that they prey on other small myrmecophiles such as the springtail *Cyphoderus albinus*, beetle larvae, mites, and other small arthropods.

The rove beetle *Stenus aterrimus*, which was never observed to prey on ant brood, also appears to specialize in hunting springtails and mites inside the ant nest. The rove beetle *Quedius brevis*, which is a facultative predator on ant brood, also preys on larvae of other myrmecophiles. These trophic interactions are very well illustrated in the diagram summarizing the results obtained by Thomas Parmentier and his coworkers (Parmentier et al. 2016a) (Figure 9-12).

To recapitulate, the brood chambers are not the main localities where these "non-specialist" myrmecophiles, studied by Parmentier, can be found. Yet many of them sneak into the ants' brood chambers and prey on eggs and larvae opportunistically. Based on these observations Parmentier and colleagues (2016b) divided the myrmecophiles into three groups: (1) species attracted to dense brood chambers; (2) species randomly distributed throughout the nest; and (3) species rarely or never present in the brood chambers. For example, to group 1 belong the staphylinid beetle *Thiasophila angulata*, or the larvae of the chrysomelid beetle *Clytra quadripunctata* (see Chapter 3). Of course, the behavioral means that enable these two species to prey on the hosts' brood are very different. Nevertheless, they both successfully exploit the same resource. Most of the investigated species belong to group 2, randomly distributed through the nest. Their appearance in the brood chambers varied from an average of 10% to about 25% of the sightings, with considerable variation between species. A few species, such as

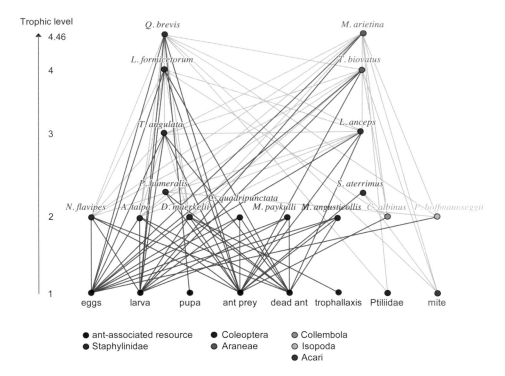

9-12 Representation of the trophic interactions in the community of myrmecophiles in a nest of red mound-building wood ants (*Formica polyctena* and *F. rufa*). The trophic level is based on average chain length, which is one plus the average chain length of all paths from each node to basal species. Black links refer to trophic pathways where the source was associated with the host ants. Gray links refer to predator-prey interactions between myrmecophile species. The following species are involved in this interaction network: *Mastigusa arietina, Thyreosthenius biovatus* (spiders); *Quedius brevis, Dinarda maerkelii, Pella humeralis, Thiasophila angulata, Notothecta flavipes, Lyprocorrhe anceps, Amidobia talpa, Leptacinus formicetorum, Stenus aterrimus* (rove beetles); *Clytra quadripunctata* (chrysomelid beetle); *Monotoma angusticollis* (monotomid beetle); *Myrmetes paykulli* (histerid beetle); *Cyphoderus albinus* (springtail); *Platyarthrus hoffmannseggii* (isopod). (Courtesy of Thomas Parmentier).

the histerid beetle *Dendrophilus pygmaeus* (Figure 9-13), the staphylinid beetle *Quedius brevis,* the spiders *Mastigusa arietina* and *Thyreosthenius biovatus,* and the isopod *Porcellio scaber* appear to be "repulsed" when approaching the dense brood chambers and were hardly ever seen inside these niches (group 3).

Most importantly, however, the observational data did not reveal a correlation between the tendency of parasites to prey on brood and aggressive treatment by the ants. In other words, red wood ants did not react with more hos-

9-13　The histerid beetle *Dendrophilus pygmaeus* with its host ant *Formica polyctena*. The right picture is a close-up of the beetle, whose size is about the same as the head of the ant. (Courtesy of Pavel Krásenský).

tility toward "species that have a higher tendency to prey on brood compared to species that are less likely or do not prey on brood."

As we have learned, aleocharine myrmecophiles investigated to date employ a "gentle" appeasement defense when encountering aggressive host ants, and we dare to speculate that the abovementioned, less-integrated aleocharine myrmecophiles may also be endowed with appeasement glands at the tip of the abdomen. Whatever the nature of these appeasement secretions might be, we assume they contain something the ants are enticed to lick, perhaps of a proteinaceous or sugary nature. This appears to briefly deflect the ant's aggressive behavior, thus enabling the beetle to swiftly escape. Most likely such appeasement behavior is absent in the staphylinine and stenine species, which are endowed with a repellent defense pygidial gland at the tip of the abdomen (Schierling and Dettner 2013). Except for *Dinarda,* no particular behavioral interactions of these staphylinid co-inhabitants in *Formica* nests have been described, and therefore our musings concerning the behavioral mechanisms employed by these beetles can only be speculation.

Miroslaw Zagaja and his coworkers (2017) published a study on the defensive secretions and general biology of the myrmecophilous aleocharine *Thiasophila*

　MYRMECOPHILES IN THE ECOSYSTEM OF ANT NESTS

angulata (the species belong to the guild of myrmecophiles of the *F. rufa* group analyzed by Parmentier and his coworkers). They report that this rather small beetle (1.9–4.3 mm in body length) was found in nests of *Formica aquilonia, F. lugubris, F. polyctena, F. pratensis, F. rufa, F. sanguinea, F. uralensis, F. truncorum, Lasius brunneus,* and *L. fuliginosus* (Päivinen et al. 2002, 2003, 2004; Staniec and Zagaja 2008), and it has also occasionally been seen outside of ant nests in leaf litter and sand dunes (Tenenbaum 1913, cited by Zagaja et al. 2017). These observations indicate that *T. angulata* does not exhibit a strict host specificity, and this is most likely also true for many of these "non-specialized" myrmecophiles. They seem to live as scavengers and opportunistic brood predators, preferring "dense brood chambers," where they mainly feed on ant eggs. One of us (B. H., unpublished observations) observed two specimens of *T. angulata* in a laboratory nest of *Formica sanguinea*. Although no predation of brood was recorded, the beetles were seen in and out of the brood chambers. They were mostly ignored by the ants, but should an ant touch them with its antennae, the ant sometimes jerked aggressively toward the beetle. In response, the myrmecophile would swiftly escape, usually accompanied by a brief lifting of its abdominal tip toward the larger ant. We hesitate to consider the integration of this myrmecophile at the same level as that of the genus *Dinarda,* but it is certainly better integrated than, for example, the genera *Stenus* or *Quedius,* which are rather generalized scavengers and predators.

The trophic network among the myrmecophiles and their host ants is a major factor shaping the ecosystem in the ant colony. But extrinsic factors also influence the community of myrmecophiles inside the *Formica* nests. Parmentier et al. (2015a) studied how the size of the nest mounds and the spacing of neighboring mounds in populations of *F. polyctena* and *F. rufa* affect the community structure of myrmecophiles. They found that the community of myrmecophiles within nests of both species were very similar, and that "species richness per unit (nest) volume is negatively correlated with increasing nest mound isolation." This suggests that a few isolated nest mounds support fewer species of myrmecophiles—that is, a minimum of neighboring nests is needed to maintain species-rich metacommunities of myrmecophiles.

Finally, Parmentier et al. (2017a) scrutinized the question of whether arthropods associated with the red wood ants match their cuticular hydrocarbon patterns with those of their host ants. They compared the hydrocarbon profiles of the guild of species they had previously studied with respect to trophic interactions in *Formica* nests. Thus, they compared a total of twenty-two arthropod species, most of them beetles (eighteen species), two spiders, two isopods, and one springtail. In all but one beetle species, the differences in the cuticular hydrocarbon profiles of the myrmecophiles and host ants were highly significant. In one test, with the staphylinine species *Xantholinus linearis,* a facultative myrmecophile, the difference was not significant (however, only one specimen was available for testing). Thus, we can conclude, at least for these twenty-one myrmecophiles, a resemblance of their hydrocarbon profiles with that of the ants is not relevant to their ability to make a living inside the nest.

Parmentier et al. (2017a) emphasized that the species tested, albeit mostly obligate myrmecophiles, are not highly specialized, as is, for example, *Lomechusa pubicollis* (which they did not test). Unfortunately, we have no information about how closely the hydrocarbons of this highly specialized myrmecophile match those of its hosts. We dare to speculate, however, that the resemblance of the hydrocarbon profiles may not be of crucial significance, because these myrmecophiles probably mimic a much more powerful chemical signal, namely a putative brood pheromone of their hosts. How they can accomplish this for two host species that even belong to different subfamilies (see Chapter 8) remains a conundrum. We dare to speculate that the adult beetles may mainly mimic the brood pheromone of the *Myrmica* hosts, and the beetle larvae mimic the brood pheromone of their *Formica* hosts. We do not deny that, over time, after the beetles have been adopted by the host ants, they may acquire some of the colony-specific mixture of cuticular hydrocarbons when they receive regurgitated food and are frequently licked by the host ants. These acquired chemical cues may facilitate their being tolerated by the hosts but are most likely not the main mechanisms underlying their myrmecophily.

Parmentier et al. (2017a) found lower concentrations of cuticular hydrocarbons in some of the less-specialized, though obligate, myrmecophiles. This may enable them to evade detection by their hosts, and indeed, the authors recognized that such myrmecophiles provoked less aggression in the ants. Other symbionts employ specific evasion tactics and are thereby able to coexist with their host ants, even though they feature different hydrocarbon profiles. In summary, one of the main results of the research by Thomas Parmentier and his coworkers is that ants of the *Formica rufa* group are not more hostile "toward species that have a high tendency to prey on brood compared to species that are less likely or do not prey on brood." They also found evidence that predation exists among the myrmecophiles, and that intra- and interspecific predation may contribute to the stabilization of these diverse associations with host ants.

Let us now return to the work by Volker Witte and Christoph von Beeren, who, together with their collaborators, pioneered this kind of analyses in the Asian army ant species of the genus *Leptogenys.*

The Case of Army Ants, Especially the *Leptogenys* "Army Ants"

As we discussed, in the Old World tropics, several ant species are commonly called "true army ants," which, together with the Neotropical army ant species, belong to the subfamily Dorylinae (Borowiec 2016). These army ant colonies house an enormous number of myrmecophiles, in particular many species of staphylinid beetles (Seevers 1965; Akre and Rettenmeyer 1966; Kistner 1966, 1979, 1982, 1993, 1997; Kistner and Jacobson 1975, 1990; Kistner et al. 2003; Gotwald 1995; Maruyama et al. 2011; Maruyama and Parker 2017). An astoundingly similar lifestyle, which features many of the traits of the so-called army ant syndrome, evolved convergently in several species of the ponerine genus *Leptogenys,* which have been studied in the Indomalayan ecozone by Maschwitz et al. (1989), Witte (2001), and Witte et al. (2010) (see Figure 5-26). Many myrmecophiles, particularly staphylinid beetles, have been found in the large colonies of these species (Kistner 1975, 1989, 2003; Kistner et al. 2008; Maruyama et al. 2010a,b).

Christoph von Beeren and his collaborators (2011) investigated whether the intensity of the host ants' aggression toward the staphylinid parasitic symbionts was correlated with the parasites' degree of predation inside the host ants' colony. They asked, to use the terminology proposed by David Kistner (1979), whether integrated myrmecophilous species ("which by their behavior and their hosts' behavior can be seen as incorporated into the hosts' social life") are less predacious than nonintegrated species ("which are not integrated into the social life of their hosts, but which are adapted to the nest as an ecological niche"). Obviously, there are many grades in between, but for a rough categorization, this dichotomy might be useful.

Specifically, two *Leptogenys* army ant species were the focus in these investigations: *L. distinguenda* and *L. borneensis*. The colonies are usually huge, with up to 50,000 workers. They conduct swarm raids with 5,000 to 10,000 workers, and almost every night the colonies emigrate to new temporary nest sites. Von Beeren et al. (2011) studied five staphylinid beetle species that are associated with two *Leptogenys* species. Three of these myrmecophiles (*Maschwitzia ulrichi, Witteia dentilabrum,* and *Togpelenys gigantea*) were found in colonies of *L. distinguenda*. Colonies of *L. borneensis* hosted two different species (*Parawroughtonilla hirsutus* and *Leptogenonia roslii*).

The comparative study of these beetles revealed distinct differences in both the field and the laboratory. *Maschwitzia ulrichi* and *Witteia dentilabrum* usually migrated at the end of the emigration column of their *L. distinguenda* hosts and avoided any contact with the ants. In the laboratory nest, they spent most of the time in the ants' refuse sites. Because of these habits, these species were considered "non-integrated." In contrast, *Togpelenys gigantea* lived in the center of the nest and were even seen together with the ant brood. During emigrations, they usually moved along with the ants in the center of their emigration columns. This species was therefore considered spatially "well-integrated." A similar

MYRMECOPHILES IN THE ECOSYSTEM OF ANT NESTS

pattern was found in the myrmecophiles of *Leptogenys borneensis.* Both myrmecophiles (*Parawroughtonilla hirsutus* and *Leptogenonia roslii*) lived in the center of the host nest and migrated with their hosts, with whom they had frequent contact. These species were also considered "well-integrated." These characterizations closely matched the behavior the host ants exhibited toward the myrmecophiles: whenever the ants encountered a member of the nonintegrated species, they behaved aggressively toward it, whereas members of the well-integrated myrmecophile species were either ignored or antennated, with some occasional mandible gaping. Feeding experiments conducted in the laboratory, during which the beetles were placed together with ant larvae, demonstrated unequivocally that only the nonintegrated beetle species feed on ant larvae. All larvae placed with integrated species survived.

Unfortunately, nothing is known about what the "integrated" and "nonintegrated" myrmecophiles feed on in natural nests. Based on observations in the laboratory, it appears that integrated species prefer to feed on insect prey brought in by host ants. This kind of kleptoparasitism is negligible, considering the small number of these myrmecophiles inside the huge host colonies. Although the nonintegrated species will prey on ant brood when exposed to ant larvae, no evidence exists to suggest that they seek out the nest area where the ant brood is housed. If they did, they would probably be attacked by the ants. Most likely they feed on insect cadavers left over by the *Leptogenys* ants or prey on other small arthropods and beetle larvae in refuse sites of the host ants, but they are certainly not specialized ant brood predators. Also, for these nonintegrated species, the number found in the *Leptogenys* nests is very small. The authors sampled twenty-one colonies of *L. distinguenda,* and the median *M. ulrichi* per nest was five (range zero to nineteen), while the median of *W. dentilabrum* was one (range zero to nine). Only three individuals in total were collected of the integrated species *T. gigantea.* From seven *L. borneensis* colonies a total of twelve individuals of the integrated species *P. hirsutus* were collected, and the median number per colony was one (range zero to six). Five individuals were found of the integrated species *L. roslii,* and the median per colony was one (range zero to two).

To reiterate, the number of individuals per colony for both groups is extremely low, and therefore it is difficult to comprehend that any of these beetles cause a serious problem for the ants. In our view, any assertions about selective forces that have shaped the ants' behavior toward the beetles, in a way that would allow one group of beetles a higher degree of integration than the other group, can only be speculative. Evolutionary arms-race scenarios lack a solid foundation, given the sparseness of currently available data. Yes, ants will attack co-inhabitants that prey on their brood, provided these predators have not evolved other deceptive means of intrusion into the brood chambers of their host. In fact, as we previously stated, the staphylinid beetles *Lomechusa* and *Lomechusoides* will even be carried into the brood chambers by their hosts, even though both adults and larvae prey on the ants' brood (see Chapter 8).

Nevertheless, it is interesting that the *Leptogenys* host ants exhibit markedly different behavior toward the so-called integrated species and nonintegrated species. This is a different result than that of Parmentier et al. (2016b), obtained with myrmecophiles in *Formica* nests. The interesting questions, from our point of view, are: What do the integrated staphylinids have that allows them to be tolerated by the ants, and how do the ants identify the nonintegrated species as potential predators, while identifying integrated species as relatively harmless? Until we have answers to these questions, it is not possible to imagine a reasonable scenario about selection for traits that support better integration of the myrmecophiles in their host ants' societies.

Having said this, we do not want to appear discouraging. In fact, we think these kinds of studies are important, although there is still a long way to go to understand the role of these myrmecophiles in the ecosystem of an ant colony. Von Beeren et al. (2011, 273) point out that the number of myrmecophile species in the two *Leptogenys* species differ strongly. They write, "Among studied taxa only three symbiont species are known to occur in *L. borneensis* in low numbers.

In contrast, *L. distinguenda* colonies are parasitized by at least 15 different species (Witte et al. 2008), some of which reach numbers more than 1000 individuals per colony. Several symbionts in *L. distinguenda* reach integration levels comparable to those observed for the integrated staphylinid beetles" (Witte et al. 2008; von Beeren et al. 2011). Obviously, these host ant species and their symbiont offer a wide terrain for insightful analyses pioneered by Volker Witte and Christoph von Beeren.

Networks and Colony-Level Censuses

Several other investigations attempted to provide as complete a census as possible of the myrmecophile diversity in colonies of host ant species. Paramount among these records is the lifelong collecting and studying of symbionts in the Neotropical army ants by Carl and Miriam Rettenmeyer and their co-workers (Rettenmeyer et al. 2011). Similar studies are those already mentioned on *Leptogenys* army ants (Witte et al. 2008); on species of the *Formica rufa* group (Päivinen et al. 2003, 2004; Robinson and Robinson 2013; Parmentier et al. 2014; Robinson et al. 2016); on some *Myrmica* species (Witek et al. 2013, 2014); on the ponerine *Neoponera villosa* (Rocha et al. 2020); on the ectatomine *Ectatomma tuberculatum* (Pérez-Lachaud et al. 2011); and on the Neotropical weaver ant *Camponotus* sp. aff. *textor* (Pérez-Lachaud and Lachaud 2014). These are mostly faunistic studies, which provide the basis for all attempts to analyze interconnections among myrmecophiles and with their hosts. Aniek Ivens, Christoph von Beeren, Nico Blüthgen, and Daniel Kronauer (2016) proposed using ecological network analysis for studying the complex communities of ants and their symbionts. This approach has been very fruitful in ecosystem studies of ant-plant interactions and trophobiotic relationships between ants and honeydew-producing Hemiptera (Blüthgen and Fiedler 2004; Blüthgen et al. 2003, 2004, 2007; Stadler and Dixon 2008; Bascompte and Jordano 2007, among others). To our knowledge, there are no comparable network studies of the myrmecophile community in ant nests and in ant populations, except perhaps the previously discussed work by Parmentier et al. (2016c) on the trophic interactions in mound-building *Formica* nests. This work covers only a small fraction of the multiple co-inhabitants in a nest of the red wood ants, but it is a very promising beginning. Indeed, ecological network studies analyzing species interactions within host societies are challenging, and they are especially difficult for ecological communities within insect societies. The most comprehensive data on the diversity of myrmecophiles exist for the army ants. For the Neotropical army ant *Eciton burchellii* alone, hundreds of myrmecophilous arthropod species are associated with the ants (Rettenmeyer et al. 2011). For several species, the behavioral mechanisms underlying the interrelations between guest and host have been analyzed, but for most very little is known about their way of life within the ant society (Kronauer 2020). In this regard, we cannot yet endorse the suggestions of Ivens et al. (2016) to employ network analysis for evaluating "the degree of interaction specificity in army ant-myrmecophile communities," when so little of the myrmecophiles' behavioral interactions with their hosts is understood: in fact, many of the collected specimens still await proper identification and likely represent cryptic species. We would like to suggest a bottom-up approach that begins with understanding the behavioral mechanisms and morphological features that enable the myrmecophiles to coexist and even exploit their host species, and then applies network analysis to explore how the complex nest community is interconnected. We believe this approach will reveal enough empirical data for a reasonable discussion concerning the selective forces within this complex network that shaped the particular myrmecophilous traits of the symbionts and the host ants' behavior toward the myrmecophiles.

Having made these somewhat critical comments, we hasten to acknowledge that in a very recent, excellent study Christoph von Beeren and his collaborators (2021b) conducted a quantitative faunistic study of myrmecophiles from six sympatric *Eciton* army ant species in a Costa Rican rainforest. "Combining DNA barcoding with morphological identification of over 2,000 specimens," they discovered 62 myrmecophilous species, including 49 beetle-, 11 fly-, one millipede-, and one silverfish-species. Several of these species were new to science.

The authors applied ecological network analysis, which was based on the incidences of collecting particular myrmecophilous species with all of the co-occurring *Eciton* species. The resulting network revealed that a majority of myrmecophiles were affiliated with a single *Eciton* host species, while only four were true host generalists. Without any question, such data are important for linking cryptic species diversity and host identity. Our friendly, critical caution concerns the absence of behavioral data in ecological networks. We are not opposed to network syntheses, but think in most cases such networks lack utility in questions related to the evolution of community interactions, because so little is known about the idiosyncratic behavioral mechanisms that sustain each symbiosis, beyond the presence or absence of partners.

Let us now briefly consider a kind of network analysis of the diversity of myrmecophiles in colonies of the Neotropical weaver ant *Camponotus* sp. aff. *textor* conducted by Gabriela Pérez-Lachaud and Jean-Paul Lachaud (2014). These are relatively small communities of parasites, parasitoids, predators, and commensals. Three *Camponotus* colonies investigated were in the apical region of the tree canopy. Each colony occupied a single tree, and each colony had only one queen. The average colony size varied (mean ± standard error: 16,734 ± 4,039 workers: 2,656 ± 556 pupal cocoons: 7,380 ± 2,242 larvae). The authors attempted a systematic survey of the "macro- and microfaunal diversity" within these weaver ant colonies. They explored internal parasites, parasitoids, myrmecophilous parasites, and commensals and identified eighteen taxa belonging to three classes inside the weaver ant nests. Three parasitoid wasp species attacked the ant larvae but emerged from the pupae. The ant workers were found to be parasitized by two endoparasites, a myrmecolacid Strepsiptera and many mermithid nematodes. A small number of ant predators were found, such as an unidentified syrphid microdontine larva that preys on ant brood, and three different spider species. One mite species and one nymph of a cockroach were detected, which probably live as scavengers in the host ant nest.

In addition, the authors registered myrmecophiles inside the nest that were themselves attacked by different parasitoids or predators. For example, the microdontine fly larva was parasitized by a eulophid wasp. Similarly, one of us (B. H.) discovered another syrphid myrmecophile inside the nests of an Australian *Polyrhachis* weaver ant (probably *P. australis*) (Hölldobler and Wilson 1990). It belongs to the subfamily of the Pipizinae and was recently described as a new species of the genus *Trichopsomyia* and named *T. formiciphila* (Downes, Skevington, and Thompson 2017). The longish, slug-like larvae probably prey on ant brood, but were often found to be infected by parasitoid wasps, which unfortunately have not been identified.

In the *Camponotus* sp. weaver ants, the parasitoid wasps were *Horismenus myrmecophagus* (Eucharitidae: Eucharitinae) and the eucharitine *Obeza* sp. A larva of a coccinellid (ladybird beetle) detected in the nests probably preyed on hemipterans (coccid scale insects) that occurred inside the nests, which most likely serve as trophobionts for the ants. The authors report that even some of these scale insects were parasitized by an unidentified parasitoid. In addition, they found several other wasps, flies, and even a weevil beetle, whose role inside the ant nests remained unknown. We conclude this particular case study with the instructive diagram assembled by Gabriela Pérez-Lachaud and Jean-Paul Lachaud, which illustrates several of these intricate interconnections of a relatively small community of myrmecophiles in the *Camponotus* weaver ant nests (Figure 9-14).

We have chosen this study because the diversity of myrmecophilous parasites, parasitoids, predators, and commensals is quite limited, and the numbers of individuals small. Nevertheless, the construction of an interaction network remains a fragmentary "snapshot." This is not meant to be a criticism, because the author's understanding of the natural history study is admirable. However, imagine a huge ant colony with hundreds of myrmecophilous species for which we lack most data about interactions with hosts and the other co-residents in the colony. To conduct a network analysis of interactions in such a complex interspecific community would be daunting, and most likely unproductive at the

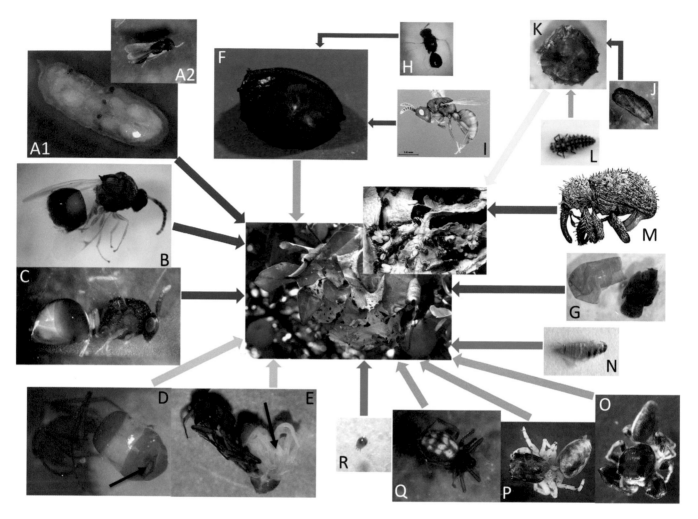

9-14 Silk nest of *Camponotus* sp. aff. *textor* (at center) and the interaction network of the *Camponotus* weaver ant, with its associates. (A) *Horismenus myrmecophagus* (Eulophidae, gregarious endoparasitoid of ant larvae / pupae): (A1) parasitized ant pupa, (A2) adult parasitoid wasp; (B) *Obeza* sp. (Eucharitidae, solitary ectoparasitoid of ant larvae / pupae); (C) *Pseudochalcura americana* (Eucharitidae, solitary ectoparasitoid of ant larvae / pupae); (D) *Caenocholax* sp. (Strepsiptera, Myrmecolacidae, endoparasite of larval, pupal, and adult ants); (E) unidentified (Mermithidae, likely endoparasite of larval, pupal, and adult ants); (F) unidentified (Microdontinae, predator of ant brood); (G) unidentified (Diptera, unknown); (H) *Horismenus microdonophagus* (Eulophidae, gregarious endoparasitoid of microdontine larvae); (I) *Camponotophilus delvarei* (Eurytomidae, gregarious ectoparasitoid of microdontine larvae); (J) unidentified (Hymenoptera, endoparasitoid of the coccid *Cryptostigma* sp.); (K) *Cryptostigma* sp. (Coccoidea, trophobiont); (L) unidentified (Coccinellidae, predator of *Cryptostigma* sp. [?]); (M) *Melexerus hispidus* (Curculionidae, unknown interactions); (N) unidentified (Blattellidae, scavenger); (O) unidentified (Salticidae, predator of ant adults or brood); (P) unidentified (Salticidae, predator of ant adults or brood); (Q) unidentified (Araneae, predator of ant adults or brood); (R) unidentified (Acari, scavenger). (Courtesy of Gabriela Pérez-Lachaud and Jean-Paul Lachaud).

present stage of our knowledge. In our opinion, the practice of lumping species into guilds based on cursory information should also be avoided for studies of myrmecophile communities.

Infestation by Myrmecophilous Parasites and Colony Traits

Finally, we briefly address the question of how far ant colonies as a whole can resist the attacks by myrmecophiles. It has been hypothesized by various scholars that genetically diverse worker populations in a colony may, through polygyny (more than one reproducing queen) and / or polyandry (queens storing sperm of more than one male), increase the colony-level resistance to pathogens and possibly also to attacks by myrmecophilous parasites and parasitoids (Hamilton 1987; Sherman et al. 1988; Schmid-Hempel 1998; van Baalen and Beekman 2006; Hughes et al. 2008). Experimental evidence in support of this hypothesis is relatively scarce, though a few studies about pathogens produced confirmatory evidence for bumblebees and honeybees (Liersch and Schmid-Hempel 1998; Baer and Schmid-Hempel 1999; Tarpy 2003; Tarpy and Seeley 2006) and for ants (Hughes and Boomsma 2004, 2006; Reber et al. 2008; Ugelvig et al. 2010). Yet, we have to ask the question: If within-colony variability is so advantageous against infestations by viruses, bacteria, microsporidia, parasitoids, and perhaps also myrmecophilous parasites (Schmid-Hempel 1994), why are many ant species monogynous and in most cases also monandrous? Although in a considerable number of ant species we find populations that are monogynous and others that are polygynous, one reasonable explanation is that the polygynous populations may be under stronger parasitism pressure.

This is particularly pronounced in the subgenus *Serviformica,* the species of which are frequently raided by so-called slave-raiding ants such as *Formica sanguinea, Formica subintegra,* or *Polyergus* spp. (Hölldobler and Wilson 1990). These raiders rob the pupae of the *Serviformica* species and transport them to their own nests. When these pupae eclose, the young *Serviformica* workers become imprinted on the slave-raiders' colony odor and eventually act and work for the fitness of unrelated queens and their mating partners. Such *Serviformica* species often have some populations with monogynous colonies and others with polygynous colonies. Apparently, the queens in monogynous colonies tend to be of larger size than queens in polygynous colonies. It has been suggested that limitation of nest sites could lead to polygyny; however, another explanation for the prevalence of facultative polygyny, particularly in *Serviformica* species, might be raiding pressure by parasitic *Formica* or *Polyergus* species that plunder pupae from the *Serviformica* colonies. Polygynous nests are usually also polydomous; thus, spreading out the chambers where pupae are housed might enable the *Serviformica* colonies to reduce the harm caused by the raiders. It would be interesting to know whether *Serviformica* populations that are more frequently exposed to raiding ant species are more frequently polygynous than monogynous. This would be another explanation for the occurrence of polygyny, but, of course, it is not suggested as an alternative explanation to the genetic variability hypothesis.

These thoughts lead us to the fascinating work by Jeremy Thomas and his collaborators. They assert that it is not self-evident that a simple relationship between genetic variability and resistance applies across all types of parasitic interactions. "Interactions between social parasites [in their terminology this includes the myrmecophilous parasites] and their hosts differ fundamentally from those of pathogens, since the former infiltrate insect societies to exploit the resources of the colony, whereas the latter parasitize individual ants (Thomas et al. 2005)" (Gardner et al. 2007, 1004). Thus, individual ants of a colony can be directly attacked by pathogens, and because of their genetic variability, some individuals may be more susceptible than others. For myrmecophilous parasites, however, it is usually the entire colony, or major sections of it, that is attacked. By preying on brood or parasitizing the food resources, myrmecophiles thereby affect the entire colony's investment in growth and reproduction

(Hölldobler and Wilson 1990; Nash and Boomsma 2008). This brings us back to a study of *Microdon* myrmecophiles we briefly addressed in Chapter 3.

M. G. Gardner and his collaborators K. C. Schönrogge, G. M. Elmes, and J. A. Thomas (2007) made an intriguing discovery. They studied the genetic variation of workers in colonies of populations of *Formica (Serviformica) lemani,* which were often infested by larvae of the syrphid fly *Microdon mutabilis.* As we learned, these myrmecophilous parasites prey on the brood of their host ants. Although Wasmann (1920) and Donisthorpe (1927) report that the larvae and pupae of *M. mutabilis* occur in nests of *Formica fusca, F. rufa, F. rufibarbis, Lasius niger, L. brunneus,* and *L. flavus,* we have to consider these accounts with caution. For example, Schönrogge et al. (2002) discovered that the putative *Microdon mutabilis* that were found in nests of the two host species, *Formica lemani* and *Myrmica scabrinodis,* in fact comprise two species: *Microdon mutabilis* with *F. lemani,* and the previously unknown species *Microdon myrmicae* with *Myrmica scabrinodis.* Thus, host specificity might be more pronounced than previously assumed. In addition, Graham Elmes and his colleagues (1999) found in their study area an "extreme host specificity," that is, *Microdon mutabilis* was found only in nests of *Formica lemani* and showed a high degree of colony specificity.

The genetic studies by Gardner et al. (2007) uncovered another remarkable peculiarity. Colonies of *F. lemani* infested with *M. mutabilis* exhibited a relatively low within-colony relatedness and a high estimated queen number. Interestingly, infested colonies with lower queen numbers usually exhibited a higher degree of polyandry per queen. In contrast, in two populations of *F. lemani* where the colonies were not infested by *M. mutabilis,* the relatedness within colonies was high. The authors argue that low within-colony relatedness increases vulnerability to *Microdon* infestation, whereas high within-colony relatedness decreases proneness to infestation. As we noted above, they state, "while high genetic variation can benefit social insects by increasing their resistance to pathogens, there may be a cost in the increased likelihood of infiltration by social parasites owing to greater variation in nest mate recognition cues."

It is a puzzling paradox, that the genetically more diverse *Formica lemani* colonies appear to have less specific collective hydrocarbon profiles than colonies with higher within-colony relatedness and, therefore, are more open to infestations by *Microdon mutabilis,* even though *Microdon* parasites exhibit an extreme colony specificity, and host workers tolerate only eggs of *Microdon* flies that developed in their colony. Thus, to our understanding, the main hurdle for the parasite penetrating the host colony is the acceptance and toleration of the *Microdon* eggs. Yet, the arguments presented by Gardner et al. (2007) seem to include also the developing parasitic larvae and pupae.

This is an intriguing hypothesis, although no experimental evidence is yet available. To our knowledge, no comparisons between the cuticular hydrocarbon profiles of monogynous and polygynous *F. lemani* colonies exist. The diversity of cuticular hydrocarbons in *F. lemani* is comparatively poor (Martin and Drijfhout 2009a), so differences between polygynous and monogynous colonies may be hard to recognize. Furthermore, colony-specific recognition cues are most likely not manufactured by *Microdon* larvae but are acquired from their hosts by "contamination" (although there are claims that at least part of the hydrocarbons that make up colony-level profiles may be manufactured by the myrmecophiles; see Chapter 3). It is difficult to explain how the myrmecophiles could produce specific nestmate recognition cues for each of the countless colonies they parasitize. Even their ant nestmates are unable to do so (see Lenoir et al. 1999, 2001).

We wonder whether the authors kept monogynous and polygynous colonies of *F. lemani* in the laboratory and made observations of the transferability of *Microdon* larvae from polygynous colonies to monogynous colonies. If the *Microdon* larvae somehow mimic the ant larva's brood pheromone, a transfer should be possible, as most ant larvae can be exchanged between the colonies. Additionally, we question whether the phenomenon reported by Gardner et al. (2007) might be coincidental with the rarity of *Microdon mutabilis,* or even its absence in the habitat of monogynous *F. lemani* colonies. Gardner et al. (2007), and previously Elmes et al. (1999), and still earlier Donisthorpe (1927) observed that the adult syrphid flies, when leaving the host nests, do not fly far away but tend to hover around the area of the nest from which they have emerged. They deposit their eggs close to the nest entrance. As we discussed in Chapter 3, the survival and adoption of eggs by the host ant colony is significantly higher if the mother fly originated from the same colony. Indeed, such viscous populations tend to be very patchy. In conclusion, we want to emphasize that we found this study and proposed hypotheses intriguing, but we do not think that the current data provide sufficient evidence for establishing the hypotheses as facts.

Let us now consider the results of a study by Matthias Fürst, Maëlle Durey, and David Nash, who analyzed intercolony aggression (or rejection of non-nestmates) in polygynous and polydomous populations of *Myrmica rubra*. It has long been known that aggression among neighboring ant societies is considerably more pronounced in populations with monogynous colonies than in polygynous populations (Hölldobler and Wilson 1977, 1990; Keller 1993). Polygynous (polydomous) colonies usually propagate by budding, and within-colony mating is more prevalent than in monogynous populations. Consequently, neighboring polygynous colonies are more closely related, even though within-colony relatedness is lower than in monogynous colonies. It has been reasoned that this organization explains why monogynous ant societies behave with more territorial aggression against neighboring colonies than do polygynous societies.

Fürst et al. (2012) provide hard evidence in partial support of these hypothetical assumptions for the myrmicine species *Myrmica rubra*. Although *M. rubra* colonies can also be monogynous, in the study sites of Matthias Fürst and his colleagues, all colonies were polygynous except one. Even though only one queen could be found in this colony, genetic work revealed several matrilines represent among the workers. The authors collected data concerning the genetic structure and the cuticular hydrocarbon profiles of workers within colonies and confirmed that, in populations with polygynous (polydomous) colonies, neighboring colonies were more closely related to each other than to colonies farther away. They demonstrated that the chemical profiles were also increasingly dissimilar the farther the geographical distance between compared colonies. The genetic diversity within a colony increases with the number of reproducing queens, and consequently the cuticular hydrocarbon profiles become broader.

As we reviewed above, in highly polygynous colonies, aggression directed toward neighboring colonies (which in fact may be "sister buds") is very low. The authors registered the number of queens by excavating the nests and used highly variable microsatellite loci to determine matrilines. Often the number of determined matrilines was higher than the number of queens found during excavations. This is not surprising, because queens can be overlooked, or may have died, workers that originated in a neighboring colony may have joined the foreign colony accidentally, or there may have been some reproductive skew, meaning some queens had more offspring than others.

In the behavioral aggression bioassays, a foreign "intruder" ant was confronted with five "defender" ants inside a neutral arena. The number of aggressive acts and kinds of aggressions with which the defenders responded to the intruder were recorded, and an aggression index was calculated. Not surprisingly, the authors confirmed that aggression intensified with increasing distance between the defenders' and intruders' colonies. These observations are closely correlated with the rise in genetic distance and increasing dissimilarity of the cuticular hydrocarbon profiles.

It has commonly been argued, as the authors do, that the discrimination between relatives and nonrelatives has been favored by kin selection, because "inclusive fitness can only increase if benefits of shared labor can be directed towards colony members or close relatives rather than towards foreigners" (Fürst et al. 2012, 516). However, a negative inclusive fitness assumption would be justified for only those ants who accidentally or forcibly joined a foreign colony. In fact, additional labor power can be beneficial for colonies, and in some species, it has been demonstrated that larger colonies conduct raids on smaller conspecific colonies and rob worker pupae (Hölldobler and Wilson 1990). These "stolen" workers will lose indirect fitness benefits but enhance the inclusive fitness of their nonrelated nestmates. That is, their work will compound the fitness of unrelated queens. We suggest that the kind of aggression Fürst et al. (2012) describe expresses colony territoriality—that is, the competition for space and resources within. The foreign ant encountered by the five "resident" ants in the test arena might be assessed as a foreign scout. Especially if this scout carries a markedly different chemical profile, it may signal a serious challenge to the integrity of the residents' territory.

But why cover this interesting study in the context of this chapter? Because it details a second remarkable discovery. As we learned in Chapter 4, *Myrmica* species are hosts of lycaenid myrmecophilous species of *Phengaris,* on which the group of David Nash has published several investigations, some of which are discussed in Chapter 4. In the above mentioned paper, Fürst and colleagues report a striking discrimination behavior exhibited by *Myrmica rubra* workers that live in colonies that are prone to housing *Phengaris alcon,* one of the "cuckoo" *Phengaris* species (see Chapter 4). Even though the host colonies are highly polygynous, and therefore exhibit high within-colony genetic diversity, the ants of such colonies are far more aggressive against foreign intruders than are ants collected in habitats where the *P. alcon* is absent. Especially puzzling is the fact that the aggressiveness is equally high independent of whether the intruder ant originated from a colony in a close neighborhood or from a geographically distant one. Fürst et al. (2012) write, "The pattern of decreasing aggression with increasing within-colony genetic diversity was not apparent in these areas" (where *P. alcon* is common). In any case, these results are striking, and puzzling, especially in light of the studies by Gardner et al. (2007), which suggest that polygynous host colonies can be more easily infected by myrmecophilous parasites because the hydrocarbon profiles of their hosts are broad and less specific. The work of Gardner et al. (2007) concerns *Microdon* larvae in polygynous *Formica lemani* colonies. In addition, Fürst et al. (2012) cite results obtained by Thomas Als (unpublished, reported in Nash and Boomsma 2008), who showed that within-nest relatedness of colonies infected by *Phengaris alcon* was lower than that of unparasitized colonies in the study area. These findings would support the hypothesis proposed by Gardener and his colleagues that increased genetic diversity would facilitate infection by myrmecophilous parasites. Fürst and colleagues correctly state that "from these results, we would thus expect to find lower aggression levels in colonies with high queen numbers facilitating the acceptance of the parasite."

Nevertheless, why are polygynous, genetically diverse *Myrmica* colonies found in the *P. alcon* habitats more aggressive toward conspecific foreigners than in areas free of *P. alcon*? Fürst and colleagues hypothesize that parasite pressure favors the continuous evolutionary change of the collective colony-specific cuticular hydrocarbon profiles that mitigate the parasites' abilities to invade the host colonies. David Nash and colleagues (2008) proposed an evolutionary arms race between parasite and potential host colonies. Fürst et al. (2012) conclude, "Additional costs of failures to recognize intruders imposed by social parasites appear from our results to have a major influence on acceptance or rejection thresholds which are adjusted in an adaptive manner."

As much as we like the study on the correlations of the decrease in aggressiveness toward foreign colony members with increasing within-colony genetic diversity, and the positive correlation between increase in aggression and geographic distance, and the parallel comparison with the changing cuticular hydrocarbon profiles, we are not convinced that parasitism by *Phengaris alcon* drives specificity of the collective cuticular hydrocarbon profiles of the colonies. Though we understand that this might work in monogynous ant colonies, we cannot conceive of a possible mechanism to accomplish this in populations of highly polygynous ant societies. Furthermore, as we discussed in Chapter 4, we do not believe that it has been demonstrated that the surface hydrocarbons of the *P. alcon* caterpillar serve as the essential key stimuli for releasing adoption behavior in the host ants. They may serve as supporting modulators, but the essential key stimuli are most likely allomones that mimic the brood pheromones of the host ants. We assume the *P. alcon* larvae can be as easily transferred from one colony of *M. rubra* to another, and will be as readily accepted, as the ant larvae can be exchanged, even though the foreign workers may treat each other aggressively. So, why are the colonies in the *Phengaris* habitats more aggressive? We do not know, but it would be desirable to make additional comparisons with other populations where *P. alcon* are present and with populations where the parasites are absent.

In this chapter we have reviewed several ecosystemic studies of myrmecophiles in different host ant colonies. They revealed that myrmecophiles, depending on the degree of integration with the hosts, can be subdivided into ecological guilds that may exhibit network interactions with host ants and myrmecophiles. These network studies are relatively new and offer a very promising approach, yet we believe we need more investigations of the behavior-physiological mechanisms of guest and host interactions before such network studies can be productive. We have also explored whether certain social organizations of host societies are more prone to infection by specific myrmecophilous parasites. In our view, these studies are indeed very intriguing, but their conclusions will require additional empirical evidence to be well supported. In Chapter 10 we will find out how far vertebrates have close interactions with ants.

10 Vertebrates and Ants

A VARIETY OF vertebrate species live with, prey on, or otherwise use ants in their daily lives. Vertebrate and arthropod myrmecophiles must overcome many of the same challenges associated with finding and entering ant nests, but vertebrates are faced with additional problems of scale. Few vertebrates approach the small size of their host ants. Accordingly, most vertebrate myrmecophiles interact with whole or large parts of ant colonies at once. We know that ants respond to visitors as large as elephants and, in the case of *Crematogaster mimosa,* can discriminate between vibrations caused by wind and those made by herbivores browsing on their mutualistic host plants (Hager and Krausa 2019). Indeed, the substantial appetites of myrmecophagous vertebrates may have driven the evolution of aposematic coloration in ants, as well as certain painful protein venoms (Schmidt and Blum 1978a), projectile acids, and group defensive, evacuation, and foraging behaviors. This is because vertebrate predators exact a measurable cost on colony biomass. Horned lizards, for instance, can take dozens of *Pogonomyrmex* foragers from a colony in one day (Sherbrooke and Schwenk 2008), while American black bears consume enough ant brood to inactivate 31% of co-occurring *Formica obscuripes* colonies each season (Grinath et al. 2015). Unlike their arthropod guests, many vertebrates are large enough to exploit numerous ant colonies across broad spatial scales and, in some cases, long lifetimes.

The relationships among reptiles, birds, fish, mammals, and ants extend beyond that of predator and prey to include some of the most visible and unexpected symbioses on record. We begin with a discussion of the birds that follow

10-1 A bicolored antbird and a barred-woodcreeper wait to capture arthropods flushed out by an advancing *Eciton burchellii* swarm raid. As the army ants depart from their bivouac, they overwhelm a tailless whip scorpion. Antbirds are parasites of army ants and can deprive a colony of 30% of its daily prey income. (Courtesy of John Dawson).

nomadic army ant colonies, and the many vertebrates that wallow in ant mounds to anoint themselves with living ants. Next, we consider the fate of tadpoles and blind snakes born in the nests of fungus-farming ants and ask whether legless lizards can hijack ant pheromone trails. We conclude with a brief discussion of the myrmecophilous behaviors of humans.

Myrmecophiles of a Feather

One of the greatest spectacles of myrmecophily can be found in the Neotropics, where dozens of bird species, primarily from the family Thamnophilidae, form associations with nomadic army ant colonies (Figure 10-1). Each *Labidus praedator* or *Eciton burchellii* army ant colony represents an enormous and dependable resource for potential predators. Yet, careful analysis of bird stomach contents by Edwin O. Willis and Yoshika Oniki (1978) revealed that the army ants are rarely, and perhaps only accidentally, consumed by their attending birds. The birds also avoid prey already subdued by multiple army ant workers, preferring instead to snap up the leaf-litter arthropods that flee in advance of the ants' swarm raid. This aversion to feeding on army ants may explain why many of the myrmecophilous beetles that accompany aboveground army ant colonies have evolved to mimic the coloration of their ant hosts, whereas subterranean myrmecophiles, who are hidden from the eyes of hungry birds, seldom do.

Ant-associated birds use *Eciton* and *Labidus* colonies as "beaters" to flush out prey in the same way that seabirds attend pods of foraging whales, and pig-birds follow peccaries (Greeney 2012). In central Panama, marauding *E. burchellii* colonies can flush out and consume an impressive 22 g of leaf-litter arthropods and 24 g of social insect brood each day (Franks 1982). But what, if any, costs do colonies incur by attracting so many large-bodied guests to their raids? Willis and Oniki (1978) suggested that ant-following birds might take only arthropods too large or fast for the ants to overwhelm, or even provide a service to army ant colonies by driving escaping arthropods back toward the advancing swarm. To clarify the nature of the relationship between the birds and army ants, Peter Wrege and colleagues (2005) monitored the number of large prey items captured by *E. burchellii* colonies as they crossed control plots with ant-following birds, and experimental plots, where mixed-species flocks of birds were excluded. To achieve bird-free zones, the researchers report "standing immediately adjacent to the area and gesticulating with our arms. The occasional persistent bird was discouraged with a shot of water from a squirt gun, or by throwing a fragment of bark into the leaf litter nearby when the bird began to approach too closely." Their experiment revealed that, in the absence of birds, *E. burchellii* colonies successfully capture the large (>1.5 cm) leaf-litter arthropods, typically favored by birds.

The most common birds observed by Wrege et al. (2005) were *Gymnopithys leucaspis* (bicolored antbirds) (see Figure 10-1), *Dendrocincla fuliginosa* (plain-brown woodcreepers), *Phaenostictus mcleannani* (ocellated antbirds), *Hylophylax naevioides* (spotted antbirds), *Neomorphus geoffroyi* (rufous-vented ground cuckoos),

Dendrocolaptes sanctithomae, and *Xiphorhynchus lachrymosus* (woodcreepers). Flocks averaging just six birds succeeded in taking a whopping 30% of each colony's daily leaf-litter arthropod income. These colony losses increased with both flock size and bird biomass. Because *Eciton burchellii* growth and reproduction (fissioning) are limited by their daily foraging efficiency, ant-following birds clearly exact a cost on the "marginal economy" of colonies. Rather than acting as mutualists or commensals that target prey too large for swarms to overwhelm, ant-following birds are true parasites of neotropical army ants (Wrege et al. 2005).

Ant-following birds in the family Thamnophilidae can be divided into roughly three levels of specialization, including the occasional followers, who opportunistically attend swarms in their territories; the regular followers, who both follow swarms and spend time foraging independently; and eighteen species of obligate followers, that strictly forage alongside army ants and must commute to ant swarms while nesting (Zimmer and Isler 2003). In a phylogenetic study of seventy Thamnophilidae species, Robb T. Brumfield and colleagues (2007) found that associations with army ants likely progressed from occasional to regular to obligate-following over evolutionary time, with three independent transitions to regular following and no known reversals from the obligate state. Willis (1972) compared the relative foraging success of one occasional follower, the spotted antbird (*H. naevioides*), when it foraged independently and when it foraged alongside the army ants *Labidus praedator* and *Eciton burchellii*. He found that birds took an average of 111.8 seconds to find prey when foraging alone, and just 32.3 seconds when foraging with army ants. Thus, there is a nearly fourfold advantage to attending army ant swarms over independent foraging. Why then do spotted antbirds spend any time foraging away from their ants, when the benefits of being a professional follower are so pronounced? The answer may be found in the composition of mixed-species flocks, where the arrival of more dominant bird species reduces the foraging efficacy of spotted antbirds (Willis 1972). Species composition in mixed flocks varies considerably across geographic areas, and in some regions, flocks are composed of heavier birds

(O'Donnell 2017). When the large, ocellated antbird (*P. mcleannani*) became lo-cally extinct on Barro Colorado Island in Panama, smaller spotted antbirds there became less territorial and shifted from occasional army ant following to more regular following outside of their territories. This compensatory response led the population of spotted antbirds on the island to double in size after twenty years (Touchton and Smith 2011).

Although ant-following birds vary considerably in morphology and body size, researchers have found little evidence that they discriminate between the different types of arthropod prey made available by army ant colonies. In a study of five obligate antbird species from Bolivia, which included *Gymno-pithys salvini, Rhegmatorhina melanosticta, Myrmeciza fortis, Phlegopsis nigromacu-lata,* and *Dendrocincla merula,* only *P. nigromaculata* and the woodcreeper *D. merula* showed some preference for prey type. In addition to the other arthro-pods, both species consumed two and a half times as many spiders as the other bird species present (Chesser 1995). Although there was limited evidence for dietary specialization among the ant-following birds studied, there is evidence that the birds form dominance hierarchies within their interspecific flocks (Willis and Oniki 1978; Chesser 1995). Larger and more aggressive birds control the productive leading edge of the advancing ant swarm and, as a result, obtain the most prey. Although their diets are broad, at least two ant-following bird species, *Myrmeciza fortis* (Thamnophilidae) and *D. merula* (Dendrocolaptida), show distinct preferences for following either *Labidus praedator* or *Eciton burchellii* and can be considered host-specialists (Willson 2004). In turn, the antbirds them-selves are pursued by three species of ithomiine butterflies, *Mechanitis polymnia isthmia, Mechanitis lysimnia doryssus,* and *Melinaea lilis imitata.* Female butterflies orient to *E. burchellii* swarm raids over large distances, presumably by odor. Up to a dozen at a time have been observed flying low over a swarm's leading edge and dipping down to feed on freshly deposited antbird droppings. As many as thirty such ant-butterflies can be captured above a swarm each hour (Ray and Andrews 1980).

The cyclical nomadic and statary brood-rearing cycles of *E. burchellii* colonies present an interesting problem for obligate ant-following birds. To maintain a reliable source of food, the birds must track multiple ant colonies as they raid from new locations on each day of their fourteen-day nomadic phase. But large army ant colonies can be cryptic and widely spaced, traveling up to 200 m each day (Schneirla 1971; Swartz 2001). Monica Beth Swartz (2001) asked how birds were able to keep track of numerous wandering superorganisms, each on its own nomadic and statary schedule. She monitored 222 birds associated with 238 army ant swarms in Costa Rica and found seven bird species that seemingly learn the location and schedules of no fewer than three colonies by sequentially visiting their bivouacs in a standard order. She writes,

> When a swarm [that the birds] are foraging with stops for the day, some species follow the trail of ants back to the bivouac to examine it, then fly on to the next bivouac in their circuit, or to their roost site if the day is over. This may reinforce spatial memory of the bivouac location and of the foraging path taken in case the bivouac moves that night. If a bivouac appears to be gone, they will carefully examine the site to see if the bivouac has shifted position and become less visible. If they cannot find it, they will search for it along the path of the previous day's raid. Occasionally, nomadic ant groups are difficult to relocate and searching birds will fly back to the former bivouac site repeatedly, perhaps for spatial orientation. (Swartz 2001, 631)

While some birds keep a close eye on the location of host colonies, others find ants by eavesdropping on the calls of talkative obligate species that have already located a swarm. This was demonstrated by Johel Chaves-Campos (2003), who recorded the calls of ant-associated bird species with varying degrees of specialization, then played them back at sites in the forest where army ants were absent. Only the recordings of two obligate antbird species, which tended to arrive first at army ant swarms, attracted other bird species. Generally, eavesdropping birds responded more to the calls of less competitive subordinate species, like the white-cheeked antbird (*Gymnopithys leucaspis*), than to the calls of dominant species, like the reddish-winged bare-eye (*Phlegopsis erythoptera*) (Batcheller 2016). Though cue prominence can differ between obligate species, experience seems to be the most important factor determining whether a facultative species attends a swarm raid (Batcheller 2016). For example, antbirds on islands respond to the calls of birds they encountered frequently, but do not attend playbacks of obligate species that are locally absent from their islands (Pollock et al. 2017). In essence, some ant-following birds learn to eavesdrop on other professional species. Although vertebrates (ourselves included) are not the only organisms that build careers on the habits of ants, the need to monitor the location and foraging status of multiple colonies may be unique to this group.

Anting

On the urban campus of Arizona State University, great-tailed grackles (*Quiscalus mexicanus*) can be observed lying prone along the margins of sidewalks, heads cocked to the side, surrounded by throngs of the dolichoderine ant *Forelius mccooki*. At first glance, one might think the worst, but the birds do show signs of life as they shift to expose new parts of themselves to the pungent-smelling ants teeming on the pavement (C. L. Kwapich personal observation). A variety of vertebrates engage in a similar self-anointing behavior, known as "anting," by allowing ants to passively wander over their bodies, or by actively pressing ants into their skin, feathers, or fur. Chemically defended ant species from the subfamilies Formicinae and Dolichoderinae are preferred over stinging myrmicines and are, in some cases, goaded into participating with tools, like twigs (Goodwin 1955; Camacho and Potti 2018). Carrion crows reportedly sit on *Formica rufa* mounds "like brooding hens," and reek of the ants' acidic spray, even days after anting (Whitaker 1957). Audubon (1831) first described anting behavior after observing wild turkey (*Meleagris gallopavo*) dust-bathing in deserted ant mounds (McAtee 1947). In birds, passive anting often coincides with dust-bathing and sun-bathing but, today, is distinguished by the presence of an active ant nest or ant foraging route (Figure 10-2) (Brackbill 1948; Whitaker 1957; Simmons 1966).

Overall, more than 200 bird species (Eisner and Aneshansley 2008) and a number of mammals and reptiles engage in anting, not limited to leech-prone snapping turtles (Burke et al. 1993), wild boar (*Sus scrofa*) that wallow in wood ant nests (Zakharov and Zakharov 2010; Robinson et al. 2016), and domestic cats who "rub themselves madly" and "roll in contortions of ecstasy" where ants have been crushed (Chisholm 1959; Hauser 1964). The latter case may be attributed to the production of iridomyrmecin by some ant species, which is a compound also found in catnip (Choe et al. 2012; Lichman et al. 2020). In one fortuitous encounter, myrmecologist Jack Longino (1984) collected *Camponotus sericeiventris,* dropped from above by capuchin monkeys (*Cebus capucinus*), as they inserted the ants into the fur around their genitalia and armpits. Grey squirrels are reported to return again and again to "leap and tumble wildly" in shallow pits near the nests of at least ten ant species, including *Camponotus nearcticus* and *Crematogaster ashmeadi* (Bagg 1952; Hauser 1964). Hauser (1964) describes squirrels lying "spread eagle" on the nest discs of *Pogonomyrmex badius* at midday, when other small ants were also present. By our assessment, the smaller ants were likely the high noon ant, *Forelius pruinosus,* a common pest of the larger harvester ant species.

Despite the pervasiveness of anting behavior among birds and other vertebrates, there is little satisfying evidence to suggest why it is performed. In addition to ants, animals use a variety of potent-smelling objects for self-anointing, including mothballs, limes, millipedes, onions, smoke, and cigarette butts. These are suggested to aid in repelling parasites or fumigating bothersome leeches, mites, ticks, and lice (Goodwin 1955; Whitaker 1957; Simmons 1966; Burke et al. 1993; Clayton and Vernon 1993). Among birds, the application of ants has also been hypothesized to help with the removal of old preening lipids from the skin and feathers (Kelso and Nice 1963; Simmons 1966), or with the initiation of seasonal molting (Potter 1970). Some researchers have suggested that birds gather

10-2   A carrion crow (*Corvus corone*) allows ants to wander over its body and applies wads of ants to its feathers (upper). (Photo: Courtesy of Marie-Lan Taÿ Pamart, CC BY4.0). A wild turkey (*Meleagris gallopavo*) rolls in an active ant mound, combining anting with a dust bath (lower). (Sue Chaplin).

and store ants in feathers as a portable food supply (Groskin 1943), or apply them to particular areas to provide autoerotic stimulation (Whitaker 1957).

One of the most convincing interpretations of anting behavior proposes that acids and other chemicals produced by ants are used medicinally, as fungicides and bactericides (Ehrlich et al. 1986; Clayton and Vernon 1993). Many of these compounds serve those very functions for ants in their own nests. For the purposes of anting, birds do seem to exhibit a bias for ant subfamilies that employ such chemical defenses (Formicinae and Dolichoderinae). Hannah C. Revis and Deborah Waller (2004) quantified the inhibitory effects of ant-associated compounds against plated cultures of bacterial and fungal parasites of feathers. While pure formic acid strongly inhibited all the bacteria and the fungal hyphae tested, natural concentrations of *Camponotus pennsylvanicus* and *Lasius flavus* defensive secretions containing formic acid did not inhibit microbial growth on plates. Likewise, polar and nonpolar extracts of a variety of ant species did not prevent the growth of bird-associated bacteria or fungus. These extracts were taken by placing groups of fifty workers each of *Camponotus pennsylvanicus, Pheidole dentata, Aphaenogaster rudis, Crematogaster lineolata,* and *Lasius flavus* in 5 mL of hexane or water. It is not clear whether these ant species are preferred by birds that suffer from the tested bacteria and fungi, but, like the other hypotheses related to anting, a solid link between the chemical defenses of ants and protection from microbes was not established.

Blue jays (*Cyanocitta cristata*) eat ants and engage in conspicuous, active anting behavior. Prior to consumption, the birds hold individual *Formica exsectoides* workers by the mesosoma and drag them across their primary flight feathers and tail feathers (Figure 10-3). Tom Eisner and Daniel Aneshansley (2008) investigated the function of this behavior by offering jays either unaltered ants, or ants with their defensive glands removed. The birds readily consumed the "deactivated" ants without sweeping them across their feathers, suggesting their feathers normally act as a sort of napkin for wiping away the ants' pungent chemical secretions in preparation for consumption. The birds are adept at processing their prey

VERTEBRATES AND ANTS

10-3　A blue jay (*Cyanocitta cristata*) holds a *Formica exsectoides* worker by the mesosoma, before wiping the ant's defensive secretions on its feathers. Jays use their feathers like a napkin to disarm the ants before eating them. (Thomas Eisner, courtesy of Maria Eisner).

and manage to remove the defensive secretions without bursting the ant's fragile crop, which can hold up to one-third of the ant's body weight and nutrient content. Jays also disarm other chemically defended prey, like bombardier beetles, by wiping them on the ground prior to consumption (Eisner et al. 2005).

Fish and Amphibious Associates

Unlike the vertebrates we have encountered so far in this chapter, a different group of associates can be found deep inside the subterranean chambers of ant nests. Nests of leaf-cutting ants in the genus *Atta* comprise enormous chambers and tunnels that can span 220 m^2 (Forti et al. 2017) (see Figure 1-13). Naturally, these giant nests also house a great variety of vertebrate guests. Nichols (1940) once found a 22 cm long eel (*Synbranchus marmoratus*) in the seasonally water-logged chambers of an *Atta sexdens* nest, more than 2 meters below the surface and 100 meters from the nearest ephemeral pond. The presence of a fish in an ant nest might seem impossible, but like many eels, *S. marmoratus* can switch from using gills to breathing through the lining of the mouth and pharynx when crossing dry land. Several amphibians are also associated with ants. Andreas Schlüter (1980) first recorded calls of male painted ant-nest frogs (*Lithodytes lineatus*) ringing from the nests of the leafcutter ant *Atta cephalotes* in the Peruvian rainforest. The male frog's short whistles entice females to mate and deposit their foam nests on roots that crisscross the ants' wet subterranean tunnels (Schlüter and Regös 1981, 2005). Here, deep below the surface, dozens of tadpoles hatch and develop while surrounded by ants (Schlüter et al. 2009) (Figure 10-4).

Regös and Schlüter (1984) hypothesized that invading *L. lineatus* must employ some chemical disguise or repellant to survive a gauntlet of biting ants. They noticed that *L. lineatus* captured at the entrance of *A. cephalotes* nests had a distinctive, spicy odor, similar to a lovage plant (*Levisticum officinale*), a popular culinary herb. When separated from ant colonies or raised in captivity, frogs failed to express the odor and were mobbed and killed by ants when returned to *Atta* nests (Regös and Schlüter 1984; Schlüter and Regös 1996). This observation was explored in greater detail by André de Lima Barros and colleagues (de Lima Barros et al. 2016), who extracted skin secretions from *L. lineatus* and applied them to a co-occurring anuran, *Rhinella major* (Bufonidae), that lacks any myrmecophilous habits. *Rhinella major* was not attacked by the ants when chemically disguised as *L. lineatus* but did elicit aggression in *Atta laevigata* and *A. sexdens* workers when dipped only in water, as a control. When half of the toad's body was painted with *L. lineatus* extract, and the other half was painted with water, only the half painted with water was seized by attacking ants. To tease apart elements of the behavioral and chemical disguise of *L. lineatus,* a reciprocal experiment might also ask whether *L. lineatus* is attacked by its own host ants if coated with the skin extracts of *R. major.*

Although the important compounds on *L. lineatus* skin have not been identified, these findings suggest that *L. lineatus* might perform unseen behaviors to acquire the colony odor of its host, or gather an inert, repellent, or appeasing compound from its environment. The toxic skin secretions of poison dart frogs (Dendrobatidae) include numerous classes of lipophilic alkaloids, derived in part from diets rich in myrmicine and formicine ants (Saporito et al. 2012; Rabeling

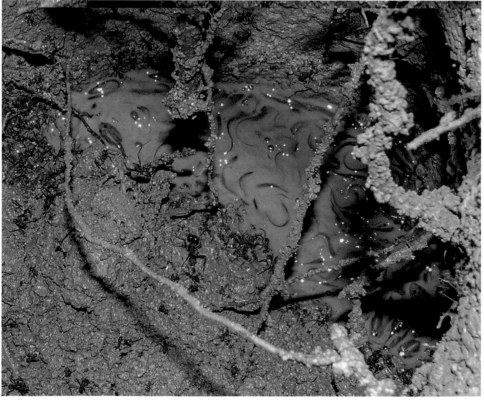

10-4 The painted ant-nest frog (*Lithodytes lineatus*) mates and lays its eggs deep inside leafcutter ant nests. Adult frogs, like this one, are not attacked by the ants (upper). Unpigmented tadpoles develop inside waterlogged chambers of an *Atta cephalotes* nest (lower). (Courtesy of Konrad Mebert).

et al. 2016). *Lithodytes lineatus* belong to a different family than the dart frogs, and inspection of stomach contents revealed that they consume crickets, earthworms, and other arthropods as well as ants (Parmelee 1999). The stomach contents of one *L. lineatus* specimen caught at the entrance of an *Atta* nest contained only two nymphs of a bug in the genus *Heza* (Reduviidae), which bears its own battery of chemical defenses (Schlüter and Regös 2005). Perhaps these bugs are the source of *L. lineatus* camouflage, and their consumption provides a chemical passport into fungus-farming ant nests. For now, the identity of the apparent chemical disguise used by eggs, tadpoles, and adult *L. lineatus* is still a mystery.

The African stink ant, *Paltothyreus tarsatus* (25 mm), is host to another amphibious guest, the rubber frog *Phrynomantis microps* (40–60 mm). Mark-Oliver Rödel and colleagues (2013) discovered that the frog inhabits stink ant nests during the dry season, presumably to take advantage of the humidity of the subterranean chambers. Although it is antennated by its hosts, *P. microps* is not attacked, and moves freely among the ant workers (Figure 10-5). *Paltothyreus tarsatus* is a capable predator and scavenger, but when presented with termites and mealworms coated in the lyophilized skin secretions of *P. microps* frogs, groups of ants did not show any aggressive or predatory behavior. Although the effects of the chemical disguise eventually wore off, the median latency to sting a coated mealworm was 286 seconds, versus 4 seconds for untreated insects. Two novel peptides were isolated from the frog's skin secretions that, when synthesized and tested, induced the same inhibitory response in the ants as the frog's natural secretions. Several lines of evidence suggest that, unlike the *L. lineatus, P. microps* synthesizes both peptides independent of its diet and exposure to ants. First, newly metamorphosed *P. microps* are not attacked when placed in an ant colony for the first time, and second, *P. microps* fed unnatural laboratory diets continue producing effective secretions (Rödel et al. 2013).

Reptile Eggs in Ant Nests

Subterranean nests provide an attractive option for animals seeking preexisting and well-defended cavities in which to house their eggs. Whether ant nests are used opportunistically or as preferred egg-laying sites by vertebrates is not well documented. Scattered reports suggest that the legless worm lizards (Amphisbaenia) *Amphisbaena alba, A. fuliginosa, A. mertensii,* and *Anops kingii* facultatively deposit their eggs in or near the nests of leaf-cutting ants (reviewed by Andrade et al. 2006). Eggs of the smallhead worm lizard, *Leposternon microcephalum,* have also been found in *Camponotus* sp. nests, though reports of associations between lizards and non-fungus-farming ant genera are rare (Andrade et al. 2006). Kwapich was lucky enough to uncover green anole (*Anolis carolinensis*) eggs and hatchlings in the top chambers of a ponerine trap-jaw ant (*Odontomachus brunneus*) nest in Florida (USA). Green anole mothers are monoallochronic and deposit a single egg each week, rather than laying a large clutch all at once (Toda et al. 2013). The observed lizards hatched on separate days beginning twenty-five

10-5   The frog *Phrynomantis microps* spends a portion of its year inside the nests of the African stink ant *Paltothyreus tarsatus*. The frog synthesizes a chemical disguise that allows it to live among the ants without being attacked. (Courtesy of Christian Brede).

days after discovery, suggesting that the nest was either revisited by the same female multiple times or was visited by several females as a communal nest (Kwapich 2021).

Several snakes also deposit their eggs inside ant nests. The eggs of *Leptodeira annulata,* a 66 cm long, frog-eating snake, were found completely embedded in the fungus gardens of both *Acromyrmex octospinosus* and *Atta colombica* leaf-cutting ants (Brandão et al. 1985; Baer et al. 2009), and Bruner et al. (2012) successfully reared the blind snake *Liotyphlops albirostris* (Figure 10-6) from eggs found in an *Apterostigma goniodes* fungus garden. In the nest, the ants actively groomed and antennated the snake eggs but never bit their leathery shells. Like many attine ants, *A. goniodes* cover their own eggs, larva, and pupae with pieces of their fungus garden, a behavior thought to protect brood from microbial infections. This same courtesy is extended to the enormous blind snake eggs (Figure 10-7). When researchers removed the fungal coverings from snake eggs, they found that they were rapidly replaced by the ants, while artificial eggs inserted in the nest were not inspected or decorated with fungus (Bruner et al. 2012). Therefore, in addition to enjoying the same physical and behavioral protection as the ants' own brood, in a carefully controlled microclimate, nesting with ants may provide guest eggs with some protection from pathogenic microbes.

Horned Lizards, Blind Snakes, and Legless Lizards

Although antbirds do not prey on the army ants that they follow, other vertebrate species are specialist predators of ants, or even predators of the myrmecophiles found with ants. These predators have special adaptations that allow them to overcome the chemical and numerical defenses of ants, including ants in the genus *Pogonomyrmex,* renowned for its powerful sting. Although each sting delivers a small dose, *Pogonomyrmex* venom has a similar LD_{50} (rodent-killing capacity) per milligram as the most potent snake venoms (Schmidt and Blum 1978b). The enzyme-rich venom of *Pogonomyrmex* workers is especially effective against vertebrates because it is hemolytic and produces a painful and long-lasting

10-6　The blind snake *Liotyphlops albirostris* is a guest of the fungus-gardening ant *Apterostigma goniodes*. (Upper: Courtesy of Paul Freed; Lower: Courtesy of Gaspar Bruner).

10-7　Eggs of the blind snake *Liotyphlops albirostris*, found in the fungus garden of *Apterostigma goniodes*. The ants cover their own brood, and the snake's eggs, in a fungus that they grow for food. (Courtesy of Gaspar Bruner).

sensation (Schmidt and Blum 1978a,b). Despite the impressive defenses of *Pogonomyrmex,* many species of horned lizards (Iguania, Phrynosomatidae, Phrynosoma) specialize in eating them. Horned lizards are squat, low-tempo lizards with short necks and faces (Figure 10-8). Sherbrooke and Schwenk (2008) demonstrated that, unlike other insectivorous lizards that process prey in the mouth, horned lizards have special morphological adaptations that allow them to swallow harvester ants whole. Immediately following capture, the lizard's tongue presses ants into mucus-secreting pharyngeal papillae and down its radially branched esophageal folds, immobilizing them and gluing them to strands of thick mucus. The ventrally curled ants are delivered to the stomach in their cocoons of mucus in a single step, presumably without ever having the opportunity to sting sensitive tissues. In addition to their feeding apparatus, horned lizards have evolved special blood plasma factor that makes them less sensitive to *Pogonomyrmex* venom and that, to some extent, detoxifies it (Schmidt et al. 1989). Pianka and Parker (1975) concluded that horned lizards possess a "unique constellation of anatomical, behavioral, physiological and ecological adaptations that facilitate efficient exploitation of ants as a food source." For their part, *Pogonomyrmex* colonies do not continue to hemorrhage workers when predators, like lizards, camp along their foraging routes. Instead, when numerous foragers are consumed by a predator, colonies either move their nest entrances, suspend foraging, or maintain a smaller foraging force until new foragers develop over time (Hölldobler 1970a; Whitford and Bryant 1979; Kwapich and Tschinkel 2013).

To access an ant nest, predatory trespassers must work quickly or rely on adaptations that allow them to either overcome the strict odor-recognition systems of ants or resist the ants' many defenses. Myrmecophagous blind snakes in the families Typhlopidae and Leptotyphlopidae, such as African *Rhinotyphlops* and *Afrotyphlops (= Typhlops)*, feed infrequently by "binging" on clumps of immobile ant brood inside nests (Figure 10-9). The ant brood are rapidly shoveled into the snake's esophagus as it protracts and retracts maxillary bones in its upper jaw. The group's unique morphology and low feeding frequency, coupled with their preference for small-bodied ant species, suggest that they are adapted to spending short periods of time in the nests of ants that cannot easily overwhelm them (Webb et al. 2001).

The flathead worm snake, *Antillotyphlops (= Typhlops) platycephalus,* is an ant generalist that feeds on the adults or brood of at least sixteen ant species available in its habitat (Torres et al. 2000). These and other ant-associated snakes are able

10-8 The Texas horned lizard, *Phrynosoma cornutum,* is a predator of the harvester ant *Pogonomyrmex rugosus* (upper). (Bert Hölldobler). The lower picture shows a juvenile greater horned lizard (*Phrynosoma hernandesi*) from southern Arizona (USA). (Christina Kwapich). Horned lizards have special morphological and physiological adaptations that allow them to eat ants with powerful venom.

10-9   The blind snake *Indotyphlops braminus* (Typhlopidae) feeds on ant brood and termites. Its eyes are covered with translucent scales. (Courtesy of Taku Shimada).

to locate and discriminate between a large pool of available ant species in their environment—but how do they do it? Jonathan K. Webb and Richard Shine (1992) designed an experiment to determine whether the Australian blind snake, *Ramphotyphlops (= Anilios) nigrescens,* could find ant nests by hijacking their chemically reinforced foraging trails. They fed wild-caught blind snakes the brood of six favored ant species over a period of several weeks. On the day of the experiment, groups of ten worker ants were placed in narrow enclosures, on top of cartridge paper in 55 × 40 cm aquaria. After twenty minutes, the ants and lane enclosures were removed, and pebbles were added on top of the paper flooring for traction. Over a six-hour period, individual snakes were introduced into the aquaria. Snakes that moved along the routes of "ant trails" for more than 4 cm were scored as having followed a trail. As controls, the researchers offered snakes lines of water, fluon (polytetrafluoroethylene, used to prevent ants from climbing walls of plastic containers), or the routes of non-prey invertebrate species. The snakes ignored trails of non-prey species like earthworms, isopods, and termites, as well as the control trails. However, they readily followed the routes of their

10-10   The Texas blind snake, *Rena dulcis* (Leptotyphlopidae), travels with army ants (*Neivamyrmex*) during their foraging raids. Like the army ants, it is a predator of ant brood. (Courtesy of Alex Wild / alexanderwild.com).

larger prey ants, including *Camponotus consobrinus, Iridomyrmex purpureus,* and two species of *Myrmecia* (four or five of six snakes). Snakes followed week-old and day-old "trails" of these larger ant species, though they never located *Rhytidoponera enigmatica* and *R. metallica* trails, which the researchers attribute to the small size and solitary nature of these ants.

Based on the reported results, there can be no doubt that the snakes are attracted to areas contacted by as few as ten *Camponotus, Iridomyrmex,* and *Myrmecia* workers, hours or days earlier. The snakes successfully followed the 40×2 cm ant routes for an average of 27 cm (SD = 10.61, N = 18, range = 10–40 cm), far exceeding the random likelihood of taking an identical path in the large aquaria. These results are puzzling, though, because *Myrmecia* are solitary huntresses that do not use pheromone trails to recruit nestmates to prey. It is also highly improbable that "ant trails" simulated in the laboratory resembled the chemically reinforced foraging routes of the other two ant species in nature. In general, ant trails comprise moderately volatile recruitment and orientation pheromones that link nests with food sources in real time. The ability of these trails to expire when they are no longer reinforced prevents continued allocation of the colony's limited foraging force to an exhausted food resource. Although stable trunk routes are frequently reinforced with more lasting orientation compounds (Hölldobler and Wilson 1970; Hölldobler 1976; Hölldobler et al. 2001; Plowes et al. 2014), and in some cases, ants do mark new terrain that they enter (Hölldobler and Wilson 1990), there can be no expectation that a randomly selected group of ten ants would lay down a pheromone trail when removed from the context of their nest. The addition and removal of ants from a paper backdrop likely included alarm behavior, but because alarm pheromones are volatile like the recruitment signals used in foraging, these are also not likely to have influenced snake behavior hours after their release.

If the snakes can truly perceive the presence of ants for hours and days after they walked across a surface, then it suggests that they do not use the more volatile recruitment pheromones of the ants, but rather detect minute quantities of heavier compounds, such as transferred cuticular hydrocarbons, hindgut

contents, or chemical footprint cues, which ants may deposit when entering new terrain. Such compounds might coincidentally be used as trunk route orientation signals by ants, or in nest site and territorial marking, in the appropriate context (Hölldobler and Wilson 1990). Given this uncertainty, we cannot yet conclude that *Ramphotyphlops nigrescens* makes use of ant foraging pheromones, though it seems likely that they can detect some odors associated with ants in a simplified environment.

Some blind snake species are reported to follow the raiding columns of army ants in the genus *Neivamyrmex*. Watkins et al. (1967) observed four Texas blind snakes, *Rena* (formerly *Leptotyphlops*) *dulcis* (Figure 10-10), traveling with *N. nigrescens* in aboveground raiding columns. The snakes were not attacked by the

10-11   The legless lizard *Amphisbaena alba* follows active and abandoned *Atta* leafcutter ant trails. It enters ant nests, where it feeds on ant larvae and the brood of myrmecophilous beetles. (Courtesy of Diogo B. Provete).

alba. The large lungworm eggs are not eaten by ants but are deposited in refuse chambers with the rest of the lizard's feces, where beetles ingest them and develop infections under laboratory conditions. In addition to feeding on these inquiline beetle larvae, the lizards readily consume *Atta* brood when offered. Although they make a living inside leafcutter ant nests, they are easily irritated by the bites of major workers, and thrash and roll to remove them.

On one occasion, *A. alba* was observed following an inactive *Atta laevigata* ant trail for 53 m before entering the colony's nest entrance. When the lizard's head crossed the trail, it flicked its tongue with increasing speed, presumably gathering chemical stimuli and delivering them to the Jacobson's organ, the chemoreception organ of lizards and snakes located in the mouth (Campos et al. 2014).

10-12   Larvae and pupae of the myrmecophilous rhinoceros beetle, *Coelosis biloba,* are preyed on by the legless lizard *Amphisbaena alba.* Beetles can become infected by the lungworm, *Raillietiella gigliolii,* when they eat the feces of myrmecophilous legless lizards. The upper photo shows male and female beetle pupae; the lower photo shows both adults. (Upper: From Pardo-Locarno et al., "Los estados inmaduros de *Coelosis biloba* [Coleoptera: Melolonthidae: Dynastinae] y notas sobre su biología," *Revista Mexicana de Biodiversidad* 77 [2006]: 215–224, Figures 17–20, CC BY-NC 3.0; lower: Courtesy of Stanislav Krejčík).

10-13   During a coming-of-age ritual, young Sateré-Mawé men from Brazil wear gloves filled with bullet ants (*Paraponera clavata*). Several hundred ants are anesthetized and sewn into each glove made of palm fronds. (Wolfgang Sauber, Wikimedia Commons / CC BY-SA 3.0).

Riley et al. (1986) were curious about the cues *A. alba* used to locate *Atta* nests, and so set out to determine whether they could orient using leafcutter ant foraging trails. To do so, they held lizards in captivity for several weeks, then returned to the forest and placed them adjacent to active *A. cephalotes* trails. The lizards correctly followed active ant trails in seventeen of twenty-four trials. They were also able to follow abandoned trails for many meters in the absence of ants, despite more frequently traveling in the incorrect direction. Artificial trails, cleared by the researchers to resemble those of the ants, were never followed. Like blind snakes, the ability of *A. alba* to travel along abandoned ant trails suggests that they may principally use substances other than the ants' volatile recruitment pheromones.

Lizards housed in captivity for many years retained their ability to follow leafcutter ant trails when returned to the forest (Riley et al. 1986), suggesting either an innate attraction to the cues associated with *Atta* trails or an exceptional memory. While it is likely that the lizards do use ant-foraging trails to locate *Atta* nests, it is also possible that the lizards detect and follow the chemical trails of conspecific lizards associated with the same colonies. As the 60 cm long lizards slither along, hardened secretions from rows of precloacal glands are abraded and deposited as granular "pheromone sachets" in the soil. Both male and female *A. alba* possess these unusual glands, which are equivalent in size between sexes and produce identical mucopolysaccharides and protein-rich secretions (Antoniazzi et al. 1993; Jared et al. 1999). More studies on learning and memory, the use of conspecific odor trails, and the use of ant trail pheromones are needed to reveal how these reptiles find ant nests.

Many other interactions between ants and vertebrates are known, including those with humans. In Europe, people use mound-building wood ants in the *Formica rufa* group for the biological control of herbivores in forests, and also illegally harvest *Formica* brood for bird feed. In Mexico, the larvae and pupae of *Liometopum* are mixed with rice to create a tangy delicacy known as escamoles,

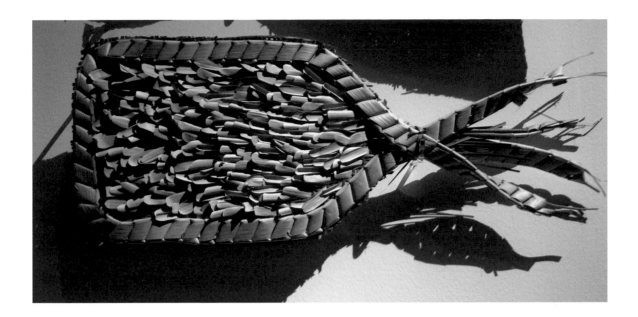

and fat, young leafcutter ant queens are gathered during the mating flight season and roasted or coated with chocolate as a delicious treat. Kevin P. Groark (1996) presented an ethnographic and toxicological review and new data about the "ritual and therapeutic use" of red harvester ants of the genus *Pogonomyrmex* by indigenous people in southern and south-central California. It has been reported that the "ants were ingested alive in massive quantities in order to induce prolonged catatonic states" accompanied by hallucinogenic visions. The harvester ant venom was also used for therapeutic purposes. In Thailand, the larvae and pupae of the weaver ant *Oecophylla smaragdina* are enjoyed as a popular culinary delicacy, and in addition, serve as medicine, bird feed, and fish bait. As early as 1000 B.C., surgeons reported using ant heads as wound sutures (Haddad 2010), and in the present day, the Sateré-Mawé people of Brazil sew bullet ants (*Paraponera clavata*) into mittens, worn during important rites of initiation (Figure 10-13). Beyond these applications, ants have been incorporated into the tapestry of human culture and imagination. Ants appear in written and oral traditions from every continent where humans co-occur (see Ellison and Gotelli 2021).

Epilogue

WE STARTED THIS book with Bert's recollection of how he was introduced as a boy to myrmecophiles by his father, Karl Hölldobler. The old, for decades unused, quarry that had been reclaimed by nature was inhabited by many ant colonies of various species. Each one housed more myrmecophiles than we have experienced in other habitats. Like erstwhile Karl Hölldobler, Bert also later collected there many specimens of a variety of ant guests for his first scientific investigations. About thirty-five years after his last visit, Bert returned to this paradise of his youth only to discover, with utmost dismay, that this former nature jewel, with the endemic population of the ant cricket (*Myrmecophilus acervorum*), and so many other myrmecophiles, had been used as a landfill. It was totally destroyed.

Many delicate nest ecosystems of host ants and their guests within the major ecosystem of this unique habitat were senselessly wiped out and replaced by a sterile seeded lawn. All the limestone rocks under which the ant nests could be found were gone, and so were the mounds of *Formica* species, the birch and black locust trees (*Robinia pseudoacacia*), which housed many honeydew-producing Hemiptera tended by the ants.

Everywhere ants are found, their species have close associations with insects that feed on plants. Aphids, scale insects, mealybugs, treehoppers, and the caterpillars of lycaenid and riodinid butterflies (the blues, coppers, and hairstreaks, and the metalmarks, respectively) excrete sugary solutions that the ants eagerly collect. In return, the ants protect these honeydew-producing insects. This mutualistic symbiosis is called trophobiosis. Except for the blues and metalmarks, our book does not cover these very intriguing trophobiotic interactions in the ants' ecosystems.

The structure of organization that the colonies of many ant species have evolved is impressive, but the basis of strength—the concatenation of simple cues—is also a source of major weakness. Ants are easily fooled. Other organisms can circumvent or break their code and exploit the social acquisitions of ant societies, for example, by imitating one or several key stimuli. A considerable number of socially parasitic ants live with and exploit colonies of other ant species. These ants are called social parasites and even some cases of intraspecific social parasitism (the parasitic exploitation of foreign colonies belonging to the same species) have been described.

Several hundred species of ants around the world are known to have become social parasites of other ants, and many more might still be discovered in the future. This is an exciting field in evolutionary myrmecology, which awaits coverage in a separate volume. But thousands of mites, silverfish, millipedes, spiders, flies, beetles, wasps, and other small creatures, and even some vertebrates, have evolved close interactions with ant species. The ant colony is "porous" to invasion, and it represents an ecological island lavishly endowed with nutrients. The colony and nest offer many kinds of niches that predators, kleptoparasites, and other symbionts can enter. Others live on or inside the bodies of their host ants. We have reported many studies of myrmecophiles that evolved special adaptations to inhabit these diverse niches. The means by which insects and other arthropods trick and rob ants are vast in number, and this book describes and critically reviews various experimental investigations that revealed some of these tricks. The study of myrmecophiles has not only become an important part of myrmecology, insect sociobiology, behavioral ecology, and population biology, but several aspects promise intriguing insights for the field of evolutionary parasitology, although more empirical studies are needed to support various proposed hypotheses in this field.

Glossary

References

Acknowledgments

Index

Glossary

active space: The space within which the concentration of a pheromone is at or above threshold concentration.

adoption substance: Secretion presented by myrmecophiles that induce the host ants to accept the parasite into their colony.

age polyethism: The regular changing of labor roles by colony members as they age.

alarm pheromone: A chemical substance that, when discharged from specific glands by ants, induces in nestmates a state of alertness or alarm and that, in higher concentrations, triggers aggressive behavior.

alate: Winged.

alitrunk: The mesosoma (section between head and waste) of higher Hymenoptera, which includes the true thorax and, fused to the thorax, the first abdominal segment.

allomone: A chemical substance or blend of substances used in communication among individuals belonging to different species. It evokes a response in the receiving individual that is favorable to the emitter, but not necessarily to the receiver.

anemotaxis: Oriented movement in response to a current of air. Positive anemotaxis means the movement is directed toward the air current (upwind anemotaxis). Often these orientation tracks are meandering movements, with the main direction upwind (positive meno-anemotaxis).

antbirds: Tropical birds that follow raiding swarms of army ants and feed on the prey stirred up by the ants.

ant-butterflies: Butterflies of the family Nymphalidae (Ithomiinae) that follow raiding swarms of army ants and feed on the droppings of antbirds.

antennal flagellum: The distal part of the insect antenna that consists of a number of antennal flagellomeres (segments).

antennal scape: The elongated basal segment of the insect antenna.

antennation: Touching with the antennae. This behavior can serve in tactile communication and sensory probing.

antennomere: One of the segments of an insect's antenna, where all of the segments are more or less uniform.

ant plant: Also known as a myrmecophyte. Species of plants closely associated with ants. Many grow domatia, specialized structures for housing ant colonies.

appeasement gland complex: Pertains to several exocrine glands in myrmecophilous rove beetles (Staphylinidae) that open near the rear end of the abdomen. The secretions of these glands are employed in defense against aggressive host ants, without causing alarm or repellent reactions in the ants.

appeasement substance: A secretion presented by insects that reduces or distracts aggression in attacking insects, thereby allowing the attacked individuals to escape without discharging repellent defense secretions. It is frequently used by social-parasitic myrmecophiles but may also be employed as "soft" defense in free-living insects against attacking ants.

appressorium: A specialized cell of pathogenic fungi. It has a special device with which it is able to penetrate the ants' integument.

apterism: Having no wings or wing-like extensions.

araneophagous: Tending to consume spiders for food.

ascoma (plural ascomata): Pertains to the fruiting body of an ascomycete fungus. It contains hundreds of thousands or millions of asci, each of which carry ascospores.

ascus (plural asci): An ascus is the sexual spore-bearing cell produced in ascomycete fungi. Each ascus usually contains eight ascospores.

assortative mating: A form of nonrandom mating in which individuals select mates with characteristics similar to their own.

bacteriocyte: Also called **mycetocyte.** Specialized cells that contain endosymbiotic organisms such as bacteria. In some cases, these microorganisms provide essential amino acids. They are usually located inside the midgut epithelium.

bivouac: The mass of army ant workers clustered together in a cavity, in the midst of which the queen and brood are sheltered by the bodies of the worker ants.

brainworm: The metacercaria of the liver fluke (*Dicrocoelium dendriticum*) that invades the suboesophageal ganglion of the ant's brain, which seems to affect the ant's behavior to the advantage of the liver fluke's propagation.

brood: The immature members of a colony: eggs, larvae, and pupae.

budding: Colony multiplication by the departure of fractions of the worker force, with one or several queens, to a new nest site in the neighborhood of the "mother colony."

callow worker: A newly eclosed, lightly colored worker whose exoskeleton is still relatively soft and not much pigmented.

caste: Basically, there are two female castes in most ant societies: the queen caste and the worker caste. In most instances, these castes can be easily differentiated based on their external morphology. The worker caste can be subdivided into several morphological subcastes such as minims, minors, medias, majors, and super-majors (the latter two are also called soldier castes). The worker caste can also be subdivided based on age groups, that is, age groups that perform special labor tasks in the colony, although this grouping might be less rigid.

caterpillar: Pertains to the larval developmental instars of butterflies and moths. They are the butterfly larvae.

cephalothorax (prosoma): The anterior section of some arthropods, including spiders. It includes the fused head and thorax and attachment points for the legs.

cibarium-pharynx: Pertains to the cavity between the base of the hypopharynx and the undersurface of the **clypeus.**

clade: An evolutionary grouping of organisms that describes an ancestor and all its descendants.

cladogram: A diagram showing the sequence in which groups of organisms have originated and diverged over the course of evolution. They are usually the basis of comparative morphological studies and / or genetic analysis.

cleptobiosis: The relation in which one species robs the food of another species.

clypeus: A sclerite that delineates the lower margin of the upper side of the head (face), to which the **labrum** (upper lip) articulates along the ventral margin.

colony: A group of individuals that collectively construct nests and rear offspring and exhibit some sort of division of labor and communication.

colony fission: The splitting of a colony into two moieties of an approximately equal size. This is the mode of colony propagation in doryline army ants.

commensalism: Symbiosis in which members of one species benefit from living close to or with ants, while the ants neither benefit nor suffer any harm.

communication: Action of one or several individuals (senders) that affects the behavior of other individuals (respondents) to the benefit of the senders or to the benefit of both senders and respondents.

crop (ingluvial):   The distensible middle proportion of the foregut in which (in many ant species) liquid food is transported, stored, and regurgitated to nestmates. It is also called the "social stomach."

dealate: An individual who has shed its wings; for example, a young freshly mated queen, after her mating flight.

diplophoresy: The use of two separate hosts during dispersal.

discriminators: Also called recognition labels. Cues that permit individuals to be classified as nestmate or foreigner.

disruptive selection (diversifying selection):   Natural selection that leads to a higher frequency of the extreme forms of a trait in a population, rather than intermediate forms.

domatium (plural domatia):   Specialized structures such as swollen leaf stems used by ant plants for housing of ant colonies. They are also called myrmecodomatia.

dorsal nectar organ (also Newcomer's gland):   Pertains to an exocrine glandular structure located on the dorsum of the seventh abdominal segment of caterpillars in species of many lycaenid subfamilies. It secretes a honeydew-like blend of substances.

Dufour's gland: Exocrine gland that empties at the base of the stinger, also known as accessory gland to the **poison or venom gland.**

eclosion: Emergence of the adult (imago) from the pupa.

ectoparasitoid: Pertains to parasitoids that live and feed on the outside of the host's body.

endosymbiosis: Pertains to symbiotic organisms that live inside the body of the host. Some of these symbionts are intracellular.

enemy-specific alarm-recruitment: Alarming and recruiting nestmates, particularly the soldier subcastes, in defense against invasion of specific enemy ant species.

epigastric furrow: A fold on the ventral side of the opisthosoma of female spiders. It separates the book lungs, the main respiratory organ in most arachnids (spiders and scorpions), from the external genital structures and contains the opening of the oviduct.

esophagus: It connects the **pharynx** with the **crop.**

ethogram: A complete description of the behavioral repertoire of an individual, a caste, or myrmecophile.

eusocial: Generally, a group of individuals that exhibit division of labor in reproduction. That is, only one or a few individuals reproduce, and the majority remain sterile and cooperatively raise the offspring. Usually two or more generations cooperate in one colony. Several degrees of eusociality exist: at one extreme, most sterile individuals have the potential to become fully functional reproductives; at the other extreme, the sterile individuals have entirely lost their reproductive organs.

exaptation: A shift in the function, or the evolution of an additional function of an existing trait.

extended phenotype: External structures or traits, the characteristics of which are the products of innate behavioral collective actions of the producers: for example, architectural constructions made by ants that modify their environment, and nest structures of solitary animals that are built based on innate behavioral actions. Furthermore, parasites that manipulate their host's behavior exploit the host as their own extended phenotype.

extrafloral nectaries: Nectar-secreting glands outside of flowers that are not involved in pollination, usually located on the leaf or petiole (foliar nectaries) and often associated with leaf venation.

exuviae: The remains of an exoskeleton and related structures that are left after the insects have molted.

femoral: Pertaining to the thigh or femur.

femur: The third segment of the leg away from the body.

gamergate: A mated, egg-laying worker.

gaster: Terminal major body part of ants, also called metasoma, the terminal part of the abdomen.

genetic drift: Pertaining to the random fluctuations of gene variants in a population (gene frequency); that is, the occurrence of variants of a gene (alleles) increase and decrease by chance over time.

Gestalt odor: A common odor is created by pooling the recognition odors of some or all individuals of a colony, which results in a unique "Gestalt odor."

gram-negative bacteria: Bacteria that do not retain the violet stain used in the gram-staining method.

gyne: Female reproductive caste, whether functioning as a reproductive or not.

harvesting ants: Ant species that collect primarily seeds and store them for food.

hemimetabolous: Gradual development that lacks the differentiation seen between larval, pupal, and adult stages. Immature stages are called nymphs.

hemolymph: Blood of insects.

holometabolous: Complete metamorphosis during development, with distinct larval, pupal, and adult stages.

homonymy: Pertains to the name of a taxon that is identical in spelling to the name of a different taxon. Such identical names of different taxa are homonyms.

honest signal: A signal that communicates the true status, intention, or state of motivation of the signaler.

honeydew: Excrement from plant-sap-feeding aphids and other insects. It contains sugars and amino acids and is a principal food of many ant species.

hypha (plural hyphae):   A branching filamentous structure of fungi. Through hyphae, the fungus achieves vegetative growth. The hyphal mass is called mycelium.

hyphal bodies: Segmented fragments of hypha, of irregular, often thickened shapes. They can propagate by budding.

inclusive fitness: The sum of an individual's (A's) direct evolutionary fitness (measured in number of personal offspring) plus the fitness increase of that individual's (A's) relatives (Bs), caused by the individual's (A's) help, weighted by the degree of relatedness between helper (A) and helped individuals (Bs).

infrabuccal pocket: A cavity in the mouth of ants in which indigestible material accumulates and is later disposed of.

instar: Any period between molts during the course of development in arthropods.

kairomone: A chemical substance emitted by one species that has an adaptive benefit to another species perceiving these chemical compounds.

kin recognition: Recognition and discrimination of close relatives and rejection of unrelated individuals.

kin selection: Evolutionary selection of helping behavior that favors the survival and reproduction of relatives who possess proportions of the genes by common descent.

labium: The lower lip in insects, located just below the mandibles. The segmented appendages arising from the labium are called labial palps.

labrum: The upper lip in insects. Located at the base of the insect's upper side of the head above the mandibles.

lateral organs: See **tentacular organs.**

macrogyne: The larger queen in species with two sizes of queens.

mandibles: The principal pair of jaws.

mandibular glands: A pair of exocrine glands, the reservoir of which opens at the base of the mandible. In many ant species this gland produces **alarm pheromones.**

maxilla: The second pair of jaws, usually folded beneath the principal pair of jaws, or mandibles. The pair of appendages originating from the maxilla are called maxillary palps.

melanization: The production and deposition of melanin, a pigment that increases resistance against pathogens, such as infectious microorganisms.

mermithergate: An aberrant worker form in ants, caused by infection with nematode parasites in the genus *Mermis.*

mesosoma: See **alitrunk.**

metacercaria: The fourth developmental stage of the liver fluke *Dicrocoelium dendriticum.*

metanotum: The dorsal portion of the posterior segment of the thorax (metathorax) of an insect.

metapleural gland: Exocrine gland peculiar to the ants located in the posteroventral angle of the metapleuron; it produces antibiotic substances and, in some cases, defensive secretions.

metasoma: See **gaster.**

metatarsus: See **tarsus.**

microgyne: The smaller queen in species with two sizes of queens.

mimicry

Batesian mimicry: The physical imitation of an unpalatable or well-defended model. Myrmecomorphs benefit from looking like ants by evading ant-averse predators.

behavioral mimicry: The imitation by myrmecophiles and myrmecomorphs of movement or tactile stimuli of associated ants.

chemical mimicry: The imitation of a chemical signal of the host ants by myrmecophiles.

topological mimicry: The imitation of cuticular structures and pubescence or shape of body parts of host ants by myrmecophiles.

Wasmannian mimicry: The imitation by myrmecophiles of the host ants' appearance, which evolved as an adaptation that favors acceptance by the host ants.

miracidium (plural miracidia): The first ciliated larval stage of the liver fluke *Dicrocoelium dendriticum.*

modulatory communication: Communication that influences the behavior of receivers, not by directly eliciting a behavioral act, but by slightly shifting the probability of performance or by lowering the response threshold necessary to react to the behavior-releasing signal.

monoallochronic ovulation: The alternating production of a single egg by the right and left ovary in some vertebrates.

monogyny: The existence of only one functional queen in an insect society.

monophasic allometry: Polymorphism wherein worker proportions are best described by the slope of a single linear regression.

mutualism: Symbiosis that benefits the members of both participating species.

mycetocyte: See **bacteriocyte.**

mycophagous: Tending to eat fungus.

myrmecoid body shape: The shape of a myrmecophile's body that resembles that of the host ant.

myrmecomorph: An organism that resembles an ant. Usually a Batesian mimic.

myrmecophagy: The act of feeding on ants.

myrmecophile: An organism that spends at least part of its life cycle with ant colonies.

myrmecophytes: Higher plants that live in a mutualistic relationship with ants.

Newcomer's gland: See **dorsal nectar organ.**

nomadic phase: The period in the activity cycle of army ants during which the colony forages more because it is also the period during which hundreds of thousands of larvae are raised. The colony moves from one bivouac to another to exploit new hunting grounds. This phase is sometimes called the migratory phase.

nuptial flight: The mating flight of young winged queens and males.

oenocytes: Glandular cells, with no duct cells, embedded in the fat body.

olfactory glomeruli (singular glomerulus):   Odor is perceived by chemo-sensilla on the antennae of insects and processed in the so-called paired antennal lobes, the olfactory center of the insect brain, which consists of spherical structures, the olfactory glomeruli.

oocyte: Female germ cell in the ovary involved in reproduction.

opisthonotal gland (oil gland):   In oribatid mites this organ consists of paired, sac-like glands that open through slit-like orifices on either side of the posterior margins of the notogaster (shield covering the dorsolateral region of the posterior body). The secretions function mainly in chemical defense and protection, and perhaps also in communication.

opisthosoma: The posterior section of some arthropods, including spiders. Similar to the abdomen of an insect.

oviposition: Laying eggs.

ovipositor: Tube-like structure through which eggs are deposited. In some ant subfamilies the ovipositor is highly modified to a stinger that is associated with venom and accessory glands. It has two functions—as ovipositor and in defense and attacks.

parasitism: Symbiosis in which members of one species exist at the cost of another species, but not necessarily killing the hosts.

parasitoid: Symbiosis in which the parasite kills the host, usually at completion of the parasite's development.

pedicel: The waist of the ant, consisting of either one segment (petiole) or two segments (petiole plus postpetiole). Also, the second segment of the antenna from the base outward is called a pedicel.

petiole: The first segment of the waist in Hymenoptera. In fact, it is the second abdominal segment. The first abdominal segment is fused with the thorax. The structure of the thorax fused with the first abdominal segment is called the **alitrunk.**

pharynx: The insect pharynx is the part of the foregut that connects with the mouth. It has a muscular structure and has the function of "sucking and swallowing" and leading food to the **esophagus.**

pheromone: A chemical substance or blend of substances, discharged from exocrine glands, that releases specific behavioral responses in members of the same species (releaser pheromone) or affects specific developmental changes (primer pheromone).

philopatric: Tending to stay near one's place of birth, rather than dispersing.

physogastry: The swelling of the gaster caused by the hypertrophy of fat bodies and / or ovaries.

planidium (plural planidia):   Pertains to the first-instar larva of parasitic and parasitoid wasps.

pleural glands: Pertain to exocrine glands that are associated with the lateral sclerites (pleura).

poison or venom gland: Pertains to the main gland associated with the stinger in many ants, bees, and wasps. It produces venomous and / or repellent secretions employed in defense.

polydomous: The occupancy of more than one nest by a single colony.

polygyny: The coexistence of two or more egg-laying queens in the same colony, where the queens do not exhibit antagonistic behavior toward each other and coexist without spacing out. A situation in which several queens are found in one colony and, even though they are accepted by the workers, do not tolerate each other is called oligogyny.

pore cupolas: Innervated, glandular structures scattered over the surface of caterpillars and pupae of many lycaenid species.

postpetiole: In certain ants, the second segment of the waist. This is in fact the third abdominal segment, since the first abdominal segment (propodeum) is fused to the thorax.

postpharyngeal glands: A paired cluster of finger-shaped exocrine glands in the head of ants that opens into the postpharynx. Among other functions, this gland plays a major role in the creation of a collective chemical colony recognition label.

prementum: Pertaining to the part of the labium (lower lip) to which the labial palps and paraglossa attach.

proboscis: Refers to the tubular feeding and sucking organ of certain insects (e.g., flies, bees, moths, butterflies) and consists of the elongated maxillae.

pronotum: The upper surface of the first segment of the thorax.

propodeal spines: The posterior segment of the **alitrunk** is called the propodeum. In several ant genera it is armed with a pair of hard spines, hooks, or teeth.

proventriculus: A muscular tubular organ posterior to the **crop** and anterior to the midgut (ventriculus), where most of the digestion occurs in insects.

pseudogynes: Ant workers with an aberrant alitrunk that somewhat resembles the alitrunk of queens.

pulvilli (singular pulvillus):   Lobes or cushion-like pads between the tarsal claws of insects that serve to attach the insect to the surface.

pupa: The last instar of the holometabolous insects, during which development into the final adult form is completed (metamorphosis).

pygidial gland: An exocrine gland, which is usually a paired cluster of glandular cells that drain the secretions into an intersegmental reservoir. The reservoir opens through the intersegmental membrane between the pygidium and the preceding tergite.

pygidium: The last externally visible tergite (upper segmental plate) of the abdomen of insects, regardless of its numerical designation.

ritualization: The evolutionary modification of a behavioral pattern, exocrine secretion, or physiological process that turns into a communication signal or improves the efficiency of a signal (modulating signal).

scolopidial sensilla: These sensory cells form large sensory organs (such as the subgenual organ) involved in proprioception and exteroception in insects.

sensillum basiconica: Olfactory sensillum of a particular shape on the antennae.

sensillum trichodea curvata: Olfactory sensillum of a particular shape on the antennae.

setae: Hair-like structures on the exterior surface of the body of insects.

social parasitism: In the strict sense, the coexistence in the same nest of two species of ants, one of which is parasitically dependent on the other. Sometimes the term is applied to the parasitic relationship between myrmecophiles and ants.

social stomach: See **crop.**

society: A group of individuals belonging to the same species that exhibit a cooperative division of labor and communication.

spermatheca (receptaculum seminis): An organ of the reproductive tract in females of ants and several other invertebrates, in which sperm from males are stored. In most ant species, only the queen caste has a functional spermatheca. In ant species where so-called gamergates (egg-laying workers) are involved in reproduction, the workers also have fully functional spermathecae.

spinneret: Refers to a silk-spinning organ of a spider.

spinulae: Small spines, often associated with the legs of insects.

spore (fungal): Fungal spores are microscopically small biological particles that serve the reproduction of fungi.

sporocyst: A sac-like structure, the second developmental stage of the liver fluke *Dicrocoelium dendriticum,* which gives rise to the motile cercaria, the third developmental stage.

statary phase: The period in the activity cycle of army ant colonies in which the colony remains in the same bivouac site. In this phase, the queen lays eggs, and new workers eclose from the pupal cocoons.

stenophagous: Feeding on one or a limited variety of foods.

sternal glands: Like tergal glands, except that they are located in the ventral region of the abdomen, where they are associated with the abdominal sternites (ventral sclerites).

sternite: The ventral sclerites, or plates of the body.

sternum: The ventral portions of the body.

stipule: In botany this is an outgrowth on either side (sometimes just one side) of the base of a leafstalk.

stomodeal trophallaxis: The exchange of liquid food mouth-to-mouth. The exchange of liquid from anus to mouth is called proctodeal trophallaxis.

stridulation: The production of sound and / or vibrations by rubbing one part of the body surface (scraper) against another that usually possesses a rippled surface.

strigilation: Scraping or licking substances from the body of another animal.

stroma: The spore-dispersal structure of pathogenic fungi that grows out of the body, usually the frontal thorax (pronotum) of the infected ant.

subgenual organ: The organ in insects that is involved in the perception of substrate vibrations, located in the lower part of the leg (tibia).

suboesophageal ganglion: Part of the central nervous system of insects, located below the **esophagus** (a tube that connects the pharynx and the crop) in the head.

super colony: A unicolonial population, in which workers move freely from one nest to another, so that the entire population is a single colony.

symbiont: An organism that lives in symbiosis with another species.

symbiosis: The close life interrelationship between organisms of different species. The symbiosis can be a commensalism, mutualism, or parasitism.

symphile: A myrmecophilous species that has a close relationship to the host ants. It is licked and groomed by the ants. The term is not used much anymore.

synechthran: A myrmecophilous species that lives as a scavenger and predator close to or in ant nests. The term is not used much anymore.

synoekete: A myrmecophilous species that is treated with indifference by the host ants. The term is not used much anymore.

tarsus: Leg of an insect; the segmented appendage attached to the **tibia**, or lower leg segment. The first tarsal segment attached to the tibia is called basitarsus or **metatarsus**.

template: Pertains to a learned pattern of odors stored in the olfactory nervous system of ants.

tentacular organs: Eversible tentacles, also called lateral organs, because they flank the dorsal nectar organ in caterpillars of many lycaenid species.

tergal: Pertaining to the upper (dorsal) surface.

tergal glands: Pertain to exocrine glands located in the dorsal region of the abdomen. They consist of either an epidermal glandular epithelium, wherein each cell drains secretions through tiny pore channels in the cuticula of the dorsal sclerites (tergites), or of clusters of glandular cells, each of which is combined with a duct cell that drains secretions into an intersegmental reservoir that opens between two tergites (dorsal sclerites).

tergite: Dorsal sclerite, or dorsal segmental plate.

thelytokous parthenogenesis: The production of females from unfertilized eggs.

threshold concentration: The concentration of pheromone molecules in the air high enough to elicit a behavioral response in the perceiving organism.

tibia: The fourth division of the leg, between femur and tarsus (foot).

topological-chemical signal: The combination of morphological features (surface structure and pubescence) with low volatile chemical signals in social communication between ants and immatures including myrmecophiles.

tracer experiment: An experiment in which food marked with a dye or radioactive compound is fed to the foraging ants, making it possible to trace the food distributed among nestmates and myrmecophiles.

trail pheromone: A substance or blend of chemical compounds originating from specific exocrine glands or the rectal bladder (hindgut) of ants that is laid down along a trail by one or few individuals and followed by other members of the same colony or species.

transformational mimic: An organism that resembles the morphology of two or more models during its development.

trichome: Tufts of yellow-golden hairs associated with exocrine glandular cells on the body surface of many myrmecophilous beetles. Often these trichome hairs are also innervated, so that they may function as mechano-receptors, perceiving the touching by ants.

trophallaxis: Exchange of liquid among colony members and myrmecophiles, either mutually or unilaterally. In stomodeal trophallaxis, the exchanged liquid originates from the mouth; in proctodeal trophallaxis, it originates from the anus.

trophic parasitism: Intrusion of one species into the social system of another to steal food.

trophobiosis: The relationship between ants and certain hemipteran organisms (and some lepidopteran species) in which they provide honeydew for the ants, and in return, the ants provide protection to the trophobionts.

tympanal organ: Hearing organ in insects. It consists of a membrane (tympanum) stretched across a frame. It is innervated by a group of mechano-sensory neurons, the so-called chordotonal organ. Airborne vibrations (sound) make the membrane vibrate, and these vibrations are sensed by the chordotonal organ.

viscous population: Pertains to populations with low genetic diversity, or high genetic relatedness among organisms. Population viscosity arises from the limited ability of organisms to move in their physical environment.

Weber's length: The diagonal length of an insect's mesosoma, in profile view. A measurement used to compare worker size.

xenobiont: Any organism found in association or close surrounding with other organisms. Ant species that regularly nest within or close to the colonies of other ant species.

References

Acharya, U, Acharya, JK. 2005. Enzymes of sphingolipid metabolism in *Drosophila melanogaster*. *Cellular and Molecular Life Sciences,* 62: 128–142.

Agrain, FA, Buffington, ML, Chaboo, CS, Chamorro, ML, Schöller, M. 2015. Leaf beetles are ant-nest beetles: The curious life of the juvenile stages of case-bearers (Coleoptera, Chrysomelidae, Cryptocephalinae). *ZooKeys,* 547: 133–164.

Akino, T. 2002. Chemical camouflage by myrmecophilous beetles *Zyras* comes (Coleoptera: Staphylinidae) and *Diaritiger fossulatus* (Coleoptera: Pselaphidae) to be integrated into the nest of *Lasius fuliginosus* (Hymenoptera: Formicidae). *Chemoecology,* 12: 83–89.

Akino, T. 2008. Chemical strategies to deal with ants: A review of mimicry, camouflage, propaganda, and phytomimesis by ants (Hymenoptera: Formicidae) and other arthropods. *Myrmecological News,* 11: 173–181.

Akino, T, Knapp, J, Thomas, J, Elmes, G. 1999. Chemical mimicry and host specificity in the butterfly *Maculinea rebeli,* a social parasite of *Myrmica* ant colonies. *Proceedings of the Royal Society of London. Series B: Biological Sciences,* 266: 1419–1426.

Akino, T, Mochizuki, R, Morimoto, M, Yamaoka, R. 1996. Chemical camouflage of myrmecophilous cricket *Myrmecophilus* sp. to be integrated with several ant species. *Japanese Journal of Applied Entomology and Zoology,* 40: 39–46.

Akino, T, Yamamura, K, Wakamura, S, Yamaoka, R. 2004. Direct behavioral evidence for hydrocarbons as nestmate recognition cues in *Formica japonica* (Hymenoptera: Formicidae). *Applied Entomology and Zoology,* 39: 381–387.

Akino, T, Yamaoka, R. 1998. Chemical mimicry in the root aphid parasitoid *Paralipsis eikoae* Yasumatsu (Hymenoptera: Aphidiidae) of the aphid-attending ant *Lasius sakagamii* Yamauchi & Hayashida (Hymenoptera: Formicidae). *Chemoecology,* 8: 153–161.

Akre, RD, Alpert, G, Alpert, T. 1973. Life cycle and behavior of *Microdon cothurnatus* in Washington (Diptera: Syrphidae). *Journal of the Kansas Entomological Society,* 39: 327–338.

Akre, RD, Garnett, WB, Zack, RS. 1988. Biology and behavior of *Microdon piperi* in the Pacific Northwest (Diptera: Syrphidae). *Journal of the Kansas Entomological Society,* 61: 441–452.

Akre, RD, Hill, WB. 1973. Behavior of *Adranes taylori,* a myrmecophilous beetle associated with *Lasius sitkaensis* in the Pacific Northwest (Coleoptera: Pselaphidae; Hymenoptera: Formicidae). *Journal of the Kansas Entomological Society,* 46: 526–536.

Akre, RD, Rettenmeyer, CW. 1966. Behavior of Staphylinidae associated with army ants (Formicidae: Ecitonini). *Journal of the Kansas Entomological Society,* 39: 745–782.

Akre, RD, Rettenmeyer, CW. 1968. Trail-following by guests of army ants (Hymenoptera: Formicidae:

Ecitonini). *Journal of the Kansas Entomological Society,* 41: 165–174.

Alexander, RD, Moore, TE, Woodruff, RE. 1963. The evolutionary differentiation of stridulatory signals in beetles (Insecta: Coleoptera). *Animal Behaviour,* 11: 111–115.

Allan, RA, Capon, RJ, Brown, WV, Elgar, MA. 2002. Mimicry of host cuticular hydrocarbons by salticid spider *Cosmophasis bitaeniata* that preys on larvae of tree ants *Oecophylla smaragdina. Journal of Chemical Ecology,* 28: 835–848.

Allan, RA, Elgar, MA. 2001. Exploitation of the green tree ant, *Oecophylla smaragdina,* by the salticid spider *Cosmophasis bitaeniata. Australian Journal of Zoology,* 49: 129–137.

Alpert, GD. 1994. A comparative study of the symbiotic relationships between beetles of the genus *Cremastocheilus* (Coleoptera: Scarabaeidae) and their host ants (Hymenoptera: Formicidae). *Sociobiology,* 25: 1–276.

Alpert, GD, Ritcher, P. 1975. Notes on the life cycle and myrmecophilous adaptations of *Cremastocheilus armatus* (Coleoptera: Scarabaeidae). *Psyche,* 82: 283–291.

Als, TD, Nash, DR, Boomsma, JJ. 2001. Adoption of parasitic *Maculinea alcon* caterpillars (Lepidoptera: Lycaenidae) by three *Myrmica* ant species. *Animal Behaviour,* 62: 99–106.

Als, TD, Nash, DR, Boomsma, JJ. 2002. Geographical variation in host-ant specificity of the parasitic butterfly *Maculinea alcon* in Denmark. *Ecological Entomology,* 27: 403–414.

Als, TD, Vila, R, Kandul, NP, Nash, DR, Yen, S-H, Hsu, Y-F, Mignault, AA, Boomsma, JJ, Pierce, NE. 2004. The evolution of alternative parasitic life histories in large blue butterflies. *Nature,* 432: 386–390.

Altson, A. 1932. On the feeding habits and breeding of *Ochromyia (Bengalia) depressa,* R.-D., and *O. peuhi. Proceedings of the Royal Entomological Society of London,* 7: 36–40.

Andersen, SB, Ferrari, M, Evans, HC, Elliot, SL, Boomsma, JJ, Hughes, DP. 2012. Disease dynamics in a specialized parasite of ant societies. *PLOS One,* 7: e36352.

Andersen, SB, Gerritsma, S, Yusah, KM, Mayntz, D, Hywel-Jones, NL, Billen, J, Boomsma, JJ, Hughes, DP. 2009. The life of a dead ant: The expression of an adaptive extended phenotype. *American Naturalist,* 174: 424–433.

Andrade, D, Nascimento, L, Abe, A. 2006. Habits hidden underground: A review on the reproduction of the Amphisbaenia with notes on four neotropical species. *Amphibia-Reptilia,* 27: 207–217.

Andries, M. 1912. Zur Systematik, Biologie und Entwicklung von *Microdon* Meigen. *Zeitschrift für Wissenschaftliche Zoologie,* 103: 300–361.

Antoniazzi, M, Jared, C, Pellegrini, C, Macha, N. 1993. Epidermal glands in Squamata: Morphology and histochemistry of the pre-cloacal glands in *Amphisbaena alba* (Amphisbaenia). *Zoomorphology,* 113: 199–203.

Araújo, JPM, Evans, HC, Kepler, R, Hughes, DP. 2018. Zombie-ant fungi across continents: 15 new species and new combinations within *Ophiocordyceps,* I: Myrmecophilous hirsutelloid species. *Studies in Mycology,* 90: 119–160.

Araújo, JPM, Hughes, DP. 2016. Diversity of entomopathogenic fungi: Which groups conquered the insect body? *Advances in Genetics,* 94: 1–39.

Atsatt, PR. 1981. Lycaenid butterflies and ants: Selection for enemy-free space. *American Naturalist,* 118: 638–654.

Audubon, J-J. 1831. *Ornithological Biography, or an Account of the Habits of the Birds of the United States of America.* Vol. 1. Edinburgh: Adam Black.

Autrum, H. 1936. Über Lautäußerungen und Schallwahrnehmung bei Arthropoden. *Zeitschrift für vergleichende Physiologie,* 23: 332–373.

Ayre, G. 1962. *Pseudometagea schwarzii* (Ashm.) (Eucharitidae: Hymenoptera), a parasite of *Lasius neoniger* Emery (Formicidae: Hymenoptera). *Canadian Journal of Zoology,* 40: 157–164.

Baer, B, Den Boer, SPA, Kronauer, D, Nash, DR, Boomsma, JJ. 2009. Fungus gardens of the leafcutter ant *Atta colombica* function as egg nurseries for the snake *Leptodeira annulata. Insectes Sociaux,* 56: 289–291.

Baer, B, Schmid-Hempel, P. 1999. Experimental variation in polyandry affects parasite loads and fitness in a bumble-bee. *Nature,* 397: 151–154.

Bagg, A. 1952. Anting not exclusively an avian trait. *American Society of Mammalogists,* 33: 243.

Bagnères, AG, Morgan, ED. 1991. The postpharyngeal glands and the cuticle of Formicidae contain the same characteristic hydrocarbons. *Experientia,* 47: 106–111.

Baker, AJ, Heraty, JM, Mottern, J, Zhang, J, Hines, HM, Lemmon, AR, Lemmon, EM. 2020. Inverse dispersal patterns in a group of ant parasitoids (Hymenoptera: Eucharitidae: Oraseminae) and their ant hosts. *Systematic Entomology,* 45: 1–19.

Barbero, F, Thomas, JA, Bonelli, S, Balletto, E, Schönrogge, K. 2009. Queen ants make distinctive sounds that are mimicked by a butterfly social parasite. *Science,* 323: 782–785.

Baroni Urbani, C, Buser, MW, Schillinger, E. 1988. Substrate vibration during recruitment in ant social organization. *Insectes Sociaux,* 35: 241–250.

Barr, B. 1995. Feeding behaviour and mouthpart structure of larvae of *Microdon eggeri* and *Microdon mutabilis* (Diptera, Syrphidae). *Dipterists Digest,* 2: 1–36.

Baruffaldi, L, Costa, FG, Rodríguez, A, González, A. 2010. Chemical communication in *Schizocosa malitiosa:* Evidence of a female contact sex pheromone and persistence in the field. *Journal of Chemical Ecology,* 36: 759–767.

Bascompte, J, Jordano, P. 2007. Plant-animal mutualistic networks: The architecture of biodiversity. *Annual Review of Ecology, Evolution, and Systematics,* 38: 567–593.

Batcheller, HJ. 2016. Interspecific information use by army-ant–following birds. *Ornithology,* 134: 247–255.

Baumann, P. 2005. Biology of bacteriocyte-associated endosymbionts of plant sap-sucking insects. *Annual Review of Microbiology,* 59: 155–189.

Baumgarten, H-T, Fiedler, K. 1998. Parasitoids of lycaenid butterfly caterpillars: Different patterns in resource use and their impact on the hosts' symbiosis with ants. *Zoologischer Anzeiger,* 236: 167–180.

Baylis, M, Pierce, N. 1992. Lack of compensation by final instar larvae of the myrmecophilous lycaenid butterfly, *Jalmenus evagoras,* for the loss of nutrients to ants. *Physiological Entomology,* 17: 107–114.

Bequaert, JC, Wheeler, WM. 1922. The predaceous enemies of ants. *Bulletin of the American Museum of Natural History,* 45: 271–331.

Berghoff, SM, Wurst, E, Ebermann, E, Sendova-Franks, AB, Rettenmeyer, CW, Franks, NR. 2009. Symbionts of societies that fission: Mites as guests or parasites of army ants. *Ecological Entomology,* 34: 684–695.

Bernardi, R, Cardani, C, Ghiringhelli, D, Selva, A, Baggini, A, Pavan, M. 1967. On the components of secretion of mandibular glands of the ant *Lasius (dendrolasius) fuliginosus. Tetrahedron Letters,* 8: 3893–3896.

Bernstein, RA. 1974. Seasonal food abundance and foraging activity in some desert ants. *American Naturalist,* 108: 490–498.

Beros, S, Jongepier, E, Hagemeier, F, Foitzik, S. 2015. The parasite's long arm: A tapeworm parasite induces behavioural changes in uninfected group members of its social host. *Proceedings of the Royal Society B: Biological Sciences,* 282: 20151473.

Bhattacharya, G. 1939. On the moulting and metamorphosis of *Myrmarachne plataleoides* Camb. *Transactions of the Bose Research Institute,* 12: 103–114.

Birer, C, Moreau, CS, Tysklind, N, Zinger, L, Duplais, C. 2020. Disentangling the assembly mechanisms of ant cuticular bacterial communities of two Amazonian ant species sharing a common arboreal nest. *Molecular Ecology,* 29: 1372–1385.

Blochmann, F. 1887. Über das Vorkommen bakterienähnlicher Gebilde in den Geweben und Eiern verschiedener Insekten. *Zentralblatt für Bakteriologie,* 11: 234–240.

Blüthgen, N, Fiedler, K. 2004. Preferences for sugars and amino acids and their conditionality in a diverse nectar-feeding ant community. *Journal of Animal Ecology,* 73: 155–166.

Blüthgen, N, Gebauer, G, Fiedler, K. 2003. Disentangling a rainforest food web using stable isotopes: Dietary diversity in a species-rich ant community. *Oecologia,* 137: 426–435.

Blüthgen, N, Menzel, F, Hovestadt, T, Fiala, B, Blüthgen, N. 2007. Specialization, constraints, and conflicting interests in mutualistic networks. *Current Biology,* 17: 341–346.

Blüthgen, N, Stork, NE, Fiedler, K. 2004. Bottom-up control and co-occurrence in complex communities: Honeydew and nectar determine a rainforest ant mosaic. *Oikos,* 106: 344–358.

Bohn, H, Nehring, V, Rodríguez, J, Klass, K-D. 2021. Revision of the genus *Attaphila* (Blattodea: Blaberoidea), myrmecophiles living in the mushroom gardens of leaf-cutting ants. *Arthropod Systematics & Phylogeny,* 79: 205.

Bolívar, I. 1905. Les blattes myrmécophiles. *Mitteilungen der Schweizerischen Entomologischen Gesellschaft,* 11: 134–141.

Bonaldo, AB. 2000. Taxonomia da subfamília Corinninae (Araneae, Corinnidae) nas regiões neotropical e neártica. *Iheringia, Série Zoologia,* 89: 3–198.

Bonaldo, AB, Brescovit, A. 2005. On new species of the Neotropical spider genus *Attacobius* Mello-Leitão, 1923 (Araneae, Corinnidae, Corinninae), with a cladistic analysis of the tribe Attacobiini. *Insect Systematics & Evolution, 36*: 35–56.

Bonavita-Cougourdan, A, Clement, J-L, Lange, C. 1993. Functional subcaste discrimination (foragers and brood-tenders) in the ant *Camponotus vagus* Scop.: Polymorphism of cuticular hydrocarbon patterns. *Journal of Chemical Ecology, 19*: 1461–1477.

Boorman, J. 1997. Book review: *The butterflies of Costa Rica and their natural history. Volume II: Riodinidae.* By Philip J. DeVries. *Bulletin of Entomological Research, 87*: 656–656.

Borges, RM, Ahmed, S, Prabhu, CV. 2007. Male ant-mimicking salticid spiders discriminate between retreat silks of sympatric females: Implications for pre-mating reproductive isolation. *Journal of Insect Behavior, 20*: 389–402.

Borgmeier, T. 1921. Zur Lebensweise von *Pseudacteon borgmeieri* Schmitz (in litt.) (Diptera: Phoridae). *Zeitschrift des deutschen Vereins für Wissenschaft und Kunst, Sao Paulo, 2*: 239–248.

Borgmeier, T. 1958. Neue Beitraege zur Kenntnis der neotropischen Phoriden (Diptera, Phoridae). *Studia Entomologica, 1*: 305–406.

Borowiec, ML. 2013. Two species of myrmecophilous Diapriidae (Hymenoptera) new to Poland. *Wiadomosci Entomologiczne, 32*: 42–48.

Borowiec, ML. 2016. Generic revision of the ant subfamily Dorylinae (Hymenoptera, Formicidae). *ZooKeys, 608*: 1–280.

Bossert, WH, Wilson, EO. 1963. The analysis of olfactory communication among animals. *Journal of Theoretical Biology, 5*: 443–469.

Bousquet, Y, Laplante, S. 2006. *Coleoptera Histeridae.* Vol. 24. Ottawa: NRC Research Press.

Boyle, JH, Kaliszewska, ZA, Espeland, M, Suderman, TR, Fleming, J, Heath, A, Pierce, NE. 2015. Phylogeny of the Aphnaeinae: Myrmecophilous African butterflies with carnivorous and herbivorous life histories. *Systematic Entomology, 40*: 169–182.

Brackbill, H. 1948. Anting by four species of birds. *Ornithology, 65*: 66–77.

Bragança, MAL, Arruda, FV, Souza, LRR, Martins, HC, Della Lucia, TMC. 2016. Phorid flies parasitizing leaf-cutting ants: Their occurrence, parasitism rates, biology and the first account of multiparasitism. *Sociobiology, 63*: 1015–1021.

Bragança, MAL, Tonhasca, A Jr, Della Lucia, TM. 1998. Reduction in the foraging activity of the leaf-cutting ant *Atta sexdens* caused by the phorid *Neodohrniphora* sp. *Entomologia Experimentalis et Applicata, 89*: 305–311.

Brake, I. 1999. *Prosaetomilichia* de Meijere: A junior subjective synonym of *Milichia* Meigen, with a phylogenetic review of the myrmecophila species-group (Diptera, Milichiidae). *Tijdschrift voor Entomologie, 142*: 31–36.

Brandão, CRF, Vanzolini, PE, Vanzolini, P. 1985. Notes on incubatory inquilinism between Squamata (Reptilia) and the neotropical fungus-growing ant genus *Acromyrmex* (Hymenoptera: Formicidae). *Papéis Avulsos de Zoologia, 36*: 31–36.

Brandstaetter, AS, Endler, A, Kleineidam, CJ. 2008. Nestmate recognition in ants is possible without tactile interaction. *Naturwissenschaften, 95*: 601–608.

Brandstaetter, AS, Kleineidam, CJ. 2011. Distributed representation of social odors indicates parallel processing in the antennal lobe of ants. *Journal of Neurophysiology, 106*: 2437–2449.

Brandstaetter, AS, Rössler, W, Kleineidam, CJ. 2010. Dummies versus air puffs: Efficient stimulus delivery for low-volatile odors. *Chemical Senses, 35*: 323–333.

Brandstaetter, AS, Rössler, W, Kleineidam, CJ. 2011. Friends and foes from an ant brain's point of view—neuronal correlates of colony odors in a social insect. *PLOS One, 6*: e21383–21392.

Brandt, M, Mahsberg, D. 2002. Bugs with a backpack: The function of nymphal camouflage in the West African assassin bugs *Paredocla* and *Acanthaspis* spp. *Animal Behaviour, 63*: 277–284.

Brian, MV. 1975. Larval recognition by workers of the ant *Myrmica*. *Animal Behaviour, 23*: 745–756.

Brown, BV. 2012. Small size no protection for acrobat ants: World's smallest fly is a parasitic phorid (Diptera: Phoridae). *Annals of the Entomological Society of America, 105*: 550–554.

Brown, BV, Feener, DH. 1998. Parasitic phorid flies (Diptera: Phoridae) associated with army ants (Hymenoptera: Formicidae: Ecitoninae, Dorylinae) and their conservation biology. *Biotropica, 30*: 482–487.

Brown, BV, Feener, DH Jr. 1991. Behavior and host location cues of *Apocephalus paraponerae* (Diptera: Phoridae), a parasitoid of the giant tropical ant, *Paraponera clavata* (Hymenoptera: Formicidae). *Biotropica, 23:* 182–187.

Brown, BV, Hash, JM, Hartop, EA, Porras, W, de Souza Amorim, D. 2017. Baby killers: Documentation and evolution of scuttle fly (Diptera: Phoridae) parasitism of ant (Hymenoptera: Formicidae) brood. *Biodiversity Data Journal, 5:* e11277.

Brown, CG, Funk, DJ. 2005. Aspects of the natural history of *Neochlamisus* (Coleoptera: Chrysomelidae): Fecal case–associated life history and behavior, with a method for studying insect constructions. *Annals of the Entomological Society of America, 98:* 711–725.

Brückner, A, Klompen, H, Bruce, AI, Hashim, R, von Beeren, C. 2018. Infection of army ant pupae by two new parasitoid mites (Mesostigmata: Uropodina). *PeerJ, 5:* e3870.

Brumfield, RT, Tello, JG, Cheviron, ZA, Carling, MD, Crochet, N, Rosenberg, KV. 2007. Phylogenetic conservatism and antiquity of a tropical specialization: Army-ant-following in the typical antbirds (Thamnophilidae). *Molecular Phylogenetics and Evolution, 45:* 1–13.

Bruner, G, Fernández-Marín, H, Touchon, JC, Wcislo, WT. 2012. Eggs of the blind snake, *Liotyphlops albirostris* are incubated in a nest of the lower fungus-growing ant, *Apterostigma* cf. *goniodes. Psyche,* 2012: 532314.

Buczkowski, G, Kumar, R, Suib, SL, Silverman, J. 2005. Diet-related modification of cuticular hydrocarbon profiles of the Argentine ant, *Linepithema humile,* diminishes intercolony aggression. *Journal of Chemical Ecology, 31:* 829–843.

Burke, VJ, Nagle, RD, Osentoski, M, Congdon, JD. 1993. Common snapping turtles associated with ant mounds. *Journal of Herpetology, 27:* 114–115.

Buschinger, A. 1973. Ameisen des Tribus Leptothoracini (Hym., Formicidae) als Zwischenwirte von Cestoden. *Zoologischer Anzeiger, 191:* 369–380.

Buschinger, A. 2009. Social parasitism among ants: A review (Hymenoptera: Formicidae). *Myrmecological News, 12:* 219–235.

Buschinger, A, Maschwitz, U. 1984. Defensive behavior and defensive mechanisms in ants. In *Defensive Mechanisms in Social Insects,* ed. RH Herman, 95–150. New York: Praeger.

Camacho, C, Potti, J. 2018. Non-foraging tool use in European honey-buzzards: An experimental test. *PLOS One,* 13: e0206843.

Camargo, RdS, Forti, LC, de Matos, CAO, Brescovit, AD. 2015. Phoretic behaviour of *Attacobius attarum* (Roewer, 1935) (Araneae: Corinnidae: Corinninae) dispersion not associated with predation? *Journal of Natural History,* 49: 1653–1658.

Cammaerts, MC, Evershed, RP, Morgan, ED. 1982. Mandibular gland secretions of workers of *Myrmica rugulosa* and M. schencki: Comparison with four other *Myrmica* species. *Physiological Entomology,* 7: 119–125.

Cammaerts, R. 1974. Le système glandulaire tégumentaire du coléoptère myrmécophile Claviger testaceus Preyssler, 1790 (Pselaphidae) / The integumentary glandular system of the myrmecophilous beetle *Claviger testaceus* Preyssler, 1790 (Pselaphidae). *Zeitschrift für Morphologie der Tiere,* 77: 187–219.

Cammaerts, R. 1991a. Interactions comportementales entre la Fourmi *Lasius flavus* (Formicidae) et de Coléoptère myrmécophile *Claviger testaceus* (Pselaphidae), I: Ethogramme et modalités des interactions avec les ouvrières. *Bulletin et annales de la Société royale belge d'entomologie,* 127: 155–190.

Cammaerts, R. 1991b. Interactions comportementales entre la fourmi Lasius flavus (Formicidae) et le coléoptère myrmécophile *Claviger testaceus* (Pselaphidae), II: Fréquence, durée et succession des comportements des ouvrières. *Bulletin et annales de la Société royale belge d'entomologie,* 127: 271–307.

Cammaerts, R. 1992. Stimuli inducing the regurgitation of the workers of *Lasius flavus* (Formicidae) upon the myrmecophilous beetle *Claviger testaceus* (Pselaphidae). *Behavioural Processes,* 28: 81–96.

Cammaerts, R. 1995. Regurgitation behaviour of the *Lasius flavus* worker (Formicidae) towards the myrmecophilous beetle *Claviger testaceus* (Pselaphidae) and other recipients. *Behavioural Processes,* 34: 241–264.

Cammaerts, R. 1996. Factors affecting the regurgitation behaviour of the ant *Lasius flavus* (Formicidae) to the guest beetle *Claviger testaceus* (Pselaphidae). *Behavioural Processes,* 38: 297–312.

Cammaerts, R. 1999a. Transport location patterns of the guest beetle *Claviger testaceus* (Pselaphidae) and other objects moved by workers of the ant, *Lasius flavus* (Formicidae). *Sociobiology, 34*: 433–475.

Cammaerts, R. 1999b. A quantitative comparison of the behavioral reactions of *Lasius flavus* ant workers (Formicidae) toward the guest beetle *Claviger testaceus* (Pselaphidae), ant larvae, intruder insects and cadavers. *Sociobiology, 33*: 145–170.

Cammaerts, R, Detrain, C, Cammaerts, M-C. 1990. Host trail following by the myrmecophilous beetle *Edaphopaussus favieri* (Fairmaire) (Carabidae Paussinae). *Insectes Sociaux, 37*: 200–211.

Campbell, DL, Brower, AV, Pierce, NE. 2000. Molecular evolution of the wingless gene and its implications for the phylogenetic placement of the butterfly family Riodinidae (Lepidoptera: Papilionoidea). *Molecular Biology and Evolution, 17*: 684–696.

Campbell, DL, Pierce, NE. 2003. Phylogenetic relationships of the Riodinidae: Implications for the evolution of ant association. In *Butterflies as Model Systems,* ed. C Boggs, P Ehrlich, W Watt, 395–408. Chicago, IL: University of Chicago Press.

Campbell, KU, Klompen, H, Crist, TO. 2013. The diversity and host specificity of mites associated with ants: The roles of ecological and life history traits of ant hosts. *Insectes Sociaux, 60*: 31–41.

Campos, VA, Dáttilo, W, Oda, FH, Piroseli, LE, Dartora, A. 2014. Detección y uso de senderos de la hormiga cortadora de hojas *Atta laevigata* (Hymenoptera: Formicidae) por *Amphisbaena alba* (Reptilia: Squamata). *Acta Zoológica Mexicana, 30*: 403–407.

Cannon, PF, Hywel-Jones, NL, Maczey, N, Norbu, L, Samdup, T, Lhendup, P. 2009. Steps towards sustainable harvest of *Ophiocordyceps sinensis* in Bhutan. *Biodiversity and Conservation, 18*: 2263–2281.

Carey, B, Visscher, K, Heraty, J. 2012. Nectary use for gaining access to an ant host by the parasitoid *Orasema simulatrix* (Hymenoptera, Eucharitidae). *Journal of Hymenoptera Research, 27*: 47–65.

Carico, JE. 1978. Predatory behavior in *Euryopis funebris* (Hentz) (Araneae: Theridiidae) and the evolutionary significance of web reduction. *Symposia of the Zoological Society London, 42*: 51–58.

Carlin, NF, Hölldobler, B. 1986. The kin recognition system of carpenter ants (*Camponotus* spp.), I: Hierarchical cues in small colonies. *Behavioral Ecology and Sociobiology, 19*: 123–134.

Carlin, NF, Hölldobler, B. 1987. The kin recognition system of carpenter ants (*Camponotus* spp.), II: Larger colonies. *Behavioral Ecology and Sociobiology, 20*: 209–217.

Carney, WP. 1969. Behavioral and morphological changes in carpenter ants harboring dicrocoeliid metacercariae. *American Midland Naturalist, 82*: 605–611.

Casacci, LP, Bonelli, S, Balletto, E, Barbero, F. 2019a. Multimodal signaling in myrmecophilous butterflies. *Frontiers in Ecology and Evolution, 7*: 454.

Casacci, LP, Schönrogge, K, Thomas, JA, Balletto, E, Bonelli, S, Barbero, F. 2019b. Host specificity pattern and chemical deception in a social parasite of ants. *Scientific Reports: 9*: 1619.

Cazier, MA, Mortenson, MA. 1965. The behavior and habits of the myrmecophilous scarab *Cremastocheilus. Journal of the Kansas Entomological Society, 38*: 19–44.

Ceccarelli, FS. 2010. Ant-mimicking spider, *Myrmarachne* species (Araneae: Salticidae), distinguishes its model, the green ant, *Oecophylla smaragdina,* from a sympatric Batesian *O. smaragdina* mimic, *Riptortus serripes* (Hemiptera: Alydidae). *Australian Journal of Zoology, 57*: 305–309.

Chapin, KJ, Hebets, EA. 2016. The behavioral ecology of amblypygids. *Journal of Arachnology, 44*: 1–14.

Charpentier, Td. 1825. *Horae Entomologicae, Adjectis Tabulis Novem Coloratis.* Wratislaviae: Apud A. Gosohorsky.

Chaves-Campos, J. 2003. Localization of army-ant swarms by ant-following birds on the Caribbean slope of Costa Rica: Following the vocalization of antbirds to find the swarms. *Ornitologia Neotropical, 14*: 289–294.

Chen, L, Fadamiro, HY. 2007. Behavioral and electroantennogram responses of phorid fly *Pseudacteon tricuspis* (Diptera: Phoridae) to red imported fire ant *Solenopsis invicta* odor and trail pheromone. *Journal of Insect Behavior, 20*: 267–287.

Chen, L, Fadamiro, HY. 2018. *Pseudacteon* phorid flies: Host specificity and impacts on *Solenopsis* fire ants. *Annual Review of Entomology, 63*: 47–67.

Chen, L, Porter, SD. 2020. Biology of Pseudacteon decapitating flies (Diptera: Phoridae) that parasitize ants of the *Solenopsis saevissima* complex (Hymenoptera: Formicidae) in South America. *Insects, 11*: 107.

Chen, Z, Corlett, RT, Jiao, X, Liu, S-J, Charles-Dominique, T, Zhang, S, Li, H, Lai, R, Long, C, Quan, R-C. 2018. Prolonged milk provisioning in a jumping spider. *Science, 362*: 1052–1055.

Chesser, RT. 1995. Comparative diets of obligate ant-following birds at a site in Northern Bolivia. *Biotropica*, 27: 382–390.

Chisholm, A. 1959. The history of anting. *Emu-Austral Ornithology*, 59: 101–130.

Choe, D-H, Villafuerte, DB, Tsutsui, ND. 2012. Trail pheromone of the Argentine ant, *Linepithema humile* (Mayr) (Hymenoptera: Formicidae). *PLOS One*, 7: e45016.

Choe, J, Perlman, D. 1997. Social conflict and cooperation among founding queens in ants (Hymenoptera: Formicidae). In *The Evolution of Social Behavior in Insects and Arachnids*, ed. JC Choe, BJ Crespi. Cambridge, UK: Cambridge University Press.

Cigliano, MM, Braun, H, Eades, DC, Otte, D. 2020. Orthoptera species file. http://Orthoptera.SpeciesFile.org, version 5.0 / 5.0.

Claassens, A, Dickson, C. 1977. A study of the myrmecophilous behaviour of the immature stages of *Aloeides thyra* (L.) (Lepidoptera: Lycaenidae) with special reference to the function of the retractile tubercules and with additional notes on the general biology of the species. *Entomologist's Record and Journal of Variation*, 89: 253–258.

Clausen, CP. 1941. The habits of the Eucharidae. *Psyche*, 48: 57–69.

Clayton, DH, Vernon, JG. 1993. Common grackle anting with lime fruit and its effect on ectoparasites. *The Auk*, 110: 951–952.

Clyne, D. 2011. Secrets of the predatory butterfly *Liphyra brassolis* exposed. *Metamorphosis Australia*, 62.

Corn, M. 1980. Polymorphism and polyethism in the neotropical ant *Cephalotes atratus* (L.). *Insectes Sociaux*, 27: 29–42.

Cottrell, C. 1984. Aphytophagy in butterflies: Its relationship to myrmecophily. *Zoological Journal of the Linnean Society*, 80: 1–57.

Couvreur, J. 1990. Le comportement de "présentation d'un leurre" chez *Zodarion rubidum* (Zodariidae). *Bulletin de la Société européenne d'Arachnologie hors serie*, 1: 75–79.

Covas, R. 2012. Evolution of reproductive life histories in island birds worldwide. *Proceedings of the Royal Society B: Biological Sciences*, 279: 1531–1537.

Crozier, RH, Dix, MW. 1979. Analysis of two genetic models for the innate components of colony odor in social Hymenoptera. *Behavioral Ecology and Sociobiology*, 4: 217–224.

Csősz, S. 2012. Nematode infection as significant source of unjustified taxonomic descriptions in ants (Hymenoptera: Formicidae). *Myrmecological News*, 17: 27–31.

Cushing, PE. 1995a. Description of the spider *Masoncus pogonophilus* (Araneae, Linyphiidae), a harvester ant myrmecophile. *Journal of Arachnology*, 23: 55–59.

Cushing, PE. 1995b. Natural history of the myrmecophilic spider, *Masoncus pogonophilus* Cushing, and its host ant, *Pogonomyrmex badius* (Latreille). PhD dissertation, University of Florida.

Cushing, PE. 1997. Myrmecomorphy and myrmecophily in spiders: A review. *Florida Entomologist*, 80: 165–193.

Cushing, PE. 1998. Population structure of the ant nest symbiont *Masoncus pogonophilus* (Araneae: Linyphiidae). *Annals of the Entomological Society of America*, 91: 626–631.

Cushing, PE. 2012. Spider-ant associations: An updated review of myrmecomorphy, myrmecophily, and myrmecophagy in spiders. *Psyche*, 2012: 151989.

Daniels, H, Gottsberger, G, Fiedler, K. 2005. Nutrient composition of larval nectar secretions from three species of myrmecophilous butterflies. *Journal of Chemical Ecology*, 31: 2805–2821.

Darling, DC. 1992. The life history and larval morphology of *Aperilampus* (Hymenoptera: Chalcidoidea: Philomidinae), with a discussion of the phylogenetic affinities of the Philomidinae. *Systematic Entomology*, 17: 331–339.

Darling, DC. 1999. Life history and immature stages of *Steffanolampus salicetum* (Hymenoptera: Chalcidoidea: Perilampidae). *Proceedings of the Entomological Society of Ontario*, 130: 3–14.

Darling, DC. 2009. A new species of *Smicromorpha* (Hymenoptera, Chalcididae) from Vietnam, with notes on the host association of the genus. *ZooKeys*, 20: 155–163.

Darling, DC, Miller, TD. 1991. Life history and larval morphology of *Chrysolampus* (Hymenoptera: Chalcidoidea: Chrysolampinae) in western North America. *Canadian Journal of Zoology*, 69: 2168–2177.

Dasch, GA, Weiss, E, Chang, KP. 1984. Endosymbiosis of insects. In *Bergey's Manual of Systematic Bacteriology*, Vol. 1, ed. JG Holt, NR Krieg. Baltimore, MD: Williams & Wilkins.

de Armas, LF, Seiter, M. 2013. *Phrynus gervaisii* (Pocock, 1894) is a junior synonym of *Phrynus barbadensis*

(Pocock, 1893) (Amblypygi: Phrynidae). *Revista Ibérica de Aracnología, 23:* 128–132.

de Bekker, C, Merrow, M, Hughes, DP. 2014. From behavior to mechanisms: An integrative approach to the manipulation by a parasitic fungus (*Ophiocordyceps unilateralis* s.l.) of its host ants (*Camponotus* spp.). *Integrative and Comparative Biology, 54:* 166–176.

de Bekker, C, Ohm, RA, Evans, HC, Brachmann, A, Hughes, DP. 2017. Ant-infecting *Ophiocordyceps* genomes reveal a high diversity of potential behavioral manipulation genes and a possible major role for enterotoxins. *Scientific Reports, 7:* 12508.

de Bekker, C, Will, I, Das, B, Adams, RMM. 2018. The ants (Hymenoptera: Formicidae) and their parasites: Effects of parasitic manipulations and host responses on ant behavioral ecology. *Myrmecological News, 28:* 1–24.

Deeleman-Reinhold, CL. 1992. A new spider genus from Thailand with a unique ant-mimicking device, with description of some other castianeirine spiders (Araneae: Corinnidae: Castianeirinae). *Natural History Bulletin of the Siam Society, 40:* 167–184.

Deeleman-Reinhold, CL. 2001. *Forest Spiders of South East Asia: With a Revision of the Sac and Ground Spiders (Araneae: Clubionidae, Corinnidae, Liocranidae, Gnaphosidae, Prodidomidae, and Trochanterriidae).* Leiden: Brill.

Degnan, PH, Lazarus, AB, Brock, CD, Wernegreen, JJ. 2004. Host-symbiont stability and fast evolutionary rates in an ant-bacterium association: Cospeciation of *Camponotus* species and their endosymbionts, *Candidatus* Blochmannia. *Systematic Biology, 53:* 95–110.

Degnan, PH, Lazarus, AB, Wernegreen, JJ. 2005. Genome sequence of *Blochmannia pennsylvanicus* indicates parallel evolutionary trends among bacterial mutualists of insects. *Genome research, 15:* 1023–1033.

Degueldre, F, Mardulyn, P, Kuhn, A, Pinel, A, Karaman, C, Lebas, C, Schifani, E, Bračko, G, Wagner, HC, Kiran, K, Borowiec, L, Passera, L, Abril, S, Espadaler, X, Aron, S. 2021. Evolutionary history of inquiline social parasitism in *Plagiolepis* ants. *Molecular Phylogenetics and Evolution, 155:* 107016.

Dejean, A, Beugnon, G. 1996. Host-ant trail following by myrmecophilous larvae of Liphyrinae (Lepidoptera, Lycaenidae). *Oecologia, 106:* 57–62.

Dejean, A, Orivel, J, Azémar, F, Hérault, B, Corbara, B. 2016. A cuckoo-like parasitic moth leads African weaver ant colonies to their ruin. *Scientific Reports, 6:* 1–9.

Dekoninck, W, Lock, K, Janssens, F. 2007. Acceptance of two native myrmecophilous species, *Platyarthrus hoffmannseggii* (Isopoda: Oniscidea) and *Cyphoderus albinus* (Collembola: Cyphoderidae) by the introduced invasive garden ant *Lasius neglectus* (Hymenoptera: Formicidae) in Belgium. *European Journal of Entomology, 104:* 159.

de Lima Barros, A, López-Lozano, JL, Lima, AP. 2016. The frog *Lithodytes lineatus* (Anura: Leptodactylidae) uses chemical recognition to live in colonies of leaf-cutting ants of the genus *Atta* (Hymenoptera: Formicidae). *Behavioral Ecology and Sociobiology, 70:* 2195–2201.

Dettner, K. 1993. Defensive secretions and exocrine glands in free-living staphylinid beetles: Their bearing on phylogeny (Coleoptera: Staphylinidae). *Biochemical Systematics and Ecology, 21:* 143–162.

Dettner, K, Liepert, C. 1994. Chemical mimicry and camouflage. *Annual Review of Entomology, 39:* 129–154.

d'Ettorre, P, Mondy, N, Lenoir, A, Errard, C. 2002. Blending in with the crowd: Social parasites integrate into their host colonies using a flexible chemical signature. *Proceedings of the Royal Society of London. Series B: Biological Sciences, 269:* 1911–1918.

DeVries, PJ. 1984. Of crazy-ants and Curetinae: Are *Curetis* butterflies tended by ants? *Zoological Journal of the Linnean Society, 80:* 59–66.

DeVries, PJ. 1988. The larval ant-organs of *Thisbe irenea* (Lepidoptera: Riodinidae) and their effects upon attending ants. *Zoological Journal of the Linnean Society, 94:* 379–393.

DeVries, PJ. 1990. Enhancement of symbioses between butterfly caterpillars and ants by vibrational communication. *Science, 248:* 1104–1106.

DeVries, PJ. 1991a. Evolutionary and ecological patterns in myrmecophilous riodinid butterflies. In *Ant-Plant Interactions,* ed. CR Huxley, DF Cutler, 143–156. Oxford: Oxford University Press.

DeVries, PJ. 1991b. Mutualism between *Thisbe irenea* butterflies and ants, and the role of ant ecology in the evolution of larval-ant associations. *Biological Journal of the Linnean Society, 43:* 179–195.

DeVries, PJ. 1991c. Call production by myrmecophilous riodinid and lycaenid butterfly caterpillars (Lepidoptera): Morphological, acoustical, functional, and evolutionary patterns. *American Museum Novitates, 3025:* 1–23.

DeVries, PJ. 1991d. Detecting and recording the calls produced by butterfly caterpillars and ants. *Journal of Research on the Lepidoptera,* 28: 258–262.

DeVries, PJ. 1992. Singing caterpillars, ants and symbiosis. *Scientific American,* 267: 76–83.

DeVries, PJ. 1997. *The Butterflies of Costa Rica and Their Natural History, Volume II: Riodinidae.* Princeton, NJ: Princeton University Press.

DeVries, PJ, Baker, I. 1989. Butterfly exploitation of an ant-plant mutualism: Adding insult to herbivory. *Journal of the New York Entomological Society,* 97: 332–340.

DeVries, PJ, Cocroft, RB, Thomas, J. 1993. Comparison of acoustical signals in *Maculinea* butterfly caterpillars and their obligate host *Myrmica* ants. *Biological Journal of the Linnean Society,* 49: 229–238.

DeVries, PJ, Harvey, DJ, Kitching, IJ. 1986. The ant associated epidermal organs on the larva of the lycaenid butterfly *Curetis regula* Evans. *Journal of Natural History,* 20: 621–633.

Di Giulio, A. 2008. Fine morphology of the myrmecophilous larva of *Paussus kannegieteri* (Coleoptera: Carabidae: Paussinae: Paussini). *Zootaxa,* 1741: 37–50.

Di Giulio, A, Fattorini, S, Kaupp, A, Vigna Taglianti, A, Nagel, P. 2003. Review of competing hypotheses of phylogenetic relationships of Paussinae (Coleoptera: Carabidae) based on larval characters. *Systematic Entomology,* 28: 508–537.

Di Giulio, A, Fattorini, S, Moore, W, Robertson, J, Maurizi, E. 2014. Form, function and evolutionary significance of stridulatory organs in ant nest beetles (Coleoptera: Carabidae: Paussini). *European Journal of Entomology,* 111: 692.

Di Giulio, A, Maruyama, M, Komatsu, T, Sakchoowong, W. 2017. Larval juice anyone? The unusual behaviour and morphology of an ant nest beetle larva (Coleoptera: Carabidae: Paussini) from Thailand. *Raffles Bulletin of Zoology,* 65: 49–59.

Di Giulio, A, Maurizi, E, Barbero, F, Sala, M, Fattorini, S, Balletto, E, Bonelli, S. 2015. The pied piper: A parasitic beetle's melodies modulate ant behaviours. *PLOS One,* 10: e0130541.

Di Giulio, A, Maurizi, E, Hlavac, P, Moore, W. 2011. The long-awaited first instar larva of *Paussus favieri* (Coleoptera: Carabidae: Paussini). *European Journal of Entomology,* 108: 127.

Di Giulio, A, Stacconi, MVR, Romani, R. 2009. Fine structure of the antennal glands of the ant nest beetle *Paussus favieri* (Coleoptera, Carabidae, Paussini). *Arthropod Structure and Development,* 38: 293–302.

Di Giulio, A, Taglianti, AV. 2001. Biological observations on *Pachyteles* larvae (Coleoptera Carabidae Paussinae). *Tropical Zoology,* 14: 157–173.

Dinter, K, Paarmann, W, Peschke, K, Arndt, E. 2002. Ecological, behavioural and chemical adaptations to ant predation in species of *Thermophilum* and *Graphipterus* (Coleoptera: Carabidae) in the Sahara Desert. *Journal of Arid Environments,* 50: 267–286.

Di Salvo, M, Calcagnile, M, Talà, A, Tredici, SM, Maffei, ME, Schönrogge, K, Barbero, F, Alifano, P. 2019. The microbiome of the *Maculinea-Myrmica* host-parasite interaction. *Scientific Reports,* 9: 8048.

Disney, RHL. 1994. *Scuttle flies: The Phoridae.* London: Chapman & Hall.

Disney, RHL. 1996. A new genus of scuttle fly (Diptera; Phoridae) whose legless, wingless, females mimic ant larvae (Hymenoptera; Formicidae). *Sociobiology* 27: 95–118.

Disney, RHL. 2000. Revision of European *Pseudacteon* Coquillett (Diptera, Phoridae). *Bonner Zoologische Beiträage,* 49: 79–92.

Disney, RHL, Schroth, M. 1989. Observations on *Megaselia persecutrix* Schmitz (Diptera: Phoridae) and the significance of ommatidial size-differentiation. *Entomologist's Monthly Magazine,* 125: 169–174.

Disney, RHL, Weissflog, A, Maschwitz, U. 1998. A second species of legless scuttle fly (Diptera: Phoridae) associated with ants (Hymenoptera: Formicidae). *Journal of Zoology,* 246: 269–274.

Dobrzańska, J. 1966. The control of the territory by *Lasius fuliginosus* Latr. *Acta Biologiae Experimentalis Sinica,* 26: 193–213.

Dodd, F. 1902. Contribution to the life-history of *Liphyra brassolis. Westwood Entomologist,* 35: 153–188.

Donisthorpe, HSJK. 1902. The life history of *Clytra quadripunctata. Transactions of the Entomological Society of London,* 50: 11–23.

Donisthorpe, HSJK. 1915. *British Ants: Their Life-History and Classification.* Plymouth: William Brendon and Son.

Donisthorpe, HSJK. 1927. *The Guests of British Ants.* London: George Routledge and Sons.

Donisthorpe, HSJK, Wilkinson, DS. 1930. Notes on the genus *Paxylomma* (Hym. Brac.), with the description of a new species taken in Britain. *Transactions of the Royal Entomological Society of London,* 78: 87–93.

Downes, J. 1958. The feeding habits of biting flies and their significance in classification. *Annual Review of Entomology,* 3: 249–266.

Downes, MF, Skevington, JH, Thompson, F. 2017. A new ant inquiline flower fly (Diptera: Syrphidae: Pipizinae) from Australia. *Australian Entomologist,* 44: 29–38.

Downey, JC, Allyn, AC. 1973. Butterfly ultrastructure. 1. Sound production and associated abdominal structures in pupae of Lycaenidae and Riodinidae. *Bulletin of the Allyn Museum,* 14: 1–47.

Downey, JC, Allyn, AC. 1978. Sounds produced in pupae of Lycaenidae. *Bulletin of the Allyn Museum,* 48: 1–14.

Downey, JC, Allyn, AC. 1979. Morphology and biology of the immature stages of *Leptotes cassius theonus* (Lucas) (Lepid: Lycaenidae). *Bulletin of the Allyn Museum,* 55: 1–27.

Drijfhout, F, Kather, R, Martin, SJ. 2009. The role of cuticular hydrocarbons in insects. In *Behavioral and Chemical Ecology,* ed. W Zhang, H Liu, 24. Hauppauge, NY: Nova Science Publishers.

Duffield, R. 1981. Biology of *Microdon fuscipennis* (Diptera: Syrphidae) with interpretations of the reproductive strategies of *Microdon* species found North of Mexico. *Proceedings of the Entomological Society of Washington,* 83: 716–724.

Dumpert, K. 1972. Alarmstoffrezeptoren auf der Antenne von *Lasius fuliginosus* (Latr.) (Hymenoptera, Formicidae). *Zeitschrift für vergleichende Physiologie,* 76: 403–425.

Dupont, ST, Zemeitat, DS, Lohman, DJ, Pierce, NE. 2016. The setae of parasitic *Liphyra brassolis* butterfly larvae form a flexible armour for resisting attack by their ant hosts (Lycaenidae: Lepidoptera). *Biological Journal of the Linnean Society,* 117: 607–619.

Durán, J-MG, van Achterberg, C. 2011. Oviposition behaviour of four ant parasitoids (Hymenoptera, Braconidae, Euphorinae, Neoneurini and Ichneumonidae, Hybrizontinae), with the description of three new European species. *ZooKeys,* 125: 59–106.

Dziekańska, I, Nowicki, P, Pirożnikow, E, Sielezniew, M. 2020. A unique population in a unique area: The alcon blue butterfly and its specific parasitoid in the Białowieża Forest. *Insects,* 11: 687.

Eastwood, R, Kongnoo, P, Reinkaw, M. 2010. Collecting and eating *Liphyra brassolis* (Lepidoptera: Lycaenidae) in southern Thailand. *Journal of Research on the Lepidoptera,* 43: 19–22.

Ebermann, E, Moser, JC. 2008. Mites (Acari: Scutacaridae) associated with the red imported fire ant, *Solenopsis invicta* Buren (Hymenoptera: Formicidae), from Louisiana and Tennessee, USA. *International Journal of Acarology,* 34: 55–69.

Edmunds, M. 1978. On the association between *Myrmarachne* spp. (Salticidae) and ants. *Bulletin of the British Arachnological Society,* 4: 149–160.

Ehrlich, PR, Dobkin, DS, Wheye, D. 1986. The adaptive significance of anting. *Auk,* 103: 835.

Eibl-Eibesfeldt, E. 1967. Das Parasitenabwehren der Minima-Arbeiterinnen der Blattschneider-Ameise (*Atta cephalotes*). *Zeitschrift für Tierpsychologie,* 24: 278–281.

Eidmann, H. 1937. Die Gäste und Gastverhältnisse der Battschneiderameise *Atta sexdens* L. *Zeitschrift für Morphologie und Ökologie der Tiere,* 32: 391–462.

Eisner, T. 2005. *For Love of Insects.* Cambridge, MA: Harvard University Press.

Eisner, T, Aneshansley, D. 2008. "Anting" in blue jays: Evidence in support of a food-preparatory function. *Chemoecology,* 18: 197–203.

Eisner, T, Eisner, M. 2000. Defensive use of a fecal thatch by a beetle larva (*Hemisphaerota cyanea*). *Proceedings of the National Academy of Sciences,* 97: 2632–2636.

Eisner, T, Eisner, M, Aneshansley, D. 2005. Pre-ingestive treatment of bombardier beetles by jays: Food preparation by "anting" and "sand-wiping." *Chemoecology,* 15: 227–233.

Eisner, T, Jones, TH, Aneshansley, DJ, Tschinkel, WR, Silberglied, RE, Meinwald, J. 1977. Chemistry of defensive secretions of bombardier beetles (Brachinini, Metriini, Ozaenini, Paussini). *Journal of Insect Physiology,* 23: 1383–1386.

Eisner, T, van Tassell, E, Carrel, JE. 1967. Defensive use of a "fecal shield" by a beetle larva. *Science,* 158: 1471–1473.

Elgar, MA, Allan, RA. 2004. Predatory spider mimics acquire colony-specific cuticular hydrocarbons from their ant model prey. *Naturwissenschaften,* 91: 143–147.

Elgar, MA, Allan, RA. 2006. Chemical mimicry of the ant *Oecophylla smaragdina* by the myrmecophilous spider

Cosmophasis bitaeniata: Is it colony-specific? *Journal of Ethology,* 24: 239–246.

Elgar, MA, Nash, DR, Pierce, NE. 2016. Eavesdropping on cooperative communication within an ant-butterfly mutualism. *The Science of Nature,* 103: 1–8.

Elizalde, L, Folgarait, PJ, Muscedere, M. 2012. Behavioral strategies of phorid parasitoids and responses of their hosts, the leaf-cutting ants. *Journal of Insect Science,* 12: 135.

Ellison, AM, Gotelli, NJ. 2021. Ants (Hymenoptera: Formicidae) and humans: from inspiration and metaphor to 21st-century symbiont. *Myrmecological News,* 31: 225–240.

Elmes, G, Akino, T, Thomas, J, Clarke, R, Knapp, J. 2002. Interspecific differences in cuticular hydrocarbon profiles of *Myrmica* ants are sufficiently consistent to explain host specificity by *Maculinea* (large blue) butterflies. *Oecologia,* 130: 525–535.

Elmes, G, Barr, B, Thomas, J, Clarke, R. 1999. Extreme host specificity by *Microdon mutabilis* (Diptera: Syrphiae), a social parasite of ants. *Proceedings of the Royal Society of London. Series B: Biological Sciences,* 266: 447–453.

Elmes, G, Wardlaw, J, Schönrogge, K, Thomas, J, Clarke, R. 2004. Food stress causes differential survival of socially parasitic caterpillars of *Maculinea rebeli* integrated in colonies of host and non-host Myrmica ant species. *Entomologia Experimentalis et Applicata,* 110: 53–63.

Elmes, GW, Wardlaw, JC, Schönrogge, K, Thomas, JA. 2019. Evidence of a fixed polymorphism of one-year and two-year larval growth in the myrmecophilous butterfly *Maculinea rebeli. Insect Conservation and Diversity,* 12: 501–510.

Elton, CS. 1927. *Animal Ecology.* London: Sidgwick & Jackson.

Elzinga, RJ. 1978. Holdfast mechanisms in certain uropodine mites (Acarina: Uropodina). *Annals of the Entomological Society of America,* 71: 896–900.

Elzinga, RJ. 1993. Larvamimidae, a new family of mites (Acari, Dermanyssoidea) associated with army ants. *Acarologia,* 34: 95–103.

Elzinga, RJ, Rettenmeyer, CW. 1974. Seven new species of *Circocylliba* (Acarina: Uropodina) found on army ants. Siete nuevas especies de *Circocylliba* (Acarina: Uropodina) encontrados en hormigas ronchadoras. *Acarologia,* 16: 595–611.

Erber, D. 1968. Bau, funktion und Bildung der Kotpresse mitteleuropäischer Clytrinen und Cryptocephalinen (Coleoptera, Chrysomelidae). *Zeitschrift für Morphologie der Tiere,* 62: 245–306.

Erber, D. 1969. Beitrag zur Entwicklungsbiologie mitteleuropäischer Clytrinen und Cryptocephalinen (Coleoptera, Chrysomelidae). *Zoologische Jahrbücher,* 96: 453–477.

Erber, D. 1988. Biology of the Camptosomata: Clytrinae, Cryptocephalinae, Chlamisinae, and Lamprosomatinae. In *Biology of Chrysomelidae,* ed. P Jolivet, E Petitpierre, T Hsiao, 513–552. Norwell, MA: Kluwer.

Erthal, M, Tonhasca, A. 2001. *Attacobius attarum* spiders (Corinnidae): Myrmecophilous predators of immature forms of the leaf-cutting ant *Atta sexdens* (Formicidae). *Biotropica,* 33: 374–376.

Escherich, K. 1898a. Zur Anatomie und Biologie von *Paussus turcicus* Friv., zugleich ein. Beitrag zur Kenntnis der Myrmekophilie. *Zoologische Jahrbücher Abteilung für Systematik, Geographie und Biologie der Tiere,* 12: 27–70.

Escherich, K. 1898b. Zur Biologie von Thorictus foreli Wasmann. *Zoologischer Anzeiger,* 21: 483–492.

Escherich, K. 1899a. Über myrmekophile Arthropoden, mit besonderer Berücksichtigung der Biologie. *Zoologisches Zentralblatt,* 6: 1–8.

Escherich, K. 1899b. Zur Naturgeschichte von *Paussus favieri* Fairm. *Verhandlungen der Zoologisch-Botanischen Gesellschaft in Wien. Vienna,* 49: 278–283.

Escherich, K. 1906. *Die Ameise. Schilderung ihrer Lebensweise.* Braunschweig: Fr. Vieweg & Sohn.

Escherich, K. 1907. Neue Beobachtungen über *Paussus* in Erythrea. *Zeitschrift für wissenschaftliche Insektenbiologie,* 3: 1–8.

Espeland, M, Breinholt, J, Willmott, KR, Warren, AD, Vila, R, Toussaint, EF, Maunsell, SC, Aduse-Poku, K, Talavera, G, Eastwood, R. 2018. A comprehensive and dated phylogenomic analysis of butterflies. *Current Biology,* 28: 770–778 (e775).

Espeland, M, Hall, JP, DeVries, PJ, Lees, DC, Cornwall, M, Hsu, Y-F, Wu, L-W, Campbell, DL, Talavera, G, Vila, R. 2015. Ancient Neotropical origin and recent recolonisation: Phylogeny, biogeography and diversification of the Riodinidae (Lepidoptera: Papilionoidea). *Molecular Phylogenetics and Evolution,* 93: 296–306.

Evans, HC. 1982. Entomogenous fungi in tropical forest ecosystems: An appraisal. *Ecological Entomology,* 7: 47–60.

Evans, HC, Araújo, JPM, Halfeld, VR, Hughes, DP. 2018. Epitypification and re-description of the zombie-ant fungus, *Ophiocordyceps unilateralis* (Ophiocordycipita-ceae). *Fungal Systematics and Evolution*, 1: 13–22.

Evans, HC, Elliot, SL, Hughes, DP. 2011. Hidden diversity behind the zombie-ant fungus *Ophiocordyceps unilateralis:* Four new species described from carpenter ants in Minas Gerais, Brazil. *PLOS One*, 6: e17024.

Evans, HC, Samson, RA. 1984. *Cordyceps* species and their anamorphs pathogenic on ants (Formicidae) in tropical forest ecosystems, II: The *Camponotus* (Formicinae) complex. *Transactions of the British Mycological Society*, 82: 127–150.

Evans, MEG, Forsythe, TG. 1985. Feeding mechanisms, and their variation in form, of some adult ground-beetles (Coleoptera: Caraboidea). *Journal of Zoology*, 206: 113–143.

Farquharson, C. 1918. *Harpagomyia* and other *Diptera* fed by *Crematogaster* ants in S. Nigeria. *Proceedings of the Entomological Society of London*, 5: 66.

Farrow, R, Dear, J. 1978. The discovery of the genus *Bengalia robineau-desvoidy* (Diptera: calliphoridae) in Australia. *Australian Journal of Entomology*, 17: 234.

Feener, DH Jr. 1981. Competition between ant species: Outcome controlled by parasitic flies. *Science*, 214: 815–817.

Feener, DH Jr. 1987. Size-selective oviposition in *Pseudacteon crawfordi* (Diptera: Phoridae), a parasite of fire ants. *Annals of the Entomological Society of America*, 80: 148–151.

Feener, DH Jr. 2000. Is the assembly of ant communities mediated by parasitoids? *Oikos*, 90: 79–88.

Feener, DH Jr, Brown, BV. 1992. Reduced foraging of *Solenopsis geminata* (Hymenoptera: Formicidae) in the presence of parasitic *Pseudacteon* spp. (Diptera: Phoridae). *Annals of the Entomological Society of America*, 85: 80–84.

Feener, DH Jr, Brown, BV. 1993. Oviposition behavior of an ant-parasitizing fly, *Neodohrniphora curvinervis* (Diptera: Phoridae), and defense behavior by its leaf-cutting ant host *Atta cephalotes* (Hymenoptera: Formicidae). *Journal of Insect Behavior*, 6: 675–688.

Feener, DH Jr, Brown, BV. 1997. Diptera as parasitoids. *Annual Review of Entomology*, 42: 73–97.

Feener, DH Jr, Moss, KAG. 1990. Defense against parasites by hitchhikers in leaf-cutting ants: A quantitative assessment. *Behavioral Ecology and Sociobiology*, 26: 17–29.

Feldhaar, H, Straka, J, Krischke, M, Berthold, K, Stoll, S, Mueller, MJ, Gross, R. 2007. Nutritional upgrading for omnivorous carpenter ants by the endosymbiont *Blochmannia. BMC Biology*, 5: 1–11.

Ferguson, ST, Park, KY, Ruff, AA, Bakis, I, Zwiebel, LJ. 2020. Odor coding of nestmate recognition in the eusocial ant *Camponotus floridanus. Journal of Experimental Biology*, 223: 10.

Fernández-Marín, H, Zimmerman, JK, Wcislo, WT. 2006. *Acanthopria* and *Mimopriella* parasitoid wasps (Diapri-idae) attack *Cyphomyrmex* fungus-growing ants (Formicidae, Attini). *Naturwissenschaften*, 93: 17–21.

Fiedler, K. 1990. New information on the biology of *Maculinea nausithous* and *M. teleius* (Lepidoptera: Lycaenidae). *Nota Lepidopterologica*, 12: 246–256.

Fiedler, K. 1991a. Systematic, evolutionary, and ecological implications of myrmecophily within the Lycaenidae (Insecta: Lepidoptera: Papilionoidea). *Bonner Zoolo-gische Monographien*, 31: 1–210.

Fiedler, K. 1991b. European and North West African Lycaenidae (Lepidoptera) and their associations with ants. *Journal of Research on the Lepidoptera*, 28: 239–257.

Fiedler, K. 1998. Lycaenid-ant interactions of the *Maculinea* type: Tracing their historical roots in a comparative framework. *Journal of Insect Conservation*, 2: 3–14.

Fiedler, K. 2006. Ant-associates of Palaearctic lycaenid butterfly larvae (Hymenoptera: Formicidae; Lepidop-tera: Lycaenidae): A review. *Myrmecologische Nach-richten*, 9: 77–87.

Fiedler, K. 2012. The host genera of ant-parasitic Lycaenidae butterflies: A review. *Psyche*, 2012: 153975.

Fiedler, K. 2021. The ant associates of Lycaenidae butterfly caterpillars—revisited. *Nota Lepidopterologica*, 44: 159–174.

Fiedler, K, Hölldobler, B. 1992. Ants and *Polyommatus icarus* immatures (Lycaenidae): Sex-related developmental benefits and costs of ant attendance. *Oecologia*, 91: 468–473.

Fiedler, K, Hölldobler, B, Seufert, P. 1996. Butterflies and ants: The communicative domain. *Experientia*, 52: 14–24.

Fiedler, K, Maschwitz, U. 1987. Functional analysis of the myrmecophious relationships between ants (Hyme-noptera: Formicidae) and lycaenids (Lepidoptera: Lycaenidae), III: New aspects of the function of the retractile tentacular organs of lycaenid larvae. *Zoologische Beiträge (Neue Folge),* 31: 409–416.

Fiedler, K, Maschwitz, U. 1988. Functional analysis of the myrmecophilous relationships between ants (Hymenoptera: Formicidae) and lycaenids (Lepidoptera: Lycaenidae), II: Lycaenid larvae as trophobiotic partners of ants—a quantitative approach. *Oecologia,* 75: 204–206.

Fiedler, K, Maschwitz, U. 1989a. Functional analysis of the myrmecophilous relationships between ants (Hymenoptera: Formicidae) and Lycaenids (Lepidoptera: Lycaenidae), I: Release of food recruitment in ants by lycaenid larvae and pupae. *Ethology,* 80: 71–80.

Fiedler, K, Maschwitz, U. 1989b. The symbiosis between the weaver ant, *Oecophylla smaragdina,* and *Anthene emolus,* an obligate myrmecophilous lycaenid butterfly. *Journal of Natural History,* 23: 833–846.

Fiedler, K, Seufert, P, Pierce, NE, Pearson, JG, Baumgarten, H-T. 1992. Exploitation of lycaenid-ant mutualisms by braconid parasitoids. *Journal of Research on the Lepidoptera,* 31: 153–168.

Fielde, AM. 1903. Artificial mixed nests of ants. *Biological Bulletin,* 5: 320–325.

Fielde, AM. 1904. Power of recognition among ants. *Biological Bulletin,* 7: 227–250.

Fischer, G, Friedman, NR, Huang, J-P, Narula, N, Knowles, LL, Fisher, BL, Mikheyev, AS, Economo, EP. 2020. Socially parasitic ants evolve a mosaic of host-matching and parasitic morphological traits. *Current Biology,* 30: 3639–3646 (e3634).

Fittkau, EJ, Klinge, H. 1973. On biomass and trophic structure of the central Amazonian rain forest ecosystem. *Biotropica,* 5: 2–14.

Folgarait, PJ. 2013. Leaf-cutter ant parasitoids: Current knowledge. *Psyche,* 2013: 539780.

Forel, A. 1874. Les fourmis de la Suisse: Systématique, notices anatomiques et physiologiques, architecture, distribution géographique, nouvelles expériences et observations de moeurs. *Neue Denkschriften der Allgemeinen Schweizerischen Gesellschaft für die gesammten Naturwissenschaften,* 26: 1–452.

Forel, A. 1894. Les formicides de la province d'Oran (Algerie). *Bulletin de la Societe Vaudoise des Sciences naturelles,* 30: 1–45.

Forti, LC, Camargo, RS, Verza, SS, Andrade, APP, Fujihara, RT, Lopes, JF. 2007. *Microdon tigrinus* Curran, 1940 (Diptera, Syrphidae): Populational fluctuation and specificity to the nest of *Acromyrmex coronatus* (Hymenoptera: Formicidae). *Sociobiology,* 50: 909–919.

Forti, LC, Protti de Andrade, AP, Camargo, RdS, Caldato, N, Moreira, AA. 2017. Discovering the giant nest architecture of grass-cutting ants, *Atta capiguara* (Hymenoptera, Formicidae). *Insects,* 8: 39.

Fowler, HG. 1992. Patterns of colonization and incipient nest survival in *Acromyrmex niger* and *Acromyrmex balzani* (Hymenoptera: Formicidae). *Insectes Sociaux,* 39: 347–350.

Fowler, HG. 1997. Morphological prediction of worker size discrimination and relative abundance of sympartic species of *Pseudacteon* (Dipt., Phoridae) parasitoids of the fire ant, *Solenopsis saevissima* (Hym., Formicidae) in Brazil. *Journal of Applied Entomology,* 121: 37–40.

Franks, NR. 1982. Ecology and population regulation in the army ant *Eciton burchellii.* In *The Ecology of a Tropical Forest: Seasonal Rhythms and Long-Term Changes,* ed. EG Leigh Jr., AS Rand, DM Windsor, 389–395. Washington, DC: Smithsonian Institution Press.

Franks, NR, Healey, KJ, Byrom, L. 1991. Studies on the relationship between the ant ectoparasite *Antennophorus grandis* (Acarina: Antennophoridae) and its host *Lasius flavus* (Hymenoptera: Formicidae). *Journal of Zoology,* 225: 59–70.

Franzl, S, Locke, M, Huie, P. 1984. Lenticles: Innervated secretory structures that are expressed at every other larval moult. *Tissue and Cell,* 16: 251–268.

Freitag, R, Lee, S. 1972. Sound producing structures in adult *Cicindela tranquebarica* (Coleoptera: Cicindelidae) including a list of tiger beetles and ground beetles with flight wing files. *Canadian Entomologist,* 104: 851–857.

Fric, Z, Wahlberg, N, Pech, P, Zrzavý, J. 2007. Phylogeny and classification of the *Phengaris-Maculinea* clade (Lepidoptera: Lycaenidae): Total evidence and phylogenetic species concepts. *Systematic Entomology,* 32: 558–567.

Fuchs, S. 1976a. The response to vibrations of the substrate and reactions to the specific drumming in colonies of carpenter ants (*Camponotus,* Formicidae, Hymenoptera). *Behavioral Ecology and Sociobiology,* 1: 155–184.

Fuchs, S. 1976b. An informational analysis of the alarm communication by drumming behavior in nests of carpenter ants (*Camponotus,* Formicidae, Hymenoptera). *Behavioral Ecology and Sociobiology,* 1: 315–336.

Funaro, CF, Böröczky, K, Vargo, EL, Schal, C. 2018. Identification of a queen and king recognition pheromone in the subterranean termite *Reticulitermes flavipes*. *Proceedings of the National Academy of Sciences,* 115: 3888–3893.

Funaro, CF, Schal, C, Vargo, EL. 2019. Queen and king recognition in the subterranean termite, *Reticulitermes flavipes:* Evidence for royal recognition pheromones. *PLOS One,* 14: e0209810.

Fürst, MA, Durey, M, Nash, DR. 2012. Testing the adjustable threshold model for intruder recognition on *Myrmica* ants in the context of a social parasite. *Proceedings of the Royal Society B: Biological Sciences,* 279: 516–522.

Gadeberg, RM, Boomsma, JJ. 1997. Genetic population structure of the large blue butterfly *Maculinea alcon* in Denmark. *Journal of Insect Conservation,* 1: 99–111.

Gaedike, R. 2019. *Tineidae II: (Myrmecozelinae, Perissomasticinae, Tineinae, Hieroxestinae, Teichobiinae, and Stathmopolitinae).* Leiden, The Netherlands: Koninklijke Brill.

Gardner, MG, Schönrogge, K, Elmes, G, Thomas, J. 2007. Increased genetic diversity as a defence against parasites is undermined by social parasites: *Microdon mutabilis* hoverflies infesting *Formica lemani* ant colonies. *Proceedings of the Royal Society B: Biological Sciences,* 274: 103–110.

Garnett, WB, Akre, RD, Sehlke, G. 1985. Cocoon mimicry and predation by myrmecophilous Diptera (Diptera: Syrphidae). *Florida Entomologist,* 68: 615–621.

Gehlbach, FR, Watkins, JF II, Kroll, JC. 1971. Pheromone trail–following studies of typhlopid, leptotyphlopid, and colubrid snakes. *Behaviour,* 40: 282–294.

Geiselhardt, SF, Peschke, K, Nagel, P. 2007. A review of myrmecophily in ant nest beetles (Coleoptera: Carabidae: Paussinae): Linking early observations with recent findings. *Naturwissenschaften,* 94: 871–894.

Gentry, JB, Stiritz, KL. 1972. The role of the Florida harvester ant, *Pogonomyrmex badius,* in old field mineral nutrient relationships. *Environmental Entomology,* 1: 39–41.

Gil, R, Sabater-Muñoz, B, Latorre, A, Silva, FJ, Moya, A. 2002. Extreme genome reduction in *Buchnera* spp.: Toward the minimal genome needed for symbiotic life. *Proceedings of the National Academy of Sciences,* 99: 4454–4458.

Gil, R, Silva, FJ, Zientz, E, Delmotte, F, González-Candelas, F, Latorre, A, Rausell, C, Kamerbeek, J, Gadau, J, Hölldobler, B, van Ham, RCHJ, Gross, R, Moya, A. 2003. The genome sequence of *Blochmannia floridanus:* Comparative analysis of reduced genomes. *Proceedings of the National Academy of Sciences,* 100: 9388–9393.

Gilbert, LE. 1976. Adult resources in butterflies: African lycaenid *Megalopalpus* feeds on larval nectary. *Biotropica,* 8: 282–283.

Gilbert, LE, Morrison, LW. 1997. Patterns of host specificity in *Pseudacteon* parasitoid flies (Diptera: Phoridae) that attack *Solenopsis* fire ants (Hymenoptera: Formicidae). *Environmental Entomology,* 26: 1149–1154.

Girault, A. 1913. Some chalcidoid Hymenoptera from Northern Queensland. *Archiv für Naturgeschichte,* 79: 70–90.

Glasier, JRN, Poore, AGB, Eldridge, DJ. 2018. Do mutualistic associations have broader host ranges than neutral or antagonistic associations? A test using myrmecophiles as model organisms. *Insectes Sociaux,* 65: 639–648.

Godfray, HCJ. 1994. *Parasitoids: Behavioral and Evolutionary Ecology.* Princeton, NJ: Princeton University Press.

Godfray, HCJ. 2007. Parasitoids. In *Encyclopedia of Biodiversity,* ed. SA Levin. Cambridge, MA: Academic Press.

Goodwin, D. 1955. Anting. *Avicultural,* 61: 21–25.

Gösswald, K. 1950. Pflege des Ameisenparasiten Tamiclea globula Meig. (Dipt.) durch den Wirt mit Bemerkungen über den Stoffwechsel in der parasitierten Ameise. *Verhandlungen der Deutschen Zoolologischen Gesellschaft* 1949: 256–264.

Gösswald, K. 1985. *Organisation und Leben der Ameisen.* Stuttgart: Wissenschaftliche Verlagsgesellschaft.

Gösswald, K. 1990. *Die Waldameise: Biologie, Ökologie und forstliche Nutzung, Band 2, Die Waldameise im Ökosystem Wald, ihr Nutzen und ihre Hege.* Wiesbaden, Germany: AULA-Verlag.

Gotwald, WH Jr. 1995. *Army Ants: The Biology of Social Predation.* Ithaca, NY: Cornell University Press / Comstock Press.

Greene, MJ, Gordon, DM. 2003. Cuticular hydrocarbons inform task decisions. *Nature,* 423: 32.

Greene, MJ, Gordon, DM. 2007. Structural complexity of chemical recognition cues affects the perception of group membership in the ants *Linephithema humile* and *Aphaenogaster cockerelli. Journal of Experimental Biology,* 210: 897–905.

Greeney, H. 2012. Antpittas and worm-feeders: A match made by evolution? Evidence for a possible commensal foraging relationship between antpittas (Grallariidae) and mammals. *Neotropical Biology and Conservation,* 7: 140–143.

Griffiths, HM, Hughes, WO. 2010. Hitchhiking and the removal of microbial contaminants by the leaf-cutting ant *Atta colombica. Ecological Entomology,* 35: 529–537.

Grinath, JB, Inouye, BD, Underwood, N. 2015. Bears benefit plants via a cascade with both antagonistic and mutualistic interactions. *Ecology Letters,* 18: 164–173.

Groark, KP. 1996. Ritual and therapeutic use of "hallucinogenic" harvester ants (*Pogonomyrmex*) in native south-central California. *Journal of Ethnobiology,* 16: 1–30.

Groskin, H. 1943. Scarlet tanagers' anting. *The Auk,* 60: 55–59.

Guerrieri, FJ, Nehring, V, Jørgensen, CG, Nielsen, J, Galizia, CG, d'Ettorre, P. 2009. Ants recognize foes and not friends. *Proceedings of the Royal Society B: Biological Sciences,* 276: 2461–2468.

Guillem, RM, Drijfhout, F, Martin, SJ. 2014. Chemical deception among ant social parasites. *Current Zoology,* 60: 62–75.

Guillem, RM, Drijfhout, FP, Martin, SJ. 2016. Species-specific cuticular hydrocarbon stability within European *Myrmica* ants. *Journal of Chemical Ecology,* 42: 1052–1062.

Haddad, FS. 2010. Suturing methods and materials with special emphasis on the jaws of giant ants (an old-new surgical instrument). *Lebanese Medical Journal,* 58: 53–56.

Hager, FA, Krausa, K. 2019. Acacia ants respond to plant-borne vibrations caused by mammalian browsers. *Current Biology,* 29: 717–725.

Haines, IH, Haines, JB. 1978. Colony structure, seasonality and food requirements of the crazy ant, *Anoplolepis longipes* (Jerd.), in the Seychelles. *Ecological Entomology,* 3: 109–118.

Hale, A, Bougie, T, Henderson, E, Sankovitz, M, West, M, Purcell, J. 2018. Notes on hunting behavior of the spider *Euryopis californica* Banks, 1904 (Araneae: Theridiidae), a novel predator of *Veromessor pergandei* (Mayr, 1886) harvester ants (Hymenoptera: Formicidae). *Pan-Pacific Entomologist,* 94: 141–145.

Haller, G. 1877. *Antennophorus uhlmanni,* ein neuer Gamaside. *Archiv für Naturgeschichte,* 46: 57–62.

Hamilton, W. 1987. Kinship, recognition, disease, and intelligence: Constraints of social evolution. In *Animal Societies: Theory and Facts,* ed. Y Ito, JL Brown, J Kikkawa, 81–102. Tokyo: Japanese Scientific.

Hashimoto, Y, Endo, T, Yamasaki, T, Hyodo, F, Itioka, T. 2020. Constraints on the jumping and prey-capture abilities of ant-mimicking spiders (Salticidae, Salticinae, *Myrmarachne*). *Scientific Reports,* 10: 18279.

Hauser, DC. 1964. Anting by gray squirrels. *Journal of Mammalogy,* 45: 136–138.

Hebard, M. 1920. A revision of the North American species of the genus *Myrmecophila* (Orthoptera; Gryllidae; Myrmecophilinae). *Transactions of the American Entomological Society (1890–),* 46: 91–111.

Hefetz, A. 2007. The evolution of hydrocarbon pheromone parsimony in ants (Hymenoptera: Formicidae)—interplay of colony odor uniformity and odor idiosyncrasy. *Myrmecological News,* 10: 59–68.

Henderson, G, Akre, RD. 1986a. Biology of the myrmecophilous cricket, *Myrmecophila manni* (Orthoptera: Gryllidae). *Journal of the Kansas Entomological Society,* 59: 454–467.

Henderson, G, Akre, RD. 1986b. Morphology of *Myrmecophila manni,* a myrmecophilous cricket (Orthoptera: Gryllidae). *Journal of the Entomological Society of British Columbia,* 83: 57–62.

Henderson, G, Akre, RD. 1986c. Dominance hierarchies in *Myrmecophila manni,* Orthoptera: Gryllidae. *Pan-Pacific Entomologist,* 62: 24–28.

Hendricks, P, Norment, G. 2015. Anting behavior by the northwestern crow (*Corvus caurinus*) and American crow (*Corvus brachyrhynchos*). *Northwestern Naturalist,* 96: 143–146.

Henning, SF. 1983a. Biological groups within the Lycaenidae (Lepidoptera). *Journal of the Entomological Society of Southern Africa,* 46: 65–85.

Henning, SF. 1983b. Chemical communication between lycaenid larvae (Lepidoptera: Lycaenidae) and ants (Hymenoptera: Formicidae). *Journal of the Entomological Society of Southern Africa,* 46: 341–366.

Henning, SF. 1984a. Life history and behaviour of the rare myrmecophilous lycaenid *Erikssonia acraeina* Trimen (Lepidoptera: Lycaenidae). *Journal of the Entomological Society of Southern Africa,* 47: 337–342.

Henning, SF. 1984b. The effect of ant association on lycaenid larval duration (Lepidoptera: Lycaenidae). *Entomologist's Record and Journal of Variation, 96*: 99–102.

Heraty, JM. 1994. Biology and importance of two eucharitid parasites of *Wasmannia* and *Solenopsis*. In *Exotic Ants: Biology, Impact and Control of Introduced Species,* ed. DF Williams, 104–120. Oxford: Westview Press.

Heraty, JM. 2000. Phylogenetic relationships of Oraseminae (Hymenoptera: Eucharitidae). *Annals of the Entomological Society of America, 93*: 374–390.

Heraty, JM. 2002. Revision of the genera of Eucharitidae (Hymenoptera: Chalcidoidea) of the world. *Memoirs of the American Entomological Institute, 68*: 1–367.

Heraty, JM, Barber, KN. 1990. Biology of *Obeza floridana* (Ashmead) and *Pseudochalcura gibbosa* (Provancher) (Hymenoptera: Eucharitidae). *Proceedings of the Entomological Society of Washington, 92*: 248–258.

Heraty, JM, Hawks, D, Kostecki, JS, Carmichael, A. 2004. Phylogeny and behaviour of the Gollumiellinae, a new subfamily of the ant-parasitic Eucharitidae (Hymenoptera: Chalcidoidea). *Systematic Entomology, 29*: 544–559.

Heraty, JM, Murray, E. 2013. The life history of *Pseudometagea schwarzii,* with a discussion of the evolution of endoparasitism and koinobiosis in Eucharitidae and Perilampidae (Chalcidoidea). *Journal of Hymenoptera Research, 35*: 1–35.

Heraty, JM, Wojcik, D, Jouvenaz, D. 1993. Species of *Orasema* parasitic on the *Solenopsis saevissima*-complex in South America (Hymenoptera: Eucharitidae, Formicidae). *Journal of Hymenoptera Research, 2*: 169–182.

Herreid, JS, Heraty, JM. 2017. Hitchhikers at the dinner table: A revisionary study of a group of ant parasitoids (Hymenoptera: Eucharitidae) specializing in the use of extrafloral nectaries for host access. *Systematic Entomology, 42*: 204–229.

Hickling, R, Brown, RL. 2000. Analysis of acoustic communication by ants. *Journal of the Acoustical Society of America, 108*: 1920–1929.

Higuchi, H. 1976. Comparative study on the breeding of mainland and island subspecies of the varied tit, *Parus varius. Japanese Journal of Ornithology, 25*: 11–20.

Hill, C. 1993. The myrmecophilous organs of *Arhopala madytusfruhstorfer* (Lepidoptera: Lycaenidae). *Australian Journal of Entomology, 32*: 283–288.

Hill, JG. 2009a. First report of the Eastern ant cricket, *Myrmecophilus pergandei* Bruner (Orthoptera: Mymecophilidae), collected from an imported fire ant colony, *Solenopsis invicta* × *richteri* (Hymenoptera: Formicidae). *Journal of Orthoptera Research, 18*: 57–58.

Hill, PS. 2009b. How do animals use substrate-borne vibrations as an information source? *Naturwissenschaften, 96*: 1355–1371.

Hill, WB, Akre, R, Huber, J. 1976. Structure of some epidermal glands in the myrmecophilous beetle *Adranes taylori* (Coleoptera: Pselaphidae). *Journal of the Kansas Entomological Society, 49*: 367–384.

Hinton, H. 1951. Myrmecophilous Lycaenidae and other Lepidoptera: A summary *Proceedings of the London Entomological Natural History Society, 1949–50*: 111–175.

Hiramatsu, M. 2003. A role for guanidino compounds in the brain. *Molecular and Cellular Biochemistry, 244*: 57–62.

Hlaváč, P. 2005. Revision of the myrmecophilous genus *Lomechusa* (Coleoptera: Staphylinidae: Aleocharinae). *Sociobiology, 46*: 203–250.

Hlaváč, P, Jászay, T. 2009. A revision of the genus *Zyras* (Zyras) Stephens, 1835 (Coleoptera, Staphylinidae, Aleocharinae), I: Current classification status and the redefinition of the genus. *ZooKeys, 29*: 49–71.

Hlaváč, P, Newton, AF, Maruyama, M. 2011. World catalogue of the species of the tribe Lomechusini (Staphylinidae: Aleocharinae). *Zootaxa, 3075*: 1–151.

Hobbs, HH III, Lawyer, R. 2003. A preliminary population study of the cave cricket, *Hadenoecus cumberlandicus* Hubbell and Norton, from a cave in Carter County, Kentucky. *Journal of Cave and Karst Studies, 65*: 174.

Hochkirch, A, Gröning, J. 2008. Sexual size dimorphism in Orthoptera (sens. str.): A review. *Journal of Orthoptera Research, 17*: 189–196.

Hohorst, W, Graefe, G. 1961. Ameisen—obligatorische Zwischenwirte des Lanzettegels (*Dicrocoelium dendriticum*). *Naturwissenschaften, 48*: 229–230.

Hojo, MK, Pierce, NE, Tsuji, K. 2015. Lycaenid caterpillar secretions manipulate attendant ant behavior. *Current Biology, 25*: 2260–2264.

Hojo, MK, Wada-Katsumata, A, Ozaki, M, Yamaguchi, S, Yamaoka, R. 2008. Gustatory synergism in ants mediates a species-specific symbiosis with lycaenid butterflies. *Journal of Comparative Physiology A*, 194: 1043–1052.

Hölldobler, B. 1966. Futterverteilung durch Männchen im Ameisenstaat. *Zeitschrift für vergleichende Physiologie*, 52: 430–455.

Hölldobler, B. 1967. Zur Physiologie der Gast-Wirt-Beziehung (Myrmecophilie) bei Ameisen, I: Das Gastverhältnis der *Atemeles*—und *Lomechusa*—Larven (Col. Staphylinidae) zu *Formica* (Hym. Formicidae). *Zeitschrift für vergleichende Physiologie*, 56: 1–21.

Hölldobler, B. 1968. Der Glanzkäfer als "Wegelagerer" an Ameisenstraßen. *Naturwissenschaften*, 55: 397.

Hölldobler, B. 1970a. *Steatoda fulva* (Theridiidae), a spider that feeds on harvester ants. *Psyche*, 77: 202–208.

Hölldobler, B. 1970b. Zur Physiologie der Gast-Wirt-Beziehungen (Myrmecophilie) bei Ameisen, II: Das Gastverhältnis des imaginalen *Atemeles pubicollis* Bris. (Col. Staphylinidae) zu *Myrmica* und *Formica* (Hym. Formicidae). *Zeitschrift für vergleichende Physiologie*, 66: 215–250.

Hölldobler, B. 1971. Communication between ants and their guests. *Scientific American*, 224: 86–93.

Hölldobler, B. 1976. Recruitment behavior, home range orientation and territoriality in harvester ants, *Pogonomyrmex*. *Behavioral Ecology and Sociobiology*, 1: 3–44.

Hölldobler, B. 1977. Communication in social Hymenoptera. In *How Animals Communicate*, ed. TA Sebeok, 418–471. Bloomington: Indiana University Press.

Hölldobler, B. 1984. The wonderfully diverse ways of the ant. *National Geographic*, 165: 779–881.

Hölldobler, B, Carlin, NF. 1987. Anonymity and specificity in the chemical communication signals of social insects. *Journal of Comparative Physiology A*, 161: 567–581.

Hölldobler, B, Inwood, M, Oldham, N, Liebig, J. 2001. Recruitment pheromone in the harvester ant genus *Pogonomyrmex*. *Journal of Insect Physiology*, 47: 369–374.

Hölldobler, B, Kwapich, CL. 2017. *Amphotis marginata* (Coleoptera: Nitidulidae) a highwayman of the ant *Lasius fuliginosus*. *PLOS One*, 12: e0180847.

Hölldobler, B, Kwapich, CL. 2019. Behavior and exocrine glands in the myrmecophilous beetle *Dinarda dentata* (Gravenhorst, 1806) (Coleoptera: Staphylinidae: Aleocharinae). *PLOS One*, 14: e0210524.

Hölldobler, B, Kwapich, CL, Haight, KL. 2018. Behavior and exocrine glands in the myrmecophilous beetle

Lomechusoides strumosus (Fabricius, 1775) (formerly called *Lomechusa strumosa*) (Coleoptera: Staphylinidae: Aleocharinae). *PLOS One*, 13: e0200309.

Hölldobler, B, Lumsden, CJ. 1980. Territorial strategies in ants. *Science*, 210: 732–739.

Hölldobler, B, Möglich, M, Maschwitz, U. 1981. Myrmecophilic relationship of *Pella* (Coleoptera: Staphylinidae) to *Lasius fuliginosus* (Hymenoptera: Formicidae). *Psyche*, 88: 347–374.

Hölldobler, B, Wilson, EO. 1970. Recruitment trails in the harvester ant *Pogonomyrmex badius*. *Psyche*, 77: 385–399.

Hölldobler, B, Wilson, EO. 1977. The number of queens: An important trait in ant evolution. *Naturwissenschaften*, 64: 8–15.

Hölldobler, B, Wilson, EO. 1990. *The Ants*. Cambridge, MA: Belknap Press of Harvard University Press.

Hölldobler, B, Wilson, EO. 2009. *The Superorganism: The Beauty, Elegance, and Strangeness of Insect Societies*. New York: W. W. Norton & Company.

Hölldobler, B, Wilson, EO. 2011. *The Leafcutter Ants: Civilization by Instinct*. New York: W. W. Norton & Company.

Hölldobler, B, Wilson, EO. 2013. *Auf den Spuren der Ameisen*. Berlin: Springer-Spektrum Verlag.

Hölldobler, K. 1928. Zur Biologie der diebischen Zwergameise (*Solenopsis fugax*) und ihrer Gäste. *Biologisches Zentralblatt*, 48: 129–142.

Hölldobler, K. 1929. Über die Entwicklung der Schwirrfliege *Xanthogramma citrofasciatum* im Neste von *Lasius alienus* und *niger*. *Zoologischer Anzeiger (Wasmann-Festband)*, 82: 171–176.

Hölldobler, K. 1941. Über das Gastverhältnis von Atemeles (Col. Staph.) zu Myrmica (Hym. Form.) *Mitteilungen der Münchner Entomologischen Gesellschaft*, 31: 1054–1059.

Hölldobler, K. 1947. Studien über die Ameisengrille (*Myrmecophila acervorum* Panzer) im mittleren Maingebiet. *Mitteilungen der Schweizerischen Entomologischen Gesellschaft*, 20: 607–648.

Hölldobler, K. 1948. Über ein parasitologisches Problem: Die Gastpflege der Ameisen und die Symphilieinstinkte. *Zeitschrift für Parasitenkunde*, 14: 3–26.

Hölldobler, K. 1951. Über eine Milbenschädigung der Roßameise (*Camponotus herculeanus*), die durch eine Fehlreaktion des Wirtes wirksam wird. *Zeitschrift für Angewandte Entomologie*, 33: 104–107.

Hölldobler, K. 1953. Gibt es in Deutschland Ameisengäste, die echte Täuscher sind? *Die Naturwissenschaften,* 40: 34–35.

Holmes, A. 2019. First observation of *Myrmarachne* species feeding on ants (Araneae: Salticidae: Myrmarachnini). *Peckhamia,* 178: 1–4.

Hopping, KA, Chignell, SM, Lambin, EF. 2018. The demise of caterpillar fungus in the Himalayan region due to climate change and overharvesting. *Proceedings of the National Academy of Sciences,* 115: 11489–11494.

Hoskins, A. 2015. *Butterflies of the World.* Cape Town, SA: New Holland Publishers.

Hovestadt, T, Thomas, JA, Mitesser, O, Elmes, GW, Schönrogge, K. 2012. Unexpected benefit of a social parasite for a key fitness component of its ant host. *American Naturalist,* 179: 110–123.

Howard, DF, Tschinkel, WR. 1976. Aspects of necrophoric behavior in the red imported fire ant, *Solenopsis invicta. Behaviour,* 56: 157–178.

Howard, RW, Akre, RD, Garnett, WB. 1990. Chemical mimicry in an obligate predator of carpenter ants (Hymenoptera: Formicidae). *Annals of the Entomological Society of America,* 83: 607–616.

Howard, RW, Blomquist, GJ. 2005. Ecological, behavioral, and biochemical aspects of insect hydrocarbons. *Annual Review of Entomology,* 50: 371–393.

Howard, RW, McDaniel, C, Blomquist, GJ. 1980. Chemical mimicry as an integrating mechanism: Cuticular hydrocarbons of a termitophile and its host. *Science,* 210: 431–433.

Howard, RW, Stanley-Samuelson, DW, Akre, RD. 1990. Biosynthesis and chemical mimicry of cuticular hydrocarbons from the obligate predator, *Microdon albicomatus* Novak (Diptera: Syrphidae) and its ant prey, *Myrmica incompleta* Provancher (Hymenoptera: Formicidae). *Journal of the Kansas Entomological Society,* 63: 437–443.

Hsieh, H-Y, Perfecto, I. 2012. Trait-mediated indirect effects of phorid flies on ants. *Psyche,* 2012: 380474.

Hsu, P-W, Hugel, S, Wetterer, JK, Tseng, S-P, Ooi, C-SM, Lee, C-Y, Yang, C-CS. 2020. Ant crickets (Orthoptera: Myrmecophilidae) associated with the invasive yellow crazy ant *Anoplolepis gracilipes* (Hymenoptera: Formicidae): Evidence for cryptic species and potential co-introduction with hosts. *Myrmecological News,* 30: 103–129.

Hu, Y, Sanders, JG, Łukasik, P, D'Amelio, CL, Millar, JS, Vann, DR, Lan, Y, Newton, JA, Schotanus, M, Kronauer, DJ. 2018. Herbivorous turtle ants obtain essential nutrients from a conserved nitrogen-recycling gut microbiome. *Nature Communications,* 9: 1–14.

Huang, J-N, Cheng, R-C, Li, D, Tso, I-M. 2011. Salticid predation as one potential driving force of ant mimicry in jumping spiders. *Proceedings of the Royal Society B: Biological Sciences,* 278: 1356–1364.

Huddleston, T. 1976. A revision of *Elasmosoma* Ruthe (Hymenoptera, Braconidae) with two new species from Mongolia. *Annales Historico-Naturales Musei Nationalis Hungarici,* 68: 215–225.

Hugel, S, Blard, F. 2005. Présence de Grillons du genre *Myrmecophilus* à l'île de la Réunion (Orthoptera, Myrmecophilinae). *Bulletin de la Société entomologique de France,* 110: 387–389.

Huggert, L, Masner, L. 1983. A review of myrmecophilic-symphilic diapriid wasps in the Holarctic realm, with descriptions of new taxa and a key to genera (Hymenoptera: Proctotrupoidea: Diapriidae). *Contributions of the American Entomological Institute,* 20: 63–89.

Hughes, DP. 2013. Pathways to understanding the extended phenotype of parasites in their hosts. *Journal of Experimental Biology,* 216: 142–147.

Hughes, DP, Andersen, SB, Hywel-Jones, NL, Himaman, W, Billen, J, Boomsma, JJ. 2011. Behavioral mechanisms and morphological symptoms of zombie ants dying from fungal infection. *BMC Ecology,* 11: 1–10.

Hughes, DP, Araújo, JP, Loreto, RG, Quevillon, L, de Bekker, C, Evans, HC. 2016. From so simple a beginning: The evolution of behavioral manipulation by fungi. *Advances in Genetics,* 94: 437–469.

Hughes, DP, Evans, HC, Hywel-Jones, N, Boomsma, JJ, Armitage, SA. 2009. Novel fungal disease in complex leaf-cutting ant societies. *Ecological Entomology,* 34: 214–220.

Hughes, DP, Pierce, NE, Boomsma, JJ. 2008. Social insect symbionts: Evolution in homeostatic fortresses. *Trends in Ecology & Evolution,* 23: 672–677.

Hughes, WOH, Boomsma, JJ. 2004. Genetic diversity and disease resistance in leaf-cutting ant societies. *Evolution,* 58: 1251–1260.

Hughes, WOH, Boomsma, J. 2006. Does genetic diversity hinder parasite evolution in social insect colonies? *Journal of Evolutionary Biology,* 19: 132–143.

Huhta, V. 2016. Catalogue of the Mesostigmata mites in Finland. *Memoranda Societatis pro Fauna et Flora Fennica,* 92: 129–148.

Hunt, GL Jr. 1977. Low preferred foraging temperatures and nocturnal foraging in a desert harvester ant. *American Naturalist,* 111: 589–591.

Hunt, J, Richard, F-J. 2013. Intracolony vibroacoustic communication in social insects. *Insectes Sociaux,* 60: 403–417.

Hutchinson, GE. 1959. Homage to Santa Rosalia or why are there so many kinds of animals? *American Naturalist,* 93: 145–159.

Huxley, J. 1966. A discussion on ritualization of behaviour in animals and man. *Philosophical Transactions of the Royal Society (London) B,* 251: 249–271.

Ichinose, K, Rinaldi, I, Forti, LC. 2004. Winged leaf-cutting ants on nuptial flights used as transport by *Attacobius* for dispersal. *Ecological Entomology,* 29.

Ikeshita, Y, Taniguchi, K, Kitagawa, Y, Maruyama, M, Ito, F. 2017. The rove beetle *Drusilla sparsa* (Coleoptera: Staphylinidae) is a myrmecophilous species associated with a myrmicine ant, *Crematogaster osakensis* (Hymenoptera: Formicidae). *Entomological Science,* 20: 437–442.

Ingrisch, S. 1995. Eine neue Ameisengrille aus Borneo (Ensifera: Grylloidea). *Entomologische Zeitschrift,* 105: 421–427.

Inui, Y, Shimizu-Kaya, U, Okubo, T, Yamsaki, E, Itioka, T. 2015. Various chemical strategies to deceive ants in three *Arhopala* species (Lepidoptera: Lycaenidae) exploiting *Macaranga* myrmecophytes. *PLOS One,* 10: e0120652.

Iorgu, IŞ, Iorgu, EI, Stalling, T, Puskás, G, Chobanov, D, Szövényi, G, Moscaliuc, LA, Motoc, R, Tăuşan, I, Fusu, L. 2021. Ant crickets and their secrets: *Myrmecophilus acervorum* is not always parthenogenetic (Insecta: Orthoptera: Myrmecophilidae). *Zoological Journal of the Linnean Society:* zlab084.

Ivens, AB, von Beeren, C, Blüthgen, N, Kronauer, DJ. 2016. Studying the complex communities of ants and their symbionts using ecological network analysis. *Annual Review of Entomology,* 61: 353–371.

Jackson, RR. 1986. Communal jumping spiders (Araneae: Salticidae) from Kenya: Interspecific nest complexes, cohabitation with web-building spiders, and intraspecific interactions. *New Zealand Journal of Zoology,* 13: 13–26.

Jackson, RR, Nelson, XJ. 2012. Specialized exploitation of ants (Hymenoptera: Formicidae) by spiders (Araneae). *Myrmecological News,* 17: 33–49.

Jackson, RR, Nelson, XJ, Salm, K. 2008. The natural history of *Myrmarachne melanotarsa,* a social ant-mimicking jumping spider. *New Zealand Journal of Zoology,* 35: 225–235.

Jackson, RR, Pollard, SD. 2007. Bugs with backpacks deter vision-guided predation by jumping spiders. *Journal of Zoology,* 273: 358–363.

Jackson, RR, Pollard, S, Nelson, X, Edwards, G, Barrion, A. 2001. Jumping spiders (Araneae: Salticidae) that feed on nectar. *Journal of Zoology,* 255: 25–29.

Jacobson, EE. 1909. Ein Moskito als Gast und diebischer Schmarotzer der *Crematogaster difformis* Smith und eine andere schmarotzende Fliege. *Tjidschrift voor Entomologie,* 52: 158–164.

Jacobson, EE. 1910. *Pheidologeton diversus* Jerdon und eine myrmecophile Fliegenart. *Tijdschrift voor Entomologie,* 53: 328–335.

Jacobson, EE. 1911. Nähere Mitteilungen über die myrmecophile Culicide *Harpagomyia splendens* de Meij. *Tjidschrift voor Entomolgie,* 54.

Jacobson, HR, Kistner, DH. 1991. Cladistic study, taxonomic restructuring, and revision of the myrmecophilous tribe Leptanillophilini with comments on its evolution and host relationships (Coleoptera: Staphylinidae; Hymenoptera: Formicidae). *Sociobiology,* 18: 1–150.

Jacobson, HR, Kistner, DH. 1992. Cladistic study, taxonomic restructuring, and revision of the myrmecophilous tribe Crematoxenini with comments on its evolution and host relationships (Coleoptera: Staphylinidae: Hymenoptera: Formicidae). *Sociobiology,* 20: 91–198.

Janet, C. 1897a. Études sur les fourmis les Guêpes et les Abeilles: Note 14. Rapports des animaux myrmécophiles avec les fourmis. *Annals and Magazine of Natural History,* 6: 620–623.

Janet, C. 1897b. Sur les rapports de l'*Antennophorus uklniunni* Haller avec le *Lasius mixius* Nyl. *Comptes Rendus de l'Académie des Science, Paris,* 124: 583–585.

Janzen, DH, Carroll, CRC. 1983. *Paraponera clavata* (bala, giant tropical ant). In *Costa Rican Natural History,* ed. DH Janzen, 752–753. Chicago, IL: University of Chicago Press.

Jared, C, Antoniazzi, MM, Silva, JRMC, Freymüller, E. 1999. Epidermal glands in Squamata: Microscopical examination of precloacal glands in *Amphisbaena alba* (Amphisbaenia, Amphisbaenidae). *Journal of Morphology,* 241: 197–206.

Jenkins, M. 1957. The morphology and anatomy of the pygidial glands of *Dianous coerulescens* Gyllenhal (Coleoptera: Staphylinidae). *Proceedings of the Royal Entomological Society, London (A),* 32: 159–167.

Jennings, D. 1972. An overwintering aggregation of spiders (Araneae) on cottonwood in New Mexico. *Entomological News,* 83: 61–67.

Johnson, RA. 2000. Water loss in desert ants: Caste variation and the effect of cuticle abrasion. *Physiological Entomology,* 25: 48–53.

Johnson, RA. 2021. Desiccation limits recruitment in the pleometrotic desert seed-harvester ant *Veromessor pergandei. Ecology and Evolution,* 11: 294–308.

Johnson, SJ, Valentine, PS. 1986. Observations on '*Liphyra brassolis*' Westwood (Lepidoptera: Lycaenidae) in north Queensland. *Australian Entomologist,* 13: 22–26.

Jordan, KHC. 1913. Zur Morphologie und Biologie der myrmecophilen Gattungen *Lomechusa* und *Atemeles* und einiger verwandter Formen. *Zeitschrift für Wissenschaftliche Zoologie,* 107: 346–386.

Jorgenson, C, Black, H, Hermann, H. 1984. Territorial disputes between colonies of the giant tropical ant *Paraponera clavata* (Hymenoptera: Formicidae: Ponerinae). *Journal of the Georgia Entomological Society,* 19: 156–158.

Jouvenaz, DP, Lofgren, CS, Banks, WA. 1981. Biological control of imported fire ants: A review of current knowledge. *Bulletin of the Entomological Society of America,* 27: 203–209.

Jouvenaz, DP, Wojcik, DP, Naves, MA, Lofgren, CS. 1988. Observações sobre um nematóide parasito (Tetradonematidae) da formiga lava-pé, *Solenopsis* (Formicidae), em Mato Grosso. *Pesquisa Agropecuária Brasileira,* 23: 525–528.

Junker, EA. 1997. Untersuchungen zur Lebensweise und Entwicklung von *Myrmecophilus acervorum* (Panzer, 1799) (Saltatoria, Myrmecophilidae). *Articulata,* 12: 93–106.

Jutsum, A, Quinlan, R. 1978. Flight and substrate utilisation in laboratory-reared males of *Atta sexdens. Journal of Insect Physiology,* 24: 821–825.

Kaib, M, Eisermann, B, Schoeters, E, Billen, J, Franke, S, Francke, W. 2000. Task-related variation of postpha-ryngeal and cuticular hydrocarbon compositions in the ant *Myrmicaria eumenoides. Journal of Comparative Physiology A,* 186: 939–948.

Kaiser, H. 1986. Über Wechselbeziehungen zwischen Nematoden (Mermithidae) und Ameisen. *Zoologischer Anzeiger,* 217: 156–177.

Kaiser, H. 1991. Terrestrial and semiterrestrial Mermithidae. In *Manual of Agricultural Nematology,* ed. WR Nickle, 899–965. Boca Raton, FL: CRC Press.

Kaliszewska, ZA, Lohman, DJ, Sommer, K, Adelson, G, Rand, DB, Mathew, J, Talavera, G, Pierce, NE. 2015. When caterpillars attack: Biogeography and life history evolution of the Miletinae (Lepidoptera: Lycaenidae). *Evolution,* 69: 571–588.

Kaminski, LA, Volkmann, L, Callaghan, CJ, DeVries, PJ, Vila, R. 2020. The first known riodinid "cuckoo" butterfly reveals deep-time convergence and parallelism in ant social parasites. *Zoological Journal of the Linnean Society,* 193: 860–879.

Karawajew, W. 1906. Weitere Beobachtungen über Arten der Gattung *Antennophorus. Mémoires de la Société des Naturalistes de Kiew,* 20: 209–230.

Kaston, B. 1948. Spiders of Connecticut. *Bulletin of the Connecticut State Geological and Natural History Survey,* 70: 1–874.

Kather, R, Martin, SJ. 2015. Evolution of cuticular hydrocarbons in the Hymenoptera: A meta-analysis. *Journal of Chemical Ecology,* 41: 871–883.

Kelber, C, Rössler, W, Kleineidam, CJ. 2010. Phenotypic plasticity in number of glomeruli and sensory innervation of the antennal lobe in leaf-cutting ant workers (*A. vollenweideri*). *Developmental Neurobiology,* 70: 222–234.

Keller, L. 1993. *Queen Number and Sociality in Insects.* Oxford: Oxford University Press.

Kelso, L, Nice, MM. 1963. A Russian contribution to anting and feather mites. *Wilson Bulletin,* 75: 23–26.

Kemner, NVA. 1923. Hyphaenosymphilie, eine neue, merkwürdige Art von Myrmekophilie bei einem neuen myrmekophilen Schmetterling (*Wurthia aurivillii* n. Sp.) aus Java beobachtet. *Arkiv för Zoologi,* 15: 1–28.

Khaustov, AA. 2014. A review of myrmecophilous mites of the family Microdispidae (Acari, Heterostigmatina) of Western Siberia. *ZooKeys,* 454: 13–28.

Kieffer, J. 1904. Nouveaux proctotrypides myrmécophiles. *Bulletin de la Société d'Histoire Naturelle de Metz,* 23: 31–58.

Kieffer, J, André, E. 1911. Species des Hyménopteres d'Europe et d'Algérie. *Librerie Scientifique A,* 10: 913–1015.

King, JR, Tschinkel, WR. 2016. Experimental evidence that dispersal drives ant community assembly in human-altered ecosystems. *Ecology,* 97: 236–249.

King, JR, Warren, RJ, Bradford, MA. 2013. Social insects dominate Eastern US temperate hardwood forest macroinvertebrate communities in warmer regions. *PLOS One,* 8: e75843.

Kistner, DH. 1966. A revision of the African species of the Aleocharine tribe Dorylomimini (Coleoptera: Staphylinidae), II: The genera *Dorylomimus, Dorylonannus, Dorylogaster, Dorylobactrus,* and *Mimanomma,* with notes on their behavior. *Annals of the Entomological Society of America,* 59: 320–340.

Kistner, DH. 1969. The biology of termitophiles. In *Biology of Termites,* vol. 1, ed. K Krishna, FM Weesner, 525–557. New York: Academic Press.

Kistner, DH. 1975. Myrmecophilous staphylinidae associated with *Leptogenys* roger (Coleoptera; Hymenoptera, Formicidae). *Sociobiology,* 1: 1–19.

Kistner, DH. 1979. Social and evolutionary significance of social insect symbionts. In *Social Insects,* vol. 1, ed. HR Hermann, 340–413. New York: Academic Press.

Kistner, DH. 1982. The social insects' bestiary. In *Social Insects,* vol. 3, ed. HR Hermann, 1–244. New York: Academic Press.

Kistner, DH. 1989. New genera and species of Aleocharinae associated with ants of the genus *Leptogenys* and their relationships (Coleoptera: Staphylinidae; Hymenoptera: Formicidae). *Sociobiology,* 15: 299–323.

Kistner, DH. 1993. Cladistic analysis, taxonomic restructuring and revision of the Old World genera formerly classified as Dorylomimini with comments on their evolution and behavior (Coleoptera: Staphylinidae). *Sociobiology,* 22: 151–383.

Kistner, DH. 1997. New species, new genera, and new records of myrmecophiles associated with army ants (*Aenictus* sp.) with the description of a new subtribe of Staphylinidae (Coleoptera; Formicidae, Aenictinae). *Sociobiology,* 29: 123–221.

Kistner, DH. 2003. A new species of *Trachydonia* (Coleoptera: Staphylinidae, Aleocharinae) from Malaysia with some notes on its behavior as a guest of *Leptogenys* (Hymenoptera: Formicidae). *Sociobiology,* 42: 381–389.

Kistner, DH, Berghoff, SM, Maschwitz, U. 2003. Myrmecophilous Staphylinidae (Coleoptera) associated with *Dorylus* (*Dichthadia*) *laevigatus* (Hymenoptera: Formicidae) in Malaysia with studies of their behavior. *Sociobiology,* 41: 207–268.

Kistner, DH, Blum, MS. 1971. Alarm pheromone of *Lasius* (*Dendrolasius*) *spathepus* (Hymenoptera: Formicidae) and its possible mimicry by two species of *Pella* (Coleoptera: Staphylinidae). *Annals of the Entomological Society of America,* 64: 589–594.

Kistner, DH, Hermann, HR. 1982. *Social Insects.* London: Academic Press.

Kistner, DH, Jacobson, HR. 1975. A review of the myrmecophilous Staphylinidae associated with *Aenictus* in Africa and the Orient (Coleoptera, Hymenoptera, Formicidae) with notes on their behavior and glands. *Sociobiology,* 1: 20–73.

Kistner, DH, Jacobson, HR. 1990. Cladistic analysis and taxonomic revision of the ecitophilous tribe Ecitocharini with studies of their behavior and evolution (Coleoptera, Staphylinidae, Aleocharinae). *Sociobiology,* 17: 333–480.

Kistner, DH, von Beeren, C, Witte, V. 2008. Redescription of the generitype of *Trachydonia* and a new host record for *Maschwitzia ulrichi* (Coleoptera: Staphylinidae). *Sociobiology,* 52: 497.

Kitching, RL. 1983. Myrmecophilous organs of the larvae and pupa of the lycaenid butterfly *Jalmenus evagoras* (Donovan). *Journal of Natural History,* 17: 471–481.

Kitching, RL. 1987. Aspects of the natural history of the lycaenid butterfly *Allotinus major* in Sulawesi. *Journal of Natural History,* 21: 535–544.

Kitching, RL, Luke, B. 1985. The myrmecophilous organs of the larvae of some British Lycaenidae (Lepidoptera): A comparative study. *Journal of Natural History,* 19: 259–276.

Kloft, WJ, Woodruff, RE, Kloft, ES. 1979. *Formica integra* (Hymenoptera: Formicidae), IV: Exchange of food and trichome secretions between worker ants and the inquiline beetle, *Cremastocheilus castaneus* (Coleoptera: Scarabaeidae). *Tijdschrift voor Entomologie,* 122: 47–57.

Kolb, G. 1959. Untersuchungen über die Kernverhältnisse und morphologischen Eigenschaften symbiontischer Mikroorganismen bei verschiedenen Insekten. *Zeitschrift für Morphologie und Ökologie der Tiere,* 48: 1–71.

Kolbe, W. 1971. Untersuchungen über die Bindung von *Zyras humeralis* (Coleoptera, Staphylinidae) an Waldameisen. *Entomologische Blätter,* 67: 129–136.

Komatsu, T, Maruyama, M. 2016. Taxonomic recovery of the ant cricket *Myrmecophilus albicinctus* from *M. americanus* (Orthoptera, Myrmecophilidae). *ZooKeys,* 589: 97–106.

Komatsu, T, Maruyama, M, Hattori, M, Itino, T. 2018. Morphological characteristics reflect food sources and degree of host ant specificity in four *Myrmecophilus* crickets. *Insectes Sociaux,* 65: 47–57.

Komatsu, T, Maruyama, M, Itino, T. 2009. Behavioral differences between two ant cricket species in Nansei Islands: Host-specialist versus host-generalist. *Insectes Sociaux,* 56: 389–396.

Komatsu, T, Maruyama, M, Itino, T. 2010. Differences in host specificity and behavior of two ant cricket species (Orthoptera: Myrmecophilidae) in Honshu, Japan. *Journal of Entomological Science,* 45: 227–238.

Komatsu, T, Maruyama, M, Itino, T. 2013. Nonintegrated host association of *Myrmecophilus tetramorii,* a specialist myrmecophilous ant cricket (Orthoptera: Myrmecophilidae). *Psyche,* 2013: 568536.

Komatsu, T, Maruyama, M, Ueda, S, Itino, T. 2008. mtDNA phylogeny of Japanese ant crickets (Orthoptera: Myrmecophilidae): Diversification in host specificity and habitat use. *Sociobiology,* 52: 553.

Kontschán, J, Szőcs, G, Kiss, B, Khaustov, AA. 2019. Bark beetle associated trematurid mites (Acari: Uropodina: Trematuridae) from Asian Russia with description of a new species. *Systematic and Applied Acarology,* 24: 1592–1603.

Kronauer, DJ. 2020. *Army Ants: Nature's Ultimate Social Hunters.* Cambridge, MA: Harvard University Press.

Kronauer, DJ, Pierce, NE. 2011. Myrmecophiles. *Current Biology,* 21: R208–R209.

Krull, WH, Mapes, CR. 1952. Studies on the biology of *Dicrocoelium dendriticum* (Rudolphi, 1819), Looss, 1899 (Trematoda: Dicrocoeliidae), including its relation to the intermediate host, *Cionella lubrica* (Muiller), VII: The second intermediate host of *Dicrocoelium dendriticum. Cornell Veterinarian,* 42: 253–276.

Kwapich, CL. 2021. Green anole (*Anolis carolinensis*) eggs associated with nest chambers of the trap-jaw ant *Odontomachus brunneus. Southeastern Naturalist,* 20: 119–124.

Kwapich, CL. In prep. a. Mermithid worm infections alter behavior and development of the Florida harvester ant, *Pogonomyrmex badius.*

Kwapich, CL. In prep. b. *Forelius mccooki* during self-anointing (anting) by the great-tailed grackle, *Quiscalus mexicanus.*

Kwapich, CL, Eriksson, T, Hölldobler, B. In prep. a. Host breadth and behavior of the myrmecophilous Sonoran Desert spider, *Septentrinna bicalcarata.*

Kwapich, CL, Gadau, J, Hölldobler, B. 2017. The ecological and genetic basis of annual worker production in the desert seed harvesting ant, *Veromessor pergandei. Behavioral Ecology and Sociobiology,* 71: 110.

Kwapich, CL, Hölldobler, B. 2019. Destruction of spider-webs and rescue of ensnared nestmates by a granivorous desert ant (*Veromessor pergandei*). *American Naturalist,* 194: 395–404.

Kwapich, CL, Johnston, A. In prep. A new *Hymenorus* (Coleoptera: Alleculidae) kleptoparasite of the Florida harvester ant (*Pogonomyrmex badius*).

Kwapich, CL, Sosa-Calvo, J, Johnson, RA, Hölldobler, B. In prep. b. Host-specific polyphenism in the parasitic ant cricket, *Myrmecophilus manni.*

Kwapich, CL, Tschinkel, WR. 2013. Demography, demand, death, and the seasonal allocation of labor in the Florida harvester ant (*Pogonomyrmex badius*). *Behavioral Ecology and Sociobiology,* 67: 2011–2027.

Kwapich, CL, Tschinkel, WR. 2016. Limited flexibility and unusual longevity shape forager allocation in the Florida harvester ant (*Pogonomyrmex badius*). *Behavioral Ecology and Sociobiology,* 70: 221–235.

Lachaud, J-P. 1980. Les communications tactiles interspécifique chez les diapriides myrmécophiles *Lepidopria pedestris* Kieffer et *Solenopsia imitatrix* Wasmann et leur hote *Diplorhotrum fugax* Latr. (Solenopsis fugax Latr.). *Biologie- Écologie Méditerranéenne,* 7: 183–184.

Lachaud, J-P. 1981. Les glandes tégumentaires chez deux espéces de Diapriidae: Aspects structuraux et ultrastructuraux. *Bulletin Intérieur de la Section Française de l'UIEIS, Toulouse, France:* 83–85.

Lachaud, J-P. 1982. Estudio sobre las relaciones trofalacticas entre *Lepidopria pedestris* Kieffer (Hymenoptera, Diapriidae) y su nuesped *Diplorhoptrum fugax* Latreille (Hymenoptera, Formicidae). *Folia Entomológica Mexicana,* 54: 46–47.

Lachaud, J-P, Klompen, H, Pérez-Lachaud, G. 2016. *Macrodinychus* mites as parasitoids of invasive ants: An overlooked parasitic association. *Scientific Reports, 6:* 29995.

Lachaud, J-P, Lenoir, A, Hughes, DP, eds. 2013. *Ants and Their Parasites.* London: Hindawi Press.

Lachaud, J-P, Lenoir, A, Witte, V, eds. 2012. *Ants and Their Parasites.* London: Hindawi Press.

Lachaud, J-P, Passera, L. 1982. Donnees sur la biologie de trois Diapriidae myrmecophiles: *Plagiopria passerai* Masner, *Solenopsia imitatrix* Wasmann et *Lepidopria pedestris* Kieffer. *Insectes Sociaux, 29:* 561–568.

Lachaud, J-P, Pérez-Lachaud, G. 2009. Impact of natural parasitism by two eucharitid wasps on a potential biocontrol agent ant in southeastern Mexico. *Biological Control, 48:* 92–99.

Lachaud, J-P, Pérez-Lachaud, G. 2012. Diversity of species and behavior of hymenopteran parasitoids of ants: A review. *Psyche:* 134746.

Lackner, T, Hlaváč, P. 2012. Description of a new species of *Sternocoelis* from Morocco with proposal of the *Sternocoelis marseulii* species group (Coleoptera, Histeridae). *ZooKeys, 181:* 11–21.

Lackner, T, Yélamos, T. 2001. Contribution to the knowledge of the Moroccan fauna of *Sternocoelis* Lewis, 1888 and Eretmotus Lacordaire, 1854 (Coleoptera: Histeridae). *Zapateri Revista Aragon Entomology, 9:* 99–102.

Lahav, S, Soroker, V, Hefetz, A, Vander Meer, RK. 1999. Direct behavioral evidence for hydrocarbons as ant recognition discriminators. *Naturwissenschaften, 86:* 246–249.

Lanan, MC, Rodrigues, PAP, Agellon, A, Jansma, P, Wheeler, DE. 2016. A bacterial filter protects and structures the gut microbiome of an insect. *ISME Journal, 10:* 1866–1876.

Lange, D, Calixto, ES, Del-Claro, K. 2017. Variation in extrafloral nectary productivity influences the ant foraging. *PLOS One, 12:* e0169492.

Le Breton, J, Takaku, G, Tsuji, K. 2006. Brood parasitism by mites (Uropodidae) in an invasive population of the pest-ant *Pheidole megacephala. Insectes Sociaux, 53:* 168–171.

LeClerc, MG, McClain, DC, Black, HL, Jorgensen, CD. 1987. An inquiline relationship between the tailless whip-scorpion *Phrynus gervaisii* and the giant tropical ant *Paraponera clavata. Journal of Arachnology, 15:* 129–130.

Lee, JE, Morimoto, K. 1991. Descriptions of the egg and first-instar larva of *Clytra arida* Weise (Coleoptera: Chrysomelidae). *Journal of the Faculty of Agriculture, Kyushu University, 35:* 93–99.

Lehrer, AZ. 2006a. Contributions taxonomiques et zoogéographique sur la famille des Bengaliidae (Diptera). *Bulletin de la Société Entomologique de Mulhouse, 62:* 1–11.

Lehrer, AZ. 2006b. Un autre point de vue taxonomique sur les types porte-noms. *Fragmenta dipterologica, 5:* 1–8.

Lehrer, AZ. 2008. Une nouvelle espèce thaïlandaise du genre *Afridigalia* Lehrer (Diptera, Bengaliidae). *Fragmenta dipterologica, 16:* 28–29.

Lehtinen, PT. 1987. Association of uropodid, prodinychid, polyaspidid, antennophorid, sejid, microgynid, and zerconid mites with ants. *Entomologisk tidskrift, 108:* 13–20.

Le Masne, G. 1961a. Observations sur le comportement de *Paussus favieri* Fairm., hote de la fourmi *Pheidole pallidula* Nyl. *Annales de la Faculte des Sciences de Marseille, 31:* 111–130.

Le Masne, G. 1961b. Recherches sur la biologie des animaux myrmécophiles, I: L'adoption des *Paussus favieri* Fairm. Par une nouvelle société de *Pheidole pallidula* Nyl. *Comptes rendus de l'Académie des Sciences, 253:* 1621–1623.

Le Masne, G. 1961c. Recherches sur la biologie des animaux myrmécophiles: Observations sur le régime alimentaire de *Paussus favieri* Fairm., hôte de la fourmi *Pheidole pallidula. Comptes rendus de l'Académie des Sciences, 253:* 1356–1357.

Lenoir, A, Chalon, Q, Carvajal, A, Ruel, C, Barroso, A, Lackner, T, Boulay, R. 2012. Chemical integration of myrmecophilous guests in *Aphaenogaster* ant nests. *Psyche:* 840860.

Lenoir, A, d'Ettorre, P, Errard, C, Hefetz, A. 2001. Chemical ecology and social parasitism in ants. *Annual Review of Entomology, 46:* 573–599.

Lenoir, A, Fresneau, D, Errard, C, Hefetz, A. 1999. Individuality and colonial identity in ants: The emergence of the social representation concept. In *Information Processing in Social Insects,* 219–237. Basel, Switzerland: Birkhäuser Verlag.

Liang, D, Silverman, J. 2000. "You are what you eat": Diet modifies cuticular hydrocarbons and nestmate recognition in the Argentine ant, *Linepithema humile. Naturwissenschaften, 87:* 412–416.

Lichman, BR, Godden, GT, Hamilton, JP, Palmer, L, Kamileen, MO, Zhao, D, Vaillancourt, B, Wood, JC, Sun, M, Kinser, TJ, Henry, LK, Rodriguez-Lopez, C, Dudareva, N, Soltis, DE, Soltis, PS, Buell, CR, O'Connor, SE. 2020. The evolutionary origins of the cat attractant nepetalactone in catnip. *Science Advances,* 6: eaba0721.

Liebig, J. 2010. Hydrocarbon profiles indicate fertility and dominance status in ant, bee, and wasp colonies. In *Insect Hydrocarbons: Biology, Biochemistry, and Chemical Ecology,* ed. GJ Blomquist, AG Bagnères, 254–281. Cambridge, UK: Cambridge University Press.

Liebig, J, Peeters, C, Oldham, NJ, Markstädter, C, Hölldobler, B. 2000. Are variations in cuticular hydrocarbons of queens and workers a reliable signal of fertility in the ant *Harpegnathos saltator? Proceedings of the National Academy of Sciences,* 97: 4124–4131.

Liersch, S, Schmid-Hempel, P. 1998. Genetic variation within social insect colonies reduces parasite load. *Proceedings of the Royal Society of London. Series B: Biological Sciences,* 265: 221–225.

Linksvayer, TA, McCall, AC, Jensen, RM, Marshall, CM, Miner, JW, McKone, MJ. 2002. The function of hitchhiking behavior in the leaf-cutting ant *Atta cephalotes. Biotropica,* 34: 93–100.

Líznarová, E, Pekár, S. 2016. Metabolic specialisation on preferred prey and constraints in the utilisation of alternative prey in an ant-eating spider. *Zoology,* 119: 464–470.

Lohman, DJ, Samarita, VU. 2009. The biology of carnivorous butterfly larvae (Lepidoptera: Lycaenidae: Miletinae: Miletini) and their ant-tended hemipteran prey in Thailand and the Philippines. *Journal of Natural History,* 43: 569–581.

Loiácono, M. 1985. A new diapriid (Hymenoptera) parasitoid of larvae of *Acromyrmex ambiguus* (Emery) (Hymenoptera, Formicidae) from Uruguay. *Revista de la Sociedad Entomológica Argentina,* 44: 129–136.

Loiácono, M, Margaria, C, Moreira, DD, Aquino, D. 2013. A new species of *Szelenyiopria* Fabritius (Hymenoptera: Diapriidae), larval parasitoid of *Acromyrmex subterraneus subterraneus* (Forel) (Hymenoptera: Formicidae) from Brazil. *Zootaxa,* 3646: 228–234.

Loiácono, MS, Margaría, CB, Quirán, E, Molas, BC. 2002. Revision of the myrmecophilous diapriid genus *Bruchopria* Kieffer (Hymenoptera, Proctotrupoidea,

Diapriidae). *Revista Brasileira de Entomologia,* 46: 231–235.

Longino, JT. 1984. True anting by the capuchin, *Cebus capucinus. Primates,* 25: 243–245.

Lucas, C, Pho, D, Jallon, J, Fresneau, D. 2005. Role of cuticular hydrocarbons in the chemical recognition between ant species in the *Pachycondyla villosa* species complex. *Journal of Insect Physiology,* 51: 1148–1157.

MacKay, WP. 1982. The effect of predation of western widow spiders (Araneae: Theridiidae) on harvester ants (Hymenoptera: Formicidae). *Oecologia,* 53: 406–411.

Maderspacher, F, Stensmyr, M. 2011. Myrmecomorphomania. *Current Biology,* 21: R291–R293.

Malicky, H. 1969. Versuch einer Analyse der ökologischen Beziehungen zwischen Lycaeniden (Lepidoptera) und Formiciden (Hymenoptera). *Tijdschrift voor Entomologie,* 112: 213–298.

Malicky, H. 1970. New aspects of the association between lycaenid larvae (Lycaenidae) and ants (Formicidae, Hymenoptera). *Journal of the Lepidopterists' Society,* 24: 190–202.

Markl, H. 1967. Die Verständigung durch Stridulationssignale bei Blattschneiderameisen, I: Die biologische Bedeutung der Stridulation. *Zeitschrift für vergleichende Physiologie,* 57: 299–330.

Markl, H. 1968. Die Verständigung durch Stridulationssignale bei Blattschneiderameisen, II: Erzeugung und Eigenschaften der Signale. *Zeitschrift für vergleichende Physiologie,* 60: 103–150.

Markl, H. 1970. Die Verständigung durch Stridulationssignale bei Blattschneiderameisen, III: Die Empfindlichkeit für Substratvibrationen. *Zeitschrift für vergleichende Physiologie,* 69: 6–37.

Markl, H. 1973. The evolution of stridulatory communication in ants. *Proceedings of the International Congress IUSSI, London,* 7: 258–265.

Markl, H. 1983. Vibrational communication. In *Neuroethology and Behavioral Physiology,* ed. F Huber, H Markl, 332–353. Berlin: Springer-Verlag.

Markl, H. 1985. Manipulation, modulation, information, cognition: Some of the riddles of communication. In *Experimental Behavioral Ecology and Sociobiology (Fortschritte der Zoologie, no. 31),* ed. B Hölldobler, M Lindauer, 163–194. Sutherland, MA: Sinauer Associates.

Markl, H, Fuchs, S. 1972. Klopfsignale mit Alarmfunktion bei Rossameisen (Camponotus, Formicidae, Hymenoptera). *Zeitschrift für vergleichende Physiologie,* 76: 204–225.

Markl, H, Hölldobler, B. 1978. Recruitment and food-retrieving behavior in *Novomessor* (Formicidae, Hymenoptera). *Behavioral Ecology and Sociobiology,* 4: 183–216.

Marsh, P. 1979. Hybrizontidae. In *Catalog of Hymenoptera in America North of Mexico,* ed. KV Krombein, PD Hurd, DR Smoth, BD Burks, 144–313. Washington, DC: Smithsonian Institution Press.

Marsh, P. 1989. Notes on the genus *Hybrizon* in North America (Hymenoptera: Paxylommatidae). *Proceedings of the Entomological Society of Washington,* 91: 29–34.

Marson, JE. 1946. The ant mimic *Myrmarachne plataleoides* camb. in India. *Entomologists Monthly Magazine,* 82: 52–53.

Martin, JO. 1922. Studies in the genus *Hetaerius* (Co., Hiseridae). *Entomological News,* 33: 272–277.

Martin, S, Drijfhout, F. 2009a. A review of ant cuticular hydrocarbons. *Journal of Chemical Ecology,* 35: 1151–1161.

Martin, S, Drijfhout, F. 2009b. Nestmate and task cues are influenced and encoded differently within ant cuticular hydrocarbon profiles. *Journal of Chemical Ecology,* 35: 368–374.

Martins, C, Moreau, CS. 2020. Influence of host phylogeny, geographical location and seed harvesting diet on the bacterial community of globally distributed *Pheidole* ants. *PeerJ,* 8: e8492.

Martins Neto, RG. 1991. Sistemática dos Ensifera (Insecta, Orthopteroida) da Formação Santana, Cretáceo Inferior do Nordeste do Brasil. *Acta Geológica Leopoldensia,* 32: 3–160.

Martín-Vega, D, Garbout, A, Ahmed, F, Wicklein, M, Goater, CP, Colwell, DD, Hall, MJ. 2018. 3D virtual histology at the host / parasite interface: Visualisation of the master manipulator, *Dicrocoelium dendriticum,* in the brain of its ant host. *Scientific Reports,* 8: 8587.

Maruyama, M. 2004. Four new species of *Myrmecophilus* (Orthoptera, Myrmecophilidae) from Japan. *Bulletin of the National Museum of Nature and Science,* 30: 37–44.

Maruyama, M. 2006. Revision of the Palearctic species of the myrmecophilous genus *Pella:* Coleoptera, Staphylinidae, Aleocharinae. *National Science Museum Monographs,* 32: 1–207.

Maruyama, M, Akino, T, Hashim, R, Komatsu, T. 2009. Behavior and cuticular hydrocarbons of myrmecophi-lous insects (Coleoptera: Staphylinidae; Diptera: Phoridae; Thysanura) associated with Asian *Aenictus* army ants (Hymenoptera; Formicidae). *Sociobiology,* 54: 19–35.

Maruyama, M, Disney, RHL, Hashim, R. 2008. Three new species of legless, wingless scuttle flies (Diptera: Phoridae) associated with army ants (Hymenoptera: Formicidae) in Malaysia. *Sociobiology,* 52: 485–496.

Maruyama, M, Hashim, SHYR, Ito, F. 2003. A new myrmecophilous species of *Drusilla* (Coleoptera, Staphylinidae, Aleocharinae) from peninsular Malaysia, a possible Batesian mimic associated with *Crematogaster inflata* (Hymenoptera, Formicidae Myrmicinae). *Japanese Journal of Systematic Entomology,* 9: 267–275.

Maruyama, M, Komatsu, T, Kudo, S, Shimada, T, Kino-mura, K. 2013. *The Guests of Japanese Ants.* Hadano-shi, Kanagawa, Japan: Tokai University Press.

Maruyama, M, Matsumoto, T, Itioka, T. 2011. Rove beetles (Coleoptera: Staphylinidae) associated with *Aenictus laeviceps* (Hymenoptera: Formicidae) in Sarawak, Malaysia: Strict host specificity, and first myrmecoid Aleocharini. *Zootaxa,* 3102: 1–26.

Maruyama, M, Parker, J. 2017. Deep-time convergence in rove beetle symbionts of army ants. *Current Biology,* 27: 920–926.

Maruyama, M, von Beeren, C, Hashim, R. 2010a. Aleocha-rine rove beetles (Coleoptera, Staphylinidae) associated with *Leptogenys* Roger, 1861 (Hymenoptera, Formicidae), I: Review of three genera associated with *L. distinguenda* (Emery, 1887) and *L. mutabilis* (Smith, 1861). *ZooKeys,* 59: 47–60.

Maruyama, M, von Beeren, C, Witte, V. 2010b. Aleocharine rove beetles (Coleoptera, Staphylinidae) associated with *Leptogenys* Roger, 1861 (Hymenoptera, Formi-cidae), II: Two new genera and two new species associated with *L. borneensis* Wheeler, 1919. *ZooKeys,* 59: 61–72.

Maschwitz, U. 1964. Gefahrenalarmstoffe und Ge-fahrenalarmierung bei sozialen Hymenopteren. *Zeitschrift für vergleichende Physiologie,* 47: 596–655.

Maschwitz, U. 1981. Fliegen als Wegelagerer und Parasiten bei Ameisen. *Nachrichten des Entomologischen Vereins Apollo, Frankfurt am Main, N.F.,* 2: 57–60.

Maschwitz, U, Go, C, Kaufmann, E, Buschinger, A. 2004. A unique strategy of host colony exploitation in a parasitic ant: Workers of *Polyrhachis lama* rear their brood in neighbouring host nests. *Naturwissenschaften,* 91: 40–43.

Maschwitz, U, Hölldobler, B. 1970. Der Kartonnestbau bei *Lasius fuliginosus* Latr. (Hym. Formicidae). *Zeitschrift für vergleichende Physiologie,* 66: 176–189.

Maschwitz, U, Nassig, WA, Dumpert, K, Fiedler, K. 1988. Larval carnivory and myrmecoxeny, and imaginal myrmecophily in miletine lycaenids (Lepidoptera, Lycaenidae) on the Malay Peninsula. *Lepidoptera Science,* 39: 167–181.

Maschwitz, U, Schönegge, P. 1980. Fliegen als Beute und Bruträuber bei Ameisen. *Insectes Sociaux,* 27: 1–4.

Maschwitz, U, Schroth, M, Hänel, H, Pong, TY. 1984. Lycaenids parasitizing symbiotic plant-ant partnerships. *Oecologia,* 64: 78–80.

Maschwitz, U, Steghaus-Kovac, S, Gaube, R, Hänel, H. 1989. A South East Asian ponerine ant of the genus *Leptogenys* (Hym., Form.) with army ant life habits. *Behavioral Ecology and Sociobiology,* 24: 305–316.

Maschwitz, U, Weissflog, A, Seebauer, S, Disney, RHL, Witt, V. 2008. Studies on European ant decapitating flies (Diptera: Phoridae), I: Releasers and phenology of parasitism of *Pseudacteon formicarum. Sociobiology,* 51: 127–140.

Maschwitz, U, Wüst, M, Schurian, K. 1975. Bläulingsraupen als Zuckerlieferanten für Ameisen. *Oecologia,* 18: 17–21.

Masner, L. 1959. A revision of ecitophilous diapriid-genus *Mimopria* Holmgren (Hym., Proctotrupoidea). *Insectes Sociaux,* 6: 361–367.

Masner, L. 1993. Superfamily Proctotrupoidea. In *Hymenoptera of the World: An Identification Guide to Families,* ed. H Goulet, J Huber, 537–557. Ottawa, Canada: Agriculture Canada Publications.

Masner, L, García, R., Luis, J. 2002. The genera of Diapriinae (Hymenoptera: Diapriidae) in the New World. *Bulletin of the American Museum of Natural History,* 268: 1–138.

Masters, WM. 1979. Insect disturbance stridulation: Its defensive role. *Behavioral Ecology and Sociobiology,* 5: 187–200.

Masters, WM. 1980. Insect disturbance stridulation: Characterization of airborne and vibrational components of the sound. *Journal of Comparative Physiology,* 135: 259–268.

Masters, WM, Tautz, J, Fletcher, NH, Markl, H. 1983. Body vibration and sound production in an insect (*Atta sexdens*) without specialized radiating structures. *Journal of Comparative Physiology,* 150: 239–249.

Mathew, A. 1934. The life-history of the spider *Myrmarachne plataleoides* (Cambr.) a mimic of the Indian red ant. *Journal of the Bombay Natural History Society,* 37: 369–374.

Mathew, A. 1954. Observations on the habits of two spider mimics of the red ant, *Oecophylla smaragdina* (Fabr.). *Journal of the Bombay Natural History Society,* 52: 249–263.

Maurizi, E, Fattorini, S, Moore, W, Di Giulio, A. 2012. Behavior of *Paussus favieri* (Coleoptera, Carabidae, Paussini): A myrmecophilous beetle associated with *Pheidole pallidula* (Hymenoptera, Formicidae). *Psyche:* 940315.

McAtee, WL. 1947. Wild turkey anting. *Ornithology,* 64: 130–131.

McInnes, DA, Tschinkel, WR. 1996. Mermithid nematode parasitism of *Solenopsis* ants (Hymenoptera: Formicidae) of Northern Florida. *Annals of the Entomological Society of America,* 89: 231–237.

McMahan, EA. 1983. Adaptations, feeding preferences, and biometrics of a termite-baiting assassin bug (Hemiptera: Reduviidae). *Annals of the Entomological Society of America,* 76: 483–486.

Mello-Leitão, CF. 1923. Sobre uma aranha parasita de saúva. *Revista do Museu Paulista,* 13: 523–525.

Mellor, J. 1922. Notes on a *"Bengalia"*-like Fly, which I have called the "Highwayman" Fly, and its behaviour towards certain species of ants. *Sudan Notes and Records,* 5: 95–100.

Mendonça, CAF, Pesquero, MA, Carvalho, RdSD, de Arruda, FV. 2019. Myrmecophily and myrmecophagy of *Attacobius lavape* (Araneae: Corinnidae) on *Solenopsis saevissima* (Hymenoptera: Myrmicinae). *Sociobiology,* 66: 545–550.

Menzel, JG, Tautz, J. 1994. Functional morphology of the subgenual organ of the carpenter ant. *Tissue and Cell,* 26: 735–746.

Meyer-Hozak, C. 2000. Population biology of *Maculinea rebeli* (Lepidoptera: Lycaenidae) on the chalk grasslands of Eastern Westphalia (Germany) and implications for conservation. *Journal of Insect Conservation,* 4: 63–72.

Michener, CD. 1969. Comparative social behavior of bees. *Annual Review of Entomology,* 14: 299–342.

Molero-Baltanás, R, Bach de Roca, C, Tinaut, A, Pérez, J, Gaju-Ricart, M. 2017. Symbiotic relationships between silverfish (Zygentoma: Lepismatidae, Nicoletiidae) and ants (Hymenoptera: Formicidae) in the Western Palaearctic. A quantitative analysis of data from Spain. *Myrmecological News,* 24: 107–122.

Moore, W, Di Giulio, A. 2019. Out of the burrow and into the nest: Functional anatomy of three life history stages of *Ozaena lemoulti* (Coleoptera: Carabidae) reveals an obligate life with ants. *PLOS One,* 14: e0209790.

Moreau, CS. 2020. Symbioses among ants and microbes. *Current Opinion in Insect Science,* 39: 1–5.

Morel, L, Vander Meer, RK, Lavine, BK. 1988. Ontogeny of nestmate recognition cues in the red carpenter ant (*Camponotus floridanus*). *Behavioral Ecology and Sociobiology,* 22: 175–183.

Morrill, W. 1975. An unusual predator of the Florida harvester ant. *Journal of the Georgia Entomological Society,* 10: 50–51.

Morrison, LW. 2000. Biology of Pseudacteon (Diptera: Phoridae) ant parasitoids and their potential to control imported *Solenopsis* fire ants (Hymenoptera: Formicidae). *Recent Research Developments in Entomology,* 3: 1–13.

Morrison, LW, Dall'Aglio-Holvorcem, CG, Gilbert, LE. 1997. Oviposition behavior and development of *Pseudacteon* flies (Diptera: Phoridae), parasitoids of *Solenopsis* fire ants (Hymenoptera: Formicidae). *Environmental Entomology,* 26: 716–724.

Morrison, LW, King, JR. 2004. Host location behavior in a parasitoid of imported fire ants. *Journal of Insect Behavior,* 17: 367–383.

Morrison, LW, Porter, SD. 2006. Post-release host-specificity testing of *Pseudacteon tricuspis,* a phorid parasitoid of Solenopsis invicta fire ants. *BioControl,* 51: 195–205.

Moser, JC. 1964. Inquiline roach responds to trail-marking substance of leaf-cutting ants. *Science,* 143: 1048–1049.

Moser, JC. 1967. Mating activities of *Atta texana* (Hymenoptera, Formicidae). *Insectes Sociaux,* 14: 295–312.

Moser, JC, Neff, SE. 1971. *Pholeomyia comans* (Diptera: Milichiidae) an associate of *Atta texana:* Larval anatomy and notes on biology. *Zeitschrift für Angewandte Entomologie,* 69: 343–348.

Mota, LL, Kaminski, LA, Freitas, AV. 2020. The tortoise caterpillar: Carnivory and armoured larval morphology of the metalmark butterfly *Pachythone xanthe* (Lepidoptera: Riodinidae). *Journal of Natural History,* 54: 309–319.

Mou, YC. 1938. *Morphologische und histologische Studien über Paussidendrüsen; Mit 40 Abb. im Text.* PhD dissertation, Rheinische Friedrich-Wilhelms-Universität zu Bonn.

Moulton, MJ, Song, H, Whiting, MF. 2010. Assessing the effects of primer specificity on eliminating numt coamplification in DNA barcoding: A case study from Orthoptera (Arthropoda: Insecta). *Molecular Ecology Resources,* 10: 615–627.

Moura, RR, Vasconcellos-Neto, J, de Oliveira Gonzaga, M. 2017. Extended male care in *Manogea porracea* (Araneae: Araneidae): The exceptional case of a spider with amphisexual care. *Animal Behaviour,* 123: 1–9.

Murray, EA, Carmichael, AE, Heraty, JM. 2013. Ancient host shifts followed by host conservatism in a group of ant parasitoids. *Proceedings of the Royal Society B: Biological Sciences,* 280: 20130495.

Musthak Ali, T, Baroni Urbani, C, Billen, J. 1992. Multiple jumping behaviors in the ant *Harpegnathos saltator. Naturwissenschaften,* 79: 374–376.

Muzzi, M, Di Giulio, A. 2019. The ant nest "bomber": Explosive defensive system of the flanged bombardier beetle *Paussus favieri* (Coleoptera, Carabidae). *Arthropod Structure and Development,* 50: 24–42.

Mynhardt, G. 2013. Declassifying myrmecophily in the Coleoptera to promote the study of ant-beetle symbioses. *Psyche:* 696401.

Mynhardt, G, Wenzel, JW. 2010. Phylogenetic analysis of the myrmecophilous *Cremastocheilus Knoch* (Coleoptera, Scarabaeidae, Cetoniinae), based on external adult morphology. *ZooKeys,* 34: 129–140.

Nagel, Peter. 1979. Aspects of the evolution of myrmecophilous adaptations in Paussinae (Coleoptera, Carabidae). *Miscellaneous Papers of the Agricultural University Wageningen,* 18: 15–34.

Nagel, P. 1987. *Arealsystemanalyse afrikanischer Fühlerkäfer (Coleoptera, Carabidae, Paussinae): Ein Beitrag zur Rekonstruktion der Landschaftsgenese.* Vol. 21. Wiesbaden: Franz Steiner Verlag.

Nash, DR, Als, TD, Maile, R, Jones, GR, Boomsma, JJ. 2008. A mosaic of chemical coevolution in a large blue butterfly. *Science,* 319: 88–90.

Nash, DR, Boomsma, JJ. 2008. Communication between hosts and social parasites. In *Sociobiology of Communication:*

An Interdisciplinary Perspective, vol. 55, ed. P d'Ettorre, DP Hughes, 55–79. Oxford: Oxford University Press.

Naumann, I. 1986. A revision of the Indo-Australian Smicromorphinae (Hymenoptera: Chalcididae). *Memoirs of the Queensland Museum,* 22: 169–187.

Nedeljković, Z, Ricarte, A, Zorić, LŠ, Đan, M, Obreht-Vidaković, D, Vujić, A. 2018. The genus *Xanthogramma* Schiner, 1861 (Diptera: Syrphidae) in southeastern Europe, with descriptions of two new species. *Canadian Entomologist,* 150: 440–464.

Nehring, V, Dani, R, Calamai, L, Turillazzi, S, Bohn, H, Klass, KD, d'Ettorre, P. 2016. Chemical disguise of myrmecophilous cockroaches and its implications for understanding nestmate recognition mechanisms in leaf-cutting ants. *BMC Ecology,* 16: 35.

Nehring, V, Evison, SE, Santorelli, LA, d'Ettorre, P, Hughes, WO. 2011. Kin-informative recognition cues in ants. *Proceedings of the Royal Society B: Biological Sciences,* 278: 1942–1948.

Nelson, XJ. 2010. Polymorphism in an ant mimicking jumping spider. *Journal of Arachnology,* 38: 139–141.

Nelson, XJ, Card, A. 2015. Locomotory mimicry in ant-like spiders. *Behavioral Ecology,* 27: 700–707.

Nelson, XJ, Jackson, RR. 2006. Compound mimicry and trading predators by the males of sexually dimorphic Batesian mimics. *Proceedings of the Royal Society B: Biological Sciences,* 273: 367–372.

Nelson, XJ, Jackson, RR. 2007. Complex display behaviour during the intraspecific interactions of myrmecomorphic jumping spiders (Araneae, Salticidae). *Journal of Natural History,* 41: 1659–1678.

Nelson, XJ, Jackson, RR. 2009a. Aggressive use of Batesian mimicry by an ant-like jumping spider. *Biology Letters,* 5: 755–757.

Nelson, XJ, Jackson, RR. 2009b. Collective Batesian mimicry of ant groups by aggregating spiders. *Animal Behaviour,* 78: 123–129.

Nelson, XJ, Jackson, RR. 2012. How spiders practice aggressive and Batesian mimicry. *Current Zoology,* 58: 620–629.

Nentwig, W. 1982. Why do only certain insects escape from a spider's web? *Oecologia,* 53: 412–417.

Neupert, S, Hornung, M, Grenwille Millar, J, Kleineidam, CJ. 2018. Learning distinct chemical labels of nestmates in ants. *Frontiers in Behavioral Neuroscience,* 12: 12.

Newcomer, EJ. 1912. Some observations on the relations of ants and lycaenid caterpillars, and a description of the relational organs of the latter. *Journal of the New York Entomological Society,* 20: 31–36.

Nichols, JT. 1940. Synbranch eel in ant nest. *Copeia,* 3: 202.

Noda, T, Meguri, T, Iimure, K, Ono, M, Araki, T. 2011. Potential of D-erythro-C14-sphingosine as an adjuvant for a fungal pesticide of *Nomuraea rileyi. Bioscience, Biotechnology, and Biochemistry,* 75: 373–375.

Nogueira-de-Sá, F, Trigo, JR. 2002. Do fecal shields provide physical protection to larvae of the tortoise beetles *Plagiometriona flavescens* and *Stolas chalybea* against natural enemies? *Entomologia Experimentalis et Applicata,* 104: 203–206.

Notton, DG. 1994. New eastern Palaearctic myrmecophile *Lepidopria* and *Tetramopria* (Hymenoptera, Proctotrupoidea, Diapriidae, Diapriini). *Insecta Koreana,* 11: 64–74.

O'Donnell, S. 2017. Evidence for facilitation among avian army-ant attendants: Specialization and species associations across elevations. *Biotropica,* 49: 665–674.

Okubo, T, Yago, M, Itioka, T. 2009. Immature stages and biology of Bornean *Arhopala* butterflies (Lepidoptera, Lycaenidae) feeding on myrmecophytic *Macaranga. Lepidoptera Science,* 60: 37–51.

Oliveira, PS. 1985. On the mimetic association between nymphs of *Hyalymenus* spp. (Hemiptera: Alydidae) and ants. *Zoological Journal of the Linnean Society,* 83: 371–384.

Oliveira, PS, Sazima, I. 1984. The adaptive bases of ant-mimicry in a neotropical aphantochilid spider (Araneae: Aphantochilidae). *Biological Journal of the Linnean Society,* 22: 145–155.

Olmstead, KL. 1994. Waste products as chrysomelid defenses. In *Novel Aspects of the Biology of Chrysomelidae,* ed. E Petitpierre, 311–318. The Netherlands: Kluwer Academic Publishers.

Olmstead, KL, Denno, RF. 1993. Effectiveness of tortoise beetle larval shields against different predator species. *Ecology,* 74: 1394–1405.

Orivel, J, Servigne, P, Cerdan, P, Dejean, A, Corbara, B. 2004. The ladybird *Thalassa saginata,* an obligatory myrmecophile of *Dolichoderus bidens* ant colonies. *Naturwissenschaften,* 91: 97–100.

Orr, M, Seike, S, Benson, W, Gilbert, L. 1995. Flies suppress fire ants. *Nature,* 373: 292–293.

Ortega-Morales, AI, Rodríguez, QKS, Garza-Hernández, JA, Adeniran, AA, Hernández-Triana, LM, Rodríguez-Pérez, MA. 2017. First record of the ant cricket *Myrmecophilus* (Myrmecophilina) *americanus* (Orthoptera: Myrmecophilidae) in Mexico. *Zootaxa,* 4258: 195–200.

Ott, R, von Beeren, C, Hashim, R, Witte, V, Harvey, M. 2015. *Sicariomorpha,* a new myrmecophilous goblin spider genus (Araneae, Oonopidae) associated with Asian army ants. *American Museum Novitates,* 3843: 1–14.

Ozaki, M, Wada-Katsumata, A, Fujikawa, K, Iwasaki, M, Yokohari, F, Satoji, Y, Nisimura, T, Yamaoka, R. 2005. Ant nestmate and non-nestmate discrimination by a chemosensory sensillum. *Science,* 309: 311–314.

Paarmann, W, Erbeling, L, Spinnler, K. 1986. Ant and ant brood preying larvae: An adaptation of carabid beetles to arid environments. In *Carabid Beetles: Their Adaptations and Dynamics: XVIIth International Congress of Entomology,* ed. PJD Boer, ML Luff, D Mossakowski, F Weber, 79–90. Stuttgart: Gustav Fischer Verlag.

Painting, CJ, Nicholson, CC, Bulbert, MW, Norma-Rashid, Y, Li, D. 2017. Nectary feeding and guarding behavior by a tropical jumping spider. *Frontiers in Ecology and the Environment,* 15: 469–470.

Päivinen, J, Ahlroth, P, Kaitala, V. 2002. Ant-associated beetles of Fennoscandia and Denmark. *Entomologica Fennica,* 13: 20–40.

Päivinen, J, Ahlroth, P, Kaitala, V, Kotiaho, JS, Suhonen, J, Virola, T. 2003. Species richness and regional distribution of myrmecophilous beetles. *Oecologia,* 134: 587–595.

Päivinen, J, Ahlroth, P, Kaitala, V, Suhonen, J. 2004. Species richness, abundance and distribution of myrmecophilous beetles in nests of *Formica aquilonia* ants. *Annales Zoologici Fennici,* 41: 442–454.

Panzer, GWF. 1799. *Faunae Insectorum Germanicae initia oder Deutschlands Insecten.* Vol. 68. Nürnberg: Felseckerschen Buchhandlung.

Pardo-Locarno, L, Morón, M, Gaigl, A. 2006. Los estados inmaduros de *Coelosis biloba* (Coleoptera: Melolonthidae: Dynastinae) y notas sobre su biología. *Revista Mexicana de Biodiversidad,* 77: 215–224.

Park, O. 1932. The myrmecocoles of *Lasius umbratus* mixtus aphidicola Walsh. *Annals of the Entomological Society of America,* 25: 77–88.

Park, O. 1964. Observations upon the behavior of myrmecophilous pselaphid beetles. *Pedobiologia,* 4: 129–137.

Parker, J. 2016. Myrmecophily in beetles (Coleoptera): Evolutionary patterns and biological mechanisms. *Myrmecological News,* 22: 65–108.

Parker, J, Grimaldi, DA. 2014. Specialized myrmecophily at the ecological dawn of modern ants. *Current Biology,* 24: 2428–2434.

Parker, J, Owens, B. 2018. *Batriscydmaenus* Parker and Owens, new genus, and convergent evolution of a "reductive" ecomorph in socially symbiotic Pselaphinae (Coleoptera: Staphylinidae). *Coleopterists Bulletin,* 72: 219–229.

Parmelee, JR. 1999. Trophic ecology of a tropical anuran assemblage. *Scientific Papers of the Natural History Museum, the University of Kansas,* 11: 1–59.

Parmentier, T. 2019. Host following of an ant associate during nest relocation. *Insectes Sociaux,* 66: 329–332.

Parmentier, T, Bouillon, S, Dekoninck, W, Wenseleers, T. 2016a. Trophic interactions in an ant nest microcosm: A combined experimental and stable isotope (δ13C / δ15N) approach. *Oikos,* 125: 1182–1192.

Parmentier, T, Claus, R, De Laender, F, Bonte, D. 2021. Moving apart together: Co-movement of a symbiont community and their ant host, and its importance for community assembly. *Movement Ecology,* 9: 1–15.

Parmentier, T, Dekoninck, W, Wenseleers, T. 2014. A highly diverse microcosm in a hostile world: A review on the associates of red wood ants (*Formica rufa* group). *Insectes Sociaux,* 61: 229–237.

Parmentier, T, Dekoninck, W, Wenseleers, T. 2015a. Metapopulation processes affecting diversity and distribution of myrmecophiles associated with red wood ants. *Basic and Applied Ecology,* 16: 553–562.

Parmentier, T, Dekoninck, W, Wenseleers, T. 2015b. Context-dependent specialization in colony defence in the red wood ant *Formica rufa. Animal Behaviour,* 103: 161–167.

Parmentier, T, Dekoninck, W, Wenseleers, T. 2016b. Do well-integrated species of an inquiline community have a lower brood predation tendency? A test using red wood ant myrmecophiles. *BMC Evolutionary Biology,* 16: 1–12.

Parmentier, T, Dekoninck, W, Wenseleers, T. 2016c. Survival of persecuted myrmecophiles in laboratory nests of different ant species can explain patterns of host use in the field. *Myrmecological News,* 23: 71–79.

Parmentier, T, Dekoninck, W, Wenseleers, T. 2017a. Arthropods associate with their red wood ant host without matching nestmate recognition cues. *Journal of Chemical Ecology,* 43: 644–661.

Parmentier, T, De Laender, F, Wenseleers, T, Bonte, D. 2018. Prudent behavior rather than chemical deception enables a parasite to exploit its ant host. *Behavioral Ecology,* 29: 1225–1233.

Parmentier, T, Vanderheyden, A, Dekoninck, W, Wenseleers, T. 2017b. Body size in the ant-associated isopod *Platyarthrus hoffmannseggii* is host-dependent. *Biological Journal of the Linnean Society,* 121: 305–311.

Pasteels, J. 1968. Le système glandulaire tégumentaire des Aleocharinae (Coleoptera, Staphylinidae) et son évolution chez les espèces termitophiles du genre *Termitella. Archives de Biologie,* 79: 381–469.

Pearcy, M, Goodisman, MA, Keller, L. 2011. Sib mating without inbreeding in the longhorn crazy ant. *Proceedings of the Royal Society B: Biological Sciences,* 278: 2677–2681.

Pech, P, Fric, Z, Konvicka, M. 2007. Species-specificity of the *Phengaris* (Maculinea)—*Myrmica* host system: Fact or myth? (Lepidoptera: Lycaenidae; Hymenoptera: Formicidae). *Sociobiology,* 50: 983–1004.

Peeters, C, Heraty, J, Wiwatwitaya, D. 2015. Eucharitid wasp parasitoids in cocoons of the ponerine ant *Diacamma scalpratum* from Thailand. *Halteres,* 6: 90–94.

Peeters, C, Hölldobler, B, Moffett, M, Ali, TM. 1994. "Wall-papering" and elaborate nest architecture in the ponerine ant *Harpegnathos saltator. Insectes Sociaux,* 41: 211–218.

Peeters, C, Liebig, J. 2009. Fertility signaling as a general mechanism of regulating reproductive division of labor in ants. In *Organization of Insect Societies: From Genome to Socio-Complexity,* ed. J Gadau, J Fewell, 220–242. Cambridge, MA: Harvard University Press.

Pekár, S, Cárdenas, M. 2015. Innate prey preference overridden by familiarisation with detrimental prey in a specialised myrmecophagous predator. *Naturwissenschaften,* 102: 1257.

Pekár, S, Jarab, M. 2011. Assessment of color and behavioral resemblance to models by inaccurate myrmecomorphic spiders (Araneae). *Invertebrate Biology,* 130: 83–90.

Pekár, S, Jiroš, P. 2011. Do ant mimics imitate cuticular hydrocarbons of their models? *Animal Behaviour,* 82: 1193–1199.

Pekár, S, Král, J. 2002. Mimicry complex in two central European zodariid spiders (Araneae: Zodariidae): How *Zodarion* deceives ants. *Biological Journal of the Linnean Society,* 75: 517–532.

Pekár, S, Mayntz, D, Ribeiro, T, Herberstein, ME. 2010. Specialist ant-eating spiders selectively feed on different body parts to balance nutrient intake. *Animal Behaviour,* 79: 1301–1306.

Pekár, S, Sobotník, J. 2007. Comparative study of the femoral organ in *Zodarion* spiders (Araneae: Zodariidae). *Arthropod Structure and Development,* 36: 105–112.

Pereira-Filho, JMB, Saturnino, R, Bonaldo, AB. 2018. Five new species and novel descriptions of opposed sexes of four species of the spider genus *Attacobius* (Araneae: Corinnidae). *Zootaxa,* 4462: 211–228.

Peretti, AV. 2002. Courtship and sperm transfer in the whip spider *Phrynus gervaisii* (Amblypygi, Phrynidae): A complement to Weygoldt's 1977 paper. *Journal of Arachnology,* 30: 588–600.

Pérez, R, Condit, R, Lao, S. 1999. Distribución, mortalidad y asociación con plantas, de nidos de *Paraponera clavata* (Hymenoptera: Formicidae) en la isla de Barro Colorado, Panamá. *Revista de Biología Tropical,* 47: 697–709.

Pérez-Lachaud, G, Heraty, JM, Carmichael, A, Lachaud, J-P. 2006. Biology and behavior of *Kapala* (Hymenoptera: Eucharitidae) attacking *Ectatomma, Gnamptogenys,* and *Pachycondyla* (Formicidae: Ectatomminae and Ponerinae) in Chiapas, Mexico. *Annals of the Entomological Society of America,* 99: 567–576.

Pérez-Lachaud, G, Jervis, MA, Reemer, M, Lachaud, J-P. 2014. An unusual, but not unexpected, evolutionary step taken by syrphid flies: The first record of true primary parasitoidism of ants by Microdontinae. *Biological Journal of the Linnean Society,* 111: 462–472.

Pérez-Lachaud, G, Klompen, H, Poteaux, C, Santamaría, C, Armbrecht, I, Beugnon, G, Lachaud, J-P. 2019. Context dependent life-history shift in *Macrodinychus sellnicki* mites attacking a native ant host in Colombia. *Scientific Reports,* 9: 8394.

Pérez-Lachaud, G, Lachaud, J-P. 2014. Arboreal ant colonies as "hot-points" of cryptic diversity for myrmecophiles: The weaver ant *Camponotus* sp. aff. *textor* and its interaction network with its associates. *PLOS One,* 9: e100155.

Pérez-Lachaud, G, Valenzuela, JE, Lachaud, J-P. 2011. Is increased resistance to parasitism at the origin of polygyny in a Mexican population of the ant *Ectatomma tuberculatum* (Hymenoptera: Formicidae)? *Florida Entomologist,* 94: 677–684.

Pérez-Ortega, B, Fernández-Marín, H, Loiácono, M, Galgani, P, Wcislo, W. 2010. Biological notes on a fungus-growing ant, *Trachymyrmex* cf. *zeteki* (Hymenoptera, Formicidae, Attini) attacked by a diverse community of parasitoid wasps (Hymenoptera, Diapriidae). *Insectes Sociaux,* 57: 317–322.

Perlman, DL. 1993. Colony Founding among Ants. PhD dissertation, Harvard University, Department of Organismic and Evolutionary Biology, Cambridge, MA.

Phillips, ZI. 2021. Emigrating together but not establishing together: A cockroach rides ants and leaves. *American Naturalist,* 197: 138–145

Phillips, ZI, Reding, L, Farrior, CE. 2021. The early life of a leaf-cutter ant colony constrains symbiont vertical transmission and favors horizontal transmission. *Ecology and Evolution,* 11: 11718–11729.

Phillips, ZI, Zhang, MM, Mueller, UG. 2017. Dispersal of *Attaphila fungicola,* a symbiotic cockroach of leaf-cutter ants. *Insectes Sociaux,* 64: 277–284.

Pianka, ER, Parker, WS. 1975. Ecology of horned lizards: A review with special reference to *Phrynosoma platyrhinos. Copeia,* 1975: 141–162.

Pierce, NE. 1983. Associations between lycaenid butterflies and ants. *News Bulletin of the Entomological Society of Queensland,* 11: 91–97.

Pierce, NE. 1985. Lycaenid butterflies and ants: Selection for nitrogen-fixing and other protein-rich food plants. *American Naturalist,* 125: 888–895.

Pierce, NE. 1987. The evolution and biogeography of associations between lycaenid butterflies and ants. *Oxford Surveys in Evolutionary Biology,* 4: 89–116.

Pierce, NE. 1989. Butterfly-ant mutualisms. In *Toward a More Exact Ecology,* ed. PJ Grubb, JB Whittaker, 299–324. Oxford: Blackwell Science.

Pierce, NE. 1995. Predatory and parasitic Lepidoptera: Carnivores living on plants. *Journal of the Lepidopterists' Society,* 49: 412–453.

Pierce, NE, Braby, MF, Heath, A, Lohman, DJ, Mathew, J, Rand, DB, Travassos, MA. 2002. The ecology and evolution of ant association in the Lycaenidae (Lepidoptera). *Annual Review of Entomology,* 47: 733–771.

Pierce, NE, Dankowicz, E. In press. The natural history of caterpillar-ant associations. In *Caterpillars in the Middle,* ed. J Marquis, S Koptur. Dordrecht: Springer.

Pierce, NE, Easteal, S. 1986. The selective advantage of attendant ants for the larvae of a lycaenid butterfly, *Glaucopsyche lygdamus. Journal of Animal Ecology,* 55: 451–462.

Pierce, NE, Elgar, MA. 1985. The influence of ants on host plant selection by *Jalmenus evagoras,* a myrmecophilous lycaenid butterfly. *Behavioral Ecology and Sociobiology,* 16: 209–222.

Pierce, NE, Kitching, RL, Buckley, RC, Taylor, MFJ, Benbow, KF. 1987. The costs and benefits of cooperation between the Australian lycaenid butterfly, *Jalmenus evagoras,* and its attendant ants. *Behavioral Ecology and Sociobiology,* 21: 237–248.

Pierce, NE, Mead, PS. 1981. Parasitoids as selective agents in the symbiosis between lycaenid butterfly larvae and ants. *Science,* 211: 1185–1187.

Pierce, NE, Nash, DR. 1999. The imperial blue, *Jalmenus evagoras* (Lycaenidae). *Monographs on Australian Lepidoptera,* 6: 279–315.

Piza, SdT. 1937. Novas espécies de aranhas myrmecomorphas do Brazil e consideraç oes sobre o seu mimetismo. *Revista do Museu Paulista,* 23: 307–319.

Plateaux, L. 1960a. Adoptions expérimentales de larves entre des fourmis de genres différents: *Leptothorax nylanderi* Förster et *Solenopsis fugax* Latreille. *Insectes Sociaux,* 7: 163–170.

Plateaux, L. 1960b. Adoptions expérimentales de larves entre des fourmis de genres différents, II: *Myrmica laevinodis* Nylander et *Anergates atratulus* Schenck. *Insectes Sociaux,* 7: 221–226.

Plateaux, L. 1960c. Adoptions expérimentales de larves entre des fourmis de genres différents, III: *Anergates atratulus* Schenck et *Solenopsis fugax* Latreille, IV: *Leptothorax nylanderi* Förster et *Tetramorium caespitum* L. *Insectes Sociaux,* 7: 345–348.

Plateaux, L. 1972. Sur les modifications produites chez une fourmi par la présence d'un parasite cestode. *Annales des Sciences Naturelles, Zoologie,* 14: 203–220.

Platnick, NI, Baptista, RL. 1995. On the spider genus *Attacobius* (Araneae, Dionycha). *American Museum Novitates,* 3120: 1–9.

Plowes, NJR, Colella, T, Johnson, RA, Hölldobler, B. 2014. Chemical communication during foraging in the harvesting ants *Messor pergandei* and *Messor andrei*. *Journal of Comparative Physiology A,* 200: 129–137.

Plowes, NJR, Johnson, RA, Hölldobler, B. 2013. Foraging behavior in the ant genus *Messor* (Hymenoptera: Formicidae: Myrmicinae). *Myrmecological News,* 18: 33–49.

Plowes, RM, Folgarait, PJ, Gilbert, LE. 2012. The introduction of the fire ant parasitoid *Pseudacteon nocens* in North America: Challenges when establishing small populations. *BioControl,* 57: 503–514.

Plowes, RM, Lebrun, EG, Brown, BV, Gilbert, LE. 2009. A review of *Pseudacteon* (Diptera: Phoridae) that parasitize ants of the *Solenopsis geminata* complex (Hymenoptera: Formicidae). *Annals of the Entomological Society of America,* 102: 937–958.

Poinar, G. 2012. Nematode parasites and associates of ants: Past and present. *Psyche:* 192017.

Poinar, G, Yanoviak, SP. 2008. *Myrmeconema neotropicum* ng, n. sp., a new tetradonematid nematode parasitising South American populations of *Cephalotes atratus* (Hymenoptera: Formicidae), with the discovery of an apparent parasite-induced host morph. *Systematic Parasitology,* 69: 145–153.

Pollard, SD. 1994. Consequences of sexual selection on feeding in male jumping spiders (Araneae: Salticidae). *Journal of Zoology,* 234: 203–208.

Pollock, HS, Martínez, AE, Kelley, JP, Touchton, JM, Tarwater, CE. 2017. Heterospecific eavesdropping in ant-following birds of the Neotropics is a learned behaviour. *Proceedings: Biological Sciences,* 284: 20171785.

Pontoppidan, M-B, Himaman, W, Hywel-Jones, NL, Boomsma, JJ, Hughes, DP. 2009. Graveyards on the move: The spatio-temporal distribution of dead Ophiocordyceps-infected ants. *PLOS One,* 4: e4835.

Porter, SD. 1985. *Masoncus* spider: A miniature predator of Collembola in harvester ant colonies. *Psyche,* 92: 145–150.

Porter, SD. 1998a. Biology and behavior of *Pseudacteon* decapitating flies (Diptera: Phoridae) that parasitize *Solenopsis* fire ants (Hymenoptera: Formicidae). *Florida Entomologist,* 81: 292–309.

Porter, SD. 1998b. Host-specific attraction of *Pseudacteon* flies (Diptera: Phoridae) to fire ant colonies in Brazil. *Florida Entomologist,* 81: 423–429.

Porter, SD, Alonso, LE. 1999. Host specificity of fire ant decapitating flies (Diptera: Phoridae) in laboratory oviposition tests. *Journal of Economic Entomology,* 92: 110–114.

Porter, SD, Eastmond, DA. 1982. *Euryopis coki* (Theridiidae), a spider that preys on *Pogonomyrmex* ants. *Journal of Arachnology,* 10: 275–277.

Porter, SD, Gilbert, LE. 2004. Assessing host specificity and field release potential of fire ant decapitating flies (Phoridae: Pseudacteon). In *Assessing Host Ranges for Parasitoids and Predators Used for Classical Biological Control: A Guide to Best Practice,* ed. RG Van Driesche, T Murray, R Reardon, 152–176. Morgantown, WV: USDA Forest Service.

Porter, SD, Kumar, V, Calcaterra, LA, Briano, JA, Seal, DR. 2013. Release and establishment of the little decapitating fly *Pseudacteon cultellatus* (Diptera: Phoridae) on imported fire ants (Hymenoptera: Formicidae) in Florida. *Florida Entomologist,* 96: 1567–1573.

Porter, SD, Plowes, RM, Causton, CE. 2018. The fire ant decapitating fly, *Pseudacteon bifidus* (Diptera: Phoridae): Host specificity and attraction to potential food items. *Florida Entomologist,* 101: 55–60.

Potter, EF. 1970. Anting in wild birds, its frequency and probable purpose. *The Auk,* 87: 692–713.

Pringle, EG, Moreau, CS. 2017. Community analysis of microbial sharing and specialization in a Costa Rican ant-plant-hemipteran symbiosis. *Proceedings of the Royal Society B: Biological Sciences,* 284: 20162770.

Prószyński, J. 2016. Delimitation and description of 19 new genera, a subgenus and a species of Salticidae (Araneae) of the world. *Ecologica Montenegrina,* 7: 4–32.

Quinet, Y, Pasteels, J. 1991. Spatiotemporal evolution of the trail network in *Lasius fuliginosus* (Hymenoptera, Formicidae). *Belgian Journal of Zoology,* 121: 55–72.

Quinet, Y, Pasteels, J. 1995. Trail following and stowaway behaviour of the myrmecophilous staphylinid beetle, *Homoeusa acuminata,* during foraging trips of its host *Lasius fuliginosus* (Hymenoptera: Formicidae). *Insectes Sociaux,* 42: 31–44.

Rabeling, C. 2020. *Social Parasitism.* Springer Nature Switzerland AG. doi:10.1007/978-3-319-90306-4_175–1.

Rabeling, C, Sosa-Calvo, J, O'Connell, LA, Coloma, LA, Fernandez, F. 2016. *Lenomyrmex hoelldobleri:* A new ant species discovered in the stomach of the dendrobatid poison frog, *Oophaga sylvatica* (Funkhouser). *ZooKeys,* 618: 79–95.

Rafiqi, AM, Rajakumar, A, Abouheif, E. 2020. Origin and elaboration of a major evolutionary transition in individuality. *Nature*, 585: 239–244.

Ramachandra, P, Hill, DE. 2018. Predation by the weaver ant *Oecophylla smaragdina* (Hymenoptera: Formicidae: Formicinae) on its mimic jumping spider *Myrmarachne plataleoides* (Araneae: Salticidae: Astioida: Myrmarachnini). *Peckhamia*, 174: 1–8.

Ramalho, MO, Bueno, OC, Moreau, CS. 2017a. Microbial composition of spiny ants (Hymenoptera: Formicidae: *Polyrhachis*) across their geographic range. *BMC Evolutionary Biology*, 17: 96.

Ramalho, MO, Bueno, OC, Moreau, CS. 2017b. Species-specific signatures of the microbiome from *Camponotus* and *Colobopsis* ants across developmental stages. *PLOS One*, 12: e0187461.

Ramalho, MO, Moreau, CS, Bueno, OC. 2019. The potential role of environment in structuring the microbiota of *Camponotus* across parts of the body. *Advances in Entomology*, 7: 47–70.

Ramírez, MJ, Grismado, CJ, Ubick, D, Ovtsharenko, V, Cushing, PE, Platnick, NI, Wheeler, WC, Prendini, L, Crowley, LM, Horner, NV. 2019. Myrmecicultoridae, a new family of myrmecophilic spiders from the Chihuahuan Desert (Araneae: Entelegynae). *American Museum Novitates*, 2019: 1–24 (24).

Raspotnig, G, Stabentheiner, E, Föttinger, P, Schaider, M, Krisper, G, Rechberger, G, Leis, H. 2009. Opisthonotal glands in the Camisiidae (Acari, Oribatida): Evidence for a regressive evolutionary trend. *Journal of Zoological Systematics and Evolutionary Research*, 47: 77–87.

Ray, TS, Andrews, CCR. 1980. Antbutterflies: Butterflies that follow army ants to feed on antbird droppings. *Science*, 210: 1147–1148.

Reber, A, Castella, G, Christe, P, Chapuisat, M. 2008. Experimentally increased group diversity improves disease resistance in an ant species. *Ecology Letters*, 11: 682–689.

Reemer, M. 2013. Review and phylogenetic evaluation of associations between Microdontinae (Diptera: Syrphidae) and ants (Hymenoptera: Formicidae). *Psyche*, 2013: 538316.

Reemer, M, Ståhls, G. 2013a. Generic revision and species classification of the Microdontinae (Diptera, Syrphidae). *ZooKeys*, 288: 1–213.

Reemer, M, Ståhls, G. 2013b. Phylogenetic relationships of Microdontinae (Diptera: Syrphidae) based on molecular and morphological characters. *Systematic Entomology*, 38: 661–688.

Regnier, F, Wilson, E. 1971. Chemical communication and "propaganda" in slave-maker ants. *Science*, 172: 267–269.

Regös, J, Schlüter, A. 1984. Erste Ergebnisse zur Fortpflanzungsbiologie von *Lithodytes lineatus* (Schneider, 1799) (Amphibia: Leptodactylidae). *Salamandra (Frankfurt am Main)*, 20: 252–261.

Reichensperger, A. 1924. Neue südamerikanische Histeriden als Gäste von Wanderameisen und Termiten, II: Teil. *Revue Suisse de Zoologie*, 31: 117–152.

Reichensperger, A. 1948. Die Paussiden Afrikas. *Abhandlungen der Senckenberg Gesellschaft für Naturforschung*, 479: 5–31.

Renan, I, Assmann, T, Freidberg, A. 2018. Taxonomic revision of the *Graphipterus serrator* (Forskål) group (Coleoptera, Carabidae): An increase from five to 15 valid species. *ZooKeys*, 753: 23–82.

Rettenmeyer, CW. 1960. Behavior, abundance and host specificity of mites found on Neotropical army ants (Acarina; Formicidae: Dorylinae). *Proceedings of the Eleventh International Congress of Entomology, Vienna*, 1: 610–612.

Rettenmeyer, CW. 1961. Arthropods associated with neotropical army ants with a review of the behavior of these ants. PhD dissertation, University of Kansas, Lawrence.

Rettenmeyer, CW. 1962a. Notes on host specificity and behavior of myrmecophilous macrochelid mites. *Journal of the Kansas Entomological Society*, 35: 358–360.

Rettenmeyer, CW. 1962b. The diversity of arthropods found with Neotropical army ants and observations on the behavior of representative species. *Proceedings of the North Central Branch of the Entomological Society of America*, 17: 14–15.

Rettenmeyer, CW. 1963. The behavior of Thysanura found with army ants. *Annals of the Entomological Society of America*, 56: 170–174.

Rettenmeyer, CW. 1970. Insect mimicry. *Annual Review of Entomology*, 15: 43–74.

Rettenmeyer, CW, Akre, RD. 1968. Ectosynibiosis between phorid flies and army ants. *Annals of the Entomological Society of America*, 61: 1317–1326.

Rettenmeyer, CW, Rettenmeyer, ME, Joseph, J, Berghoff, SM. 2011. The largest animal association centered on one species: The army ant *Eciton burchellii* and its more than 300 associates. *Insectes Sociaux,* 58: 281–292.

Revis, HC, Waller, DA. 2004. Bactericidal and fungicidal activity of ant chemicals on feather parasites: An evaluation of anting behavior as a method of self-medication in songbirds. *Auk,* 121: 1262–1268.

Riley, J, Winch, JM, Stimson, AF, Pope, RD. 1986. The association of *Amphisbaena alba* (Reptilia: Amphisbaenia) with the leaf-cutting ant *Atta cephalotes* in Trinidad. *Journal of Natural History,* 20: 459–470.

Rink, WJ, Dunbar, JS, Tschinkel, WR, Kwapich, C, Repp, A, Stanton, W, Thulman, DK. 2013. Subterranean transport and deposition of quartz by ants in sandy sites relevant to age overestimation in optical luminescence dating. *Journal of Archaeological Science,* 40: 2217–2226.

Riva, F, Barbero, F, Bonelli, S, Balletto, E, Casacci, LP. 2017. The acoustic repertoire of lycaenid butterfly larvae. *Bioacoustics,* 26: 77–90.

Robinson, EJH, Stockan, J, Iason, GR. 2016. Wood ants and their interaction with other organisms. In *Wood Ant Ecology and Conservation,* ed. J Stockan, EJ Robinson, 177–206. Cambridge, UK: Cambridge University Press.

Robinson, NA, Robinson, EJH. 2013. Myrmecophiles and other invertebrate nest associates of the red wood ant *Formica rufa* (Hymenoptera: Formicidae) in north-west England. *British Journal of Entomology and Natural History,* 26: 67–88.

Roces, F, Hölldobler, B. 1995. Vibrational communication between hitchhikers and foragers in leaf-cutting ants (*Atta cephalotes*). *Behavioral Ecology and Sociobiology,* 37: 297–302.

Roces, F, Tautz, J. 2001. Ants are deaf. *Journal of the Acoustical Society of America,* 109: 3080–3082.

Roces, F, Tautz, J, Hölldobler, B. 1993. Stridulation in leaf-cutting ants. *Naturwissenschaften,* 80: 521–524.

Rocha, FH, Lachaud, J-P, Pérez-Lachaud, G. 2020. Myrmecophilous organisms associated with colonies of the ponerine ant *Neoponera villosa* (Hymenoptera: Formicidae) nesting in *Aechmea bracteata* bromeliads: A biodiversity hotspot. *Myrmecological News,* 30: 73–92.

Rödel, M-O, Brede, C, Hirschfeld, M, Schmitt, T, Favreau, P, Stöcklin, R, Wunder, C, Mebs, D. 2013. Chemical camouflage: A frog's strategy to co-exist with aggressive ants. *PLOS One,* 8: e81950.

Rodríguez, J, Montoya-Lerma, J, Calle, Z. 2013. Primer registro de *Attaphila fungicola* (Blattaria: Polyphagidae) en nidos de *Atta cephalotes* (Hymenoptera: Myrmicinae) en Colombia. *Boletin Cientifico Centro De Museos De Historia Natural,* 17: 219–226.

Roepke, W. 1925. Eine neue myrmekophile Tineide aus Java: *Hypophrictoides dolichoderella* n. g. n. sp. *Tijdschrift voor Entomologie,* 68: 175–194.

Rognes, K. 2009. Revision of the Oriental species of the *Bengalia peuhi* species group (Diptera, Calliphoridae). *Zootaxa,* 2251: 1–76.

Rognes, K. 2011. Revision of the *Bengalia spinifemorata* species-group (Diptera, Calliphoridae). *Zootaxa,* 2835: 1–29.

Ronque, MU, Lyra, ML, Migliorini, GH, Bacci, M, Oliveira, PS. 2020. Symbiotic bacterial communities in rainforest fungus-farming ants: Evidence for species and colony specificity. *Scientific Reports,* 10: 10172.

Ross, GN. 1964. Life history studies on Mexican butterflies, III: Early stages of *Anatole rossi,* a new myrmecophilous metalmark. *Journal of Research on the Lepidoptera,* 3: 81–94.

Ross, GN. 1966. Life-history studies on Mexican butterflies, IV: The ecology and ethology of Anatole rossi, a myrmecophilous metalmark (Lepidoptera: Riodinidae). *Annals of the Entomological Society of America,* 59: 985–1004.

Ruiz, E, Martínez, MH, Martínez, MD, Hernández, JM. 2006. Morphological study of the stridulatory organ in two species of *Crematogaster* genus: *Crematogaster scutellaris* (Olivier 1792) and *Crematogaster auberti* (Emery 1869) (Hymenoptera: Formicidae). *Annales de la Société entomologique de France,* 42: 99–105.

Russell, JA, Moreau, CS, Goldman-Huertas, B, Fujiwara, M, Lohman, DJ, Pierce, NE. 2009. Bacterial gut symbionts are tightly linked with the evolution of herbivory in ants. *Proceedings of the National Academy of Sciences,* 106: 21236–21241.

Russell, JA, Sanders, JG, Moreau, CS. 2017. Hotspots for symbiosis: Function, evolution, and specificity of ant-microbe associations from trunk to tips of the ant phylogeny (Hymenoptera: Formicidae). *Myrmecological News,* 24: 43–69.

Sakata, T, Norton, RA. 2001. Opisthonotal gland chemistry of early-derivative oribatid mites (Acari) and its relevance to systematic relationships of Astigmata. *International Journal of Acarology,* 27: 281–292.

Sala, M, Casacci, LP, Balletto, E, Bonelli, S, Barbero, F. 2014. Variation in butterfly larval acoustics as a strategy to infiltrate and exploit host ant colony resources. *PLOS One,* 9: e94341.

Sameshima, S, Hasegawa, E, Kitade, O, Minaka, N, Matsumoto, T. 1999. Phylogenetic comparison of endosymbionts with their host ants based on molecular evidence. *Zoological Science,* 16: 993–1000.

Saporito, RA, Donnelly, MA, Spande, TF, Garraffo, HM. 2012. A review of chemical ecology in poison frogs. *Chemoecology,* 22: 159–168.

Sauer, C, Stackebrandt, E, Gadau, J, Hölldobler, B, Gross, R. 2000. Systematic relationships and cospeciation of bacterial endosymbionts and their carpenter ant host species: Proposal of the new taxon Candidatus Blochmannia gen. nov. *International Journal of Systematic and Evolutionary Microbiology,* 50: 1877–1886.

Saussure, H. 1877. Gryllides. *Mémoires de la Société de Physique et d'Histoire Naturelle de Genève.,* 25: 111–341.

Savi, P. 1819. Osservazioni sopra la *Blatta acervorum* di Panzer, *Gryllus myrmecophilus* nobis. *Bibliotheka Italiana,* 15: 217–219.

Schal, C, Sevala, VL, Young, HP, Bachmann, JA. 1998. Sites of synthesis and transport pathways of insect hydrocarbons: Cuticle and ovary as target tissues. *American Zoologist,* 38: 382–393.

Scharf, I, Modlmeier, AP, Beros, S, Foitzik, S. 2012. Ant societies buffer individual-level effects of parasite infections. *American Naturalist,* 180: 671–683.

Schierling, A, Dettner, K. 2013. The pygidial defense gland system of the Steninae (Coleoptera, Staphylinidae): Morphology, ultrastructure and evolution. *Arthropod Structure and Development,* 42: 197–208.

Schimmer, F. 1909. Beitrag zu einer Monographie der Gryllodeengattung *Myrmecophila* Latr. *Zoologische Jahrbücher,* 8: 410–534.

Schlüter, A. 1980. Bio-akustische Untersuchungen an Leptodactyliden in einem begrenzten Gebiet des tropischen Regenwaldes von Peru (Amphibia: Salientia: Leptodactylidae). *Salamandra,* 16: 227–247.

Schlüter, A, Löttker, P, Mebert, K. 2009. Use of an active nest of the leaf cutter ant *Atta cephalotes* (Hymenoptera: Formicidae) as a breeding site of *Lithodytes*

lineatus (Anura: Leptodactylidae). *Herpetology Notes,* 2: 101–105.

Schlüter, A, Regös, J. 1981. *Lithodytes lineatus* (Schneider, 1799) (Amphibia: Leptodactylidae) as a dweller in nests of the leaf cutting ant *Atta cephalotes* (Linnaeus, 1758) (Hymenoptera: Attini). *Amphibia-Reptilia,* 2: 117–121.

Schlüter, A, Regös, J. 1996. The tadpole of *Lithodytes lineatus* with notes on the frogs resistance to leaf-cutting ants (Amphibia: Leptodactylidae). *Stuttgarter Beiträge zur Naturkunde: Serie A (Biologie),* 536: 1–4.

Schlüter, A, Regös, J. 2005. In der Höhle des Löwen— *Lithodytes lineatus. Amphibia,* 4: 23–27.

Schmid, VS, Morales, MN, Marinoni, L, Kamke, R, Steiner, J, Zillikens, A, Jeanne, R. 2014. Natural history and morphology of the hoverfly *Pseudomicrodon biluminiferus* and its parasitic relationship with ants nesting in bromeliads. *Journal of Insect Science,* 14: 21.

Schmid-Hempel, P. 1994. Infection and colony variability in social insects. In *Infection, Polymorphism and Evolution,* ed. WD Hamilton, JC Howard, 43–51. Dordrecht: Springer Verlag.

Schmid-Hempel, P. 1998. *Parasites in Social Insects.* Princeton, NJ: Princeton University Press.

Schmid-Hempel, P. 2021. Sociality and parasite transmission. *Behavioral Ecology and Sociobiology,* 75: 1–13.

Schmidt, JO, Blum, MS. 1978a. A harvester ant venom: Chemistry and pharmacology. *Science,* 200: 1064–1066.

Schmidt, JO, Blum, MS. 1978b. The biochemical constituents of the venom of the harvester ant, *Pogonomyrmex badius. Comparative Biochemistry and Physiology Part C: Comparative Pharmacology,* 61: 239–247.

Schmidt, PJ, Sherbrooke, WC, Schmidt, JO. 1989. The detoxification of ant (*Pogonomyrmex*) venom by a blood factor in horned lizards (*Phrynosoma*). *Copeia,* 1989: 603–607.

Schneider, G, Hohorst, W. 1971. Wanderung der Metacercarien des Lanzett-Egels in Ameisen. *Naturwissenschaften,* 58: 327–328.

Schneirla, TC. 1971. *Army Ants: A Study in Social Organization,* ed. HR Topoff. New York: W. H. Freeman & Co.

Schöller, M. 2011. Larvae of case-bearing leaf beetles (Coleoptera: Chrysomelidae: Cryptocephalinae). *Acta Entomologica Musei Nationalis Pragae,* 51: 747–748.

Schönrogge, K, Barbero, F, Casacci, LP, Settele, J, Thomas, J. 2017. Acoustic communication within ant societies and its mimicry by mutualistic and socially parasitic myrmecophiles. *Animal Behaviour,* 134: 249–256.

Schönrogge, K, Barr, B, Wardlaw, JC, Napper, E, Gardner, MG, Breen, J, Elmes, GW, Thomas, JA. 2002. When rare species become endangered: Cryptic speciation in myrmecophilous hoverflies. *Biological Journal of the Linnean Society,* 75: 291–300.

Schönrogge, K, Wardlaw, JC, Peters, AJ, Everett, S, Thomas, JA, Elmes, GW. 2004. Changes in chemical signature and host specificity from larval retrieval to full social integration in the myrmecophilous butterfly *Maculinea rebeli. Journal of Chemical Ecology,* 30: 91–107.

Schönrogge, K, Wardlaw, JC, Thomas, JA, Thomas, GW. 2000. Polymorphic growth rates in myrmecophilous insects. *Proceedings of the Royal Society of London. Series B: Biological Sciences,* 267: 771–777.

Schremmer, F. 1978. Zur Bionomie und Morphologie der myrmekophilen Raupe und Puppe der neotropischen Tagfalter-Art *Hamearis erostratus* (Lepidoptera: Riodinidae). *Entomologica Germanica,* 42: 113–121.

Schröder, D, Deppisch, H, Obermayer, M, Krohne, G, Stackebrandt, E, Hölldobler, B, Goebel, W, Gross, R. 1996. Intracellular endosymbiotic bacteria of *Camponotus* species (carpenter ants): Systematics, evolution and ultrastructural characterization. *Molecular Microbiology,* 21: 479–489.

Schroth, M, Maschwitz, U. 1984. Zur Larvalbiologie und Wirtsfindung von Maculinea teleius (Lepidoptera: Lycaenidae), eines Parasiten von Myrmica laevinodis (Hymenoptera: Formicidae). *Entomologia Generalis,* 9: 225–230.

Schurian, KG, Fiedler, K. 1991. Einfache Methoden zur Schallwahrnehmung bei Bläulings-Larven (Lepidoptera: Lycaenidae). *Entomologische Zeitschrift,* 101: 393–412.

Schurian, KG, Fiedler, K. 1994. Zur Biologie von *Polyommatus* (Lysandra) dezinus (De Freina & Witt) (Lepidoptera: Lycaenidae). *Nachrichten des Entomologischen Vereins Apollo, Frankfurt am Main, N.F.,* 14: 339–353.

Schurian, KG, Fiedler, K, Maschwitz, U. 1993. Parasitoids exploit secretions of myrmecophilous lycaenid butterfly caterpillars (Lycaenidae). *Journal of the Lepidopterists' Society,* 47: 150–154.

Schwitzke, C, Fiala, B, Linsenmair, KE, Curio, E. 2015. Eucharitid ant-parasitoid affects facultative ant-plant *Leea manillensis:* Top-down effects through three trophic levels. *Arthropod-Plant Interactions,* 9: 497–505.

Seevers, CH. 1957. A monograph on the termitophilous staphylinidae (Coleoptera). *Fieldiana Zoology,* 40: 1–334.

Seevers, CH. 1965. The systematic, evolution and zoogeography of staphylinid beetles associated with army ants (Coleoptera, Staphylinidae). *Fieldiana Zoology,* 47: 137–351.

Seraphim, N, Kaminski, LA, Devries, PJ, Penz, C, Callaghan, C, Wahlberg, N, Silva-Beandão, KL, Freitas, AV. 2018. Molecular phylogeny and higher systematics of the metalmark butterflies (Lepidoptera: Riodinidae). *Systematic Entomology,* 43: 407–425.

Shamble, PS, Hoy, RR, Cohen, I, Beatus, T. 2017. Walking like an ant: A quantitative and experimental approach to understanding locomotor mimicry in the jumping spider *Myrmarachne formicaria. Proceedings of the Royal Society B: Biological Sciences,* 284: 20170308.

Sharkey, MJ, Carpenter, JM, Vilhelmsen, L, Heraty, J, Liljeblad, J, Dowling, AP, Schulmeister, S, Murray, D, Deans, AR, Ronquist, F. 2012. Phylogenetic relationships among superfamilies of Hymenoptera. *Cladistics,* 28: 80–112.

Sharma, KR, Enzmann, BL, Schmidt, Y, Moore, D, Jones, GR, Parker, J, Berger, SL, Reinberg, D, Zwiebel, LJ, Breit, B, Liebig, J, Ray, A. 2015. Cuticular hydrocarbon pheromones for social behavior and their coding in the ant antenna. *Cell Reports,* 12: 1261–1271.

Shaw, SR. 1993. Observations on the ovipositional behavior of *Neoneurus mantis,* an ant-associated parasitoid from Wyoming (Hymenoptera: Braconidae). *Journal of Insect Behavior,* 6: 649–658.

Shenefelt, RD. 1969. Braconidae 1. *Hymenopterorum Catalogus (Nova Editio),* 4: 1–176.

Sherbrooke, WC, Schwenk, K. 2008. Horned lizards (*Phrynosoma*) incapacitate dangerous ant prey with mucus. *Journal of Experimental Zoology Part A: Ecological Genetics and Physiology,* 309: 447–459.

Sherman, PW, Seeley, TD, Reeve, HK. 1988. Parasites, pathogens, and polyandry in social Hymenoptera. *American Naturalist,* 131: 602–610.

Shimizu-kaya, U. 2014. Exploitation of food bodies on *Macaranga* myrmecophytes by larvae of a lycaenid species, *Arhopala zylda* (Lycaeninae). *Journal of the Lepidopterists' Society,* 68: 31–36.

Shimizu-kaya, U, Okubo, T, Yago, M, Inui, Y, Itioka, T. 2013. Myrmecoxeny in *Arhopala zylda* (Lepidoptera, Lycaenidae) larvae feeding on *Macaranga* myrmecophytes. *Entomological News,* 123: 63–70.

Sielezniew, M, Rutkowski, R. 2012. Population isolation rather than ecological variation explains the genetic structure of endangered myrmecophilous butterfly *Phengaris* (= *Maculinea*) *arion*. *Journal of Insect Conservation,* 16: 39–50.

Sielezniew, M, Rutkowski, R, Ponikwicka-Tyszko, D, Ratkiewicz, M, Dziekańska, I, Švitra, G. 2012. Differences in genetic variability between two ecotypes of the endangered myrmecophilous butterfly *Phengaris* (= *Maculinea*) *alcon:* The setting of conservation priorities. *Insect Conservation and Diversity,* 5: 223–236.

Silva, VM, Moreira, GF, Lopes, JM, Delabie, JH, Oliveira, AR. 2018. A new species of *Cosmolaelaps* Berlese (Acari: Laelapidae) living in the nest of the ant *Neoponera inversa* (Smith) (Hymenoptera: Formicidae) in Brazil. *Systematic and Applied Acarology,* 23: 13–24.

Simmons, KEL. 1966. Anting and the problem of self-stimulation. *Journal of Zoology,* 149: 145–162.

Singer, TL. 1998. Roles of hydrocarbons in the recognition systems of insects. *American Zoologist,* 38: 394–405.

Sinotte, VM, Freedman, SN, Ugelvig, LV, Seid, MA. 2018. *Camponotus floridanus* ants incur a trade-off between phenotypic development and pathogen susceptibility from their mutualistic endosymbiont *Blochmannia*. *Insects,* 9: 58.

Šípek, P, Fikáček, M, Bílý, S. 2008. First record of myrmecophily in buprestid beetles: Immature stages of *Habroloma myrmecophila* sp. nov. (Coleoptera: Buprestidae) associated with *Oecophylla* ants (Hymenoptera: Formicidae). *Insect Systematics & Evolution,* 39: 121–131.

Sivinski, J, Marshall, S, Petersson, E. 1999. Kleptoparasitism and phoresy in the Diptera. *Florida Entomologist,* 82: 179–197.

Skwarra, E. 1927. Über die Ernährungsweise der Larven von *Clytra quadripunctata* (L.). *Zoologischer Anzeiger,* 50: 83–96.

Smith, AA. 2019. Prey specialization and chemical mimicry between *Formica archboldi* and *Odontomachus* ants. *Insectes Sociaux,* 66: 211–222.

Smith, AA, Hölldobler, B, Liebig, J. 2008. Hydrocarbon signals explain the pattern of worker and egg policing in the ant *Aphaenogaster cockerelli*. *Journal of Chemical Ecology,* 34: 1275–1282.

Smith, AA, Hölldober, B, Liebig, J. 2009. Cuticular hydrocarbons reliably identify cheaters and allow enforcement of altruism in a social insect. *Current Biology,* 19: 78–81.

Smith, AA, Liebig, J. 2017. The evolution of cuticular fertility signals in eusocial insects. *Current Opinion in Insect Science,* 22: 79–84.

Sokolov, IM, Sokolova, YY, Fuxa, JR. 2003. Histiostomatid mites (Histiostomatidae: Astigmata: Acarina). *Journal of Entomological Science,* 38: 699–702.

Song, H, Béthoux, O, Shin, S, Donath, A, Letsch, H, Liu, S, McKenna, DD, Meng, G, Misof, B, Podsiadlowski, L, Zhou, X, Wipfler, B, Simon, S. 2020. Phylogenomic analysis sheds light on the evolutionary pathways towards acoustic communication in Orthoptera. *Nature Communications,* 11: 4939.

Song, H, Buhay, JE, Whiting, MF, Crandall, KA. 2008. Many species in one: DNA barcoding overestimates the number of species when nuclear mitochondrial pseudogenes are coamplified. *Proceedings of the National Academy of Sciences,* 105: 13486–13491.

Soroker, V, Fresneau, D, Hefetz, A. 1998. Formation of colony odor in ponerine ant *Pachycondyla apicalis*. *Journal of Chemical Ecology,* 24: 1077–1090.

Soroker, V, Hefetz, A. 2000. Hydrocarbon site of synthesis and circulation in the desert ant *Cataglyphis niger*. *Journal of Insect Physiology,* 46: 1097–1102.

Soroker, V, Hefetz, A, Cojocaru, M, Billen, J, Franke, S, Francke, W. 1995a. Structural and chemical ontogeny of the postpharyngeal gland in the desert ant *Cataglyphis niger*. *Physiological Entomology,* 20: 323–329.

Soroker, V, Vienne, C, Hefetz, A. 1995b. Hydrocarbon dynamics within and between nestmates in *Cataglyphis niger* (Hymenoptera: Formicidae). *Journal of Chemical Ecology,* 21: 365–378.

Soroker, V, Vienne, C, Hefetz, A, Nowbahari, E. 1994. The postpharyngeal gland as a "Gestalt" organ for nestmate recognition in the ant *Cataglyphis niger*. *Naturwissenschaften,* 81: 510–513.

Southern, WE. 1963. Three species observed anting on a wet lawn. *Wilson Bulletin,* 75: 275–276.

Speight, M. 2017. Species accounts of European Syrphidae (Diptera), Glasgow 2011. In *Syrph the Net, the Database of European Syrphidae,* vol. 65, ed. MCD Speight, E Castella, J-P Sarthou, C Monteil, 285. Dublin: Trinity College Department of Zoology.

Sprenger, PP, Menzel, F. 2020. Cuticular hydrocarbons in ants (Hymenoptera: Formicidae) and other insects: How and why they differ among individuals, colonies, and species. *Myrmecological News,* 30: 1–26.

Stadler, B, Dixon, AFG. 2008. *Mutualism: Ants and Their Insect Partners.* Cambridge, UK: Cambridge University Press.

Stalling, T. 2017. A new species of ant-loving cricket *Myrmecophilus* Berthold, 1827 from Cyprus (Orthoptera: Myrmecophilidae). *Zoology in the Middle East,* 49: 89–94.

Stalling, T, Iorgu, IȘ, Chobanov, DP. 2020. The ant cricket *Myrmecophilus orientalis* on the Dodecanese Islands, Greece (Orthoptera: Myrmecophilidae). *Travaux du Muséum National d'Histoire Naturelle "Grigore Antipa,"* 63: 63–67.

Staniec, B, Pietrykowska-Tudruj, E, Zagaja, M. 2017. Adaptive external larval ultrastructure of *Lomechusa* Gravenhorst, 1806 (Coleoptera: Staphylinidae: Aleocharinae), an obligate myrmecophilous genus. *Annales Zoologici,* 67: 609–626.

Staniec, B, Zagaja, M. 2008. Rove-beetles (Coleoptera, Staphylinidae) of ant nests of the vicinities of Leżajsk. *Annales UMCS, Biologia,* 63: 111–127.

Steidle, JL, Dettner, K. 1993. Chemistry and morphology of the tergal gland of freeliving adult Aleocharinae (Coleoptera: Staphylinidae) and its phylogenetic significance. *Systematic Entomology,* 18: 149–168.

Steiner, FM, Sielezniew, M, Schlick-Steiner, BC, Höttinger, H, Stankiewicz, A, Górnicki, A. 2003. Host specificity revisited: New data on *Myrmica* host ants of the lycaenid butterfly *Maculinea rebeli. Journal of Insect Conservation,* 7: 1–6.

Stiefel, VL, Margolies, DC. 1998. Is host plant choice by a clytrine leaf beetle mediated through interactions with the ant *Crematogaster lineolata? Oecologia,* 115: 434–438.

Stiefel, VL, Nechols, JR, Margolies, DC. 1995. Overwintering biology of *Anomoea flavokansiensis* (Coleoptera: Chrysomelidae). *Annals of the Entomological Society of America,* 88: 342–347.

Stoeffler, M, Boettinger, L, Tolasch, T, Steidle, JL. 2013. The tergal gland secretion of the two rare myrmecophilous species *Zyras collaris* and *Z. haworthi* (Coleoptera: Staphylinidae) and the effect on *Lasius fuliginosus. Psyche,* 2013: 601073.

Stoeffler, M, Maier, TS, Tolasch, T, Steidle, JL. 2007. Foreign-language skills in rove-beetles? Evidence for chemical mimicry of ant alarm pheromones in myrmecophilous *Pella* beetles (Coleoptera: Staphylinidae). *Journal of Chemical Ecology,* 33: 1382–1392.

Stoeffler, M, Tolasch, T, Steidle, JL. 2011. Three beetles— three concepts: Different defensive strategies of congeneric myrmecophilous beetles. *Behavioral Ecology and Sociobiology,* 65: 1605–1613.

Stoll, S, Feldhaar, H, Fraunholz, MJ, Gross, R. 2010. Bacteriocyte dynamics during development of a holometabolous insect, the carpenter ant *Camponotus floridanus. BMC Microbiology,* 10: 1–16.

Stradling, D. 1978. The influence of size on foraging in the ant, *Atta cephalotes,* and the effect of some plant defence mechanisms. *Journal of Animal Ecology,* 47: 173–188.

Stubbs, AE, Falk, SJ. 2002. *British Hoverflies: An Illustrated Identification Guide.* Reading: UK: British Entomological and Natural History Society.

Sturgis, SJ, Gordon, DM. 2012. Nestmate recognition in ants (Hymenoptera: Formicidae): A review. *Myrmecological News,* 16: 101–110.

Swann, J. 2016. Family Milichiidae. *Zootaxa,* 4122: 708–715.

Swartz, MB. 2001. Bivouac checking, a novel behavior distinguishing obligate from opportunistic species of army-ant-following birds. *The Condor,* 103: 629–633.

Szenteczki, MA, Pitteloud, C, Casacci, LP, Kešnerová, L, Whitaker, MR, Engel, P, Vila, R, Alvarez, N. 2019. Bacterial communities within *Phengaris* (*Maculinea*) *alcon* caterpillars are shifted following transition from solitary living to social parasitism of *Myrmica* ant colonies. *Ecology and Evolution,* 9: 4452–4464.

Tahami, MS, Sadeghi, S, Gorochov, AV. 2017. Cave and burrow crickets of the subfamily Bothriophylacinae (Orthoptera: Myrmecophilidae) in Iran and adjacent countries. *Zoosystematica Rossica,* 26: 241–275.

Taniguchi, K, Maruyama, M, Ichikawa, T, Ito, F. 2005. A case of Batesian mimicry between a myrmecophilous staphylinid beetle, *Pella* comes, and its host ant, *Lasius* (*Dendrolasius*) *spathepus:* An experiment using the Japanese treefrog, *Hyla japonica* as a real predator. *Insectes Sociaux,* 52: 320–322.

Tarpy, DR. 2003. Genetic diversity within honeybee colonies prevents severe infections and promotes colony growth. *Proceedings of the Royal Entomological Society of London: B,* 270: 99–103.

Tarpy, DR, Seeley, TD. 2006. Lower disease infections in honeybee (*Apis mellifera*) colonies headed by polyandrous vs monandrous queens. *Naturwissenschaften,* 93: 195–199.

Tartally, A, Thomas, JA, Anton, C, Balletto, E, Barbero, F, Bonelli, S, Bräu, M, Casacci, LP, Csősz, S, Czekes, Z. 2019. Patterns of host use by brood parasitic *Maculinea* butterflies across Europe. *Philosophical Transactions of the Royal Society B,* 374: 20180202.

Tautz, J, Fiedler, K. 1992. Mechanoreceptive properties of caterpillar hairs involved in mediation of butterfly-ant symbioses. *Naturwissenschaften,* 79: 561–563.

Tenenbaum, S. 1913. Chrząszcze (Coleoptera) zebrane w Ordynacyi Zamojskiej w gub. *Lubelskiej. Pam. Fizyogr,* 21: 1–72.

Thomas, JA. 2002. Larval niche selection and evening exposure enhance adoption of a predacious social parasite, *Maculinea arion* (large blue butterfly), by *Myrmica* ants. *Oecologia,* 132: 531–537.

Thomas, JA, Elmes, GW. 1993. Specialized searching and the hostile use of allomones by a parasitoid whose host, the butterfly *Maculinea rebeli,* inhabits ant nests. *Animal Behaviour,* 45: 593–602.

Thomas, JA, Elmes, GW, Sielezniew, M, Stankiewicz-Fiedurek, A, Simcox, DJ, Settele, J, Schönrogge, K. 2013. Mimetic host shifts in an endangered social parasite of ants. *Proceedings of the Royal Society B: Biological Sciences,* 280: 20122336.

Thomas, JA, Elmes, GW, Wardlaw, JC. 1998. Polymorphic growth in larvae of the butterfly *Maculinea rebeli,* a social parasite of *Myrmica* ant colonies. *Proceedings of the Royal Society of London. Series B: Biological Sciences,* 265: 1895–1901.

Thomas, JA, Elmes, GW, Wardlaw, JC, Woyciechowski, M. 1989. Host specificity among *Maculinea* butterflies in *Myrmica* ant nests. *Oecologia,* 79: 452–457.

Thomas, JA, Knapp, JJ, Akino, T, Gerty, S, Wakamura, S, Simcox, DJ, Wardlaw, JC, Elmes, GW. 2002. Parasitoid secretions provoke ant warfare. *Nature,* 417: 505–506.

Thomas, JA, Schönrogge, K, Elmes, GW. 2005. Specialisations and host associations of social parasites of ants. In *Insect Evolutionary Ecology,* ed. M Fellowes, G Holloway, J Rolff, 479–518. Wallingford, UK: CAB International Publishing.

Thomas, JA, Wardlaw, JC. 1992. The capacity of a *Myrmica* ant nest to support a predacious species of *Maculinea* butterfly. *Oecologia,* 91: 101–109.

Timuş, N, Constantineanu, R, Rákosy, L. 2013. *Ichneumon balteatus* (Hymenoptera: Ichneumonidae): A new parasitoid species of *Maculinea alcon* butterflies (Lepidoptera: Lycaenidae). *Entomologica Romanica,* 18: 31–35.

Tishechkin, AK, Kronauer, DJ, Von Beeren, C. 2017. Taxonomic review and natural history notes of the army ant-associated beetle genus *Ecclisister Reichensperger* (Coleoptera: Histeridae: Haeteriinae). *Coleopterists Bulletin,* 71: 279–288.

Toda, M, Komatsu, N, Takahashi, H, Nakagawa, N, Sukigara, N. 2013. Fecundity in captivity of the green anoles, *Anolis carolinensis,* established on the Ogasawara islands. *Current Herpetology,* 32: 82–88.

Torres, JA, Thomas, R, Leal, M, Gush, T. 2000. Ant and termite predation by the tropical blindsnake *Typhlops platycephalus. Insectes Sociaux,* 47: 1–6.

Touchton, JM, Smith, JNM. 2011. Species loss, delayed numerical responses, and functional compensation in an antbird guild. *Ecology,* 92: 1126–1136.

Trabalon, M, Plateaux, L, Péru, L, Bagnères, A-G, Hartmann, N. 2000. Modification of morphological characters and cuticular compounds in worker ants *Leptothorax nylanderi* induced by endoparasites *Anomotaenia brevis. Journal of Insect Physiology,* 46: 169–178.

Trach, VA, Bobylev, AN. 2018. Description of the female of the myrmecophilous mite *Antennophorus goesswaldi* Wiśniewski et Hirschmann, 1992 (Acari: Mesostigmata: Antennophoridae). *Acarina,* 26: 227–235.

Travassos, MA, DeVries, PJ, Pierce, NE. 2008. A novel organ and mechanism for larval sound production in butterfly caterpillars: *Eurybia elvina* (Lepidoptera: Riodinidae). *Tropical Lepidoptera Research,* 18: 20–23.

Travassos, MA, Pierce, NE. 2000. Acoustics, context and function of vibrational signalling in a lycaenid butterfly-ant mutualism. *Animal Behaviour,* 60: 13–26.

Tschinkel, WR. 1992. Brood raiding the fire ant, *Solenopsis invicta* (Hymenoptera: Formicidae): Laboratory and field observations. *Annals of the Entomological Society of America,* 85: 638–646.

Tschinkel, WR. 2006. *The Fire Ants.* Cambridge, MA: Belknap Press of Harvard University Press.

Tschinkel, WR. 2014. Nest relocation and excavation in the Florida harvester ant, *Pogonomyrmex badius. PLOS One,* 9: e112981.

Tschinkel, WR. 2017. Lifespan, age, size-specific mortality and dispersion of colonies of the Florida harvester ant, *Pogonomyrmex badius. Insectes Sociaux,* 64: 285–296.

Tschinkel, WR. 2021. *Ant Architecture: The Wonder, Beauty, and Science of Underground Nests.* Princeton, NJ: Princeton University Press.

Tschinkel, WR, Kwapich, CL. 2016. The Florida harvester ant, *Pogonomyrmex badius,* relies on germination to consume large seeds. *PLOS One,* 11: e0166907.

Tschinkel, WR, Rink, WJ, Kwapich, CL. 2015. Sequential subterranean transport of excavated sand and foraged seeds in nests of the harvester ant, *Pogonomyrmex badius. PLOS One,* 10: e0139922.

Tseng, S-P, Hsu, P-W, Lee, C-C, Wetterer, JK, Hugel, S, Wu, L-H, Lee, C-Y, Yoshimura, T, Yang, C-CS. 2020. Evidence for common horizontal transmission of *Wolbachia* among ants and ant crickets: Kleptoparasitism added to the list. *Microorganisms,* 8: 805.

Ueda, S, Okubo, T, Itioka, T, Shimizu-kaya, U, Yago, M, Inui, Y, Itino, T. 2012. Timing of butterfly parasitization of a plant-ant-scale symbiosis. *Ecological Research,* 27: 437–443.

Ugelvig, LV, Kronauer, DJ, Schrempf, A, Heinze, J, Cremer, S. 2010. Rapid anti-pathogen response in ant societies relies on high genetic diversity. *Proceedings of the Royal Society B: Biological Sciences,* 277: 2821–2828.

Ugelvig, LV, Vila, R, Pierce, NE, Nash, DR. 2011. A phylogenetic revision of the *Glaucopsyche* section (Lepidoptera: Lycaenidae), with special focus on the *Phengaris-Maculinea* clade. *Molecular Phylogenetics and Evolution,* 61: 237–243.

Uppstrom, KA, Klompen, H. 2011. Mites (Acari) associated with the desert seed harvester ant, *Messor pergandei* (Mayr). *Psyche:* 974646.

Uy, FMK, Adcock, JD, Jeffries, SF, Pepere, E. 2019. Intercolony distance predicts the decision to rescue or attack conspecifics in weaver ants. *Insectes Sociaux,* 66: 185–192.

van Achterberg, C. 1999. The West Palaearctic species of the subfamily Paxylommatinae (Hymenoptera: Ichneumonidae), with special reference to the genus *Hybrizon* Fallén. *Zoologische Mededelingen Leiden,* 73: 11–26.

van Achterberg, C, Argaman, Q. 1993. *Kollasmosoma* gen. nov. and a key to the genera of the subfamily Neoneurinae (Hymenoptera: Braconidae). *Zoologische Mededelingen Leiden,* 67: 63–74.

van Baalen, M, Beekman, M. 2006. The costs and benefits of genetic heterogeneity in resistance against parasites in social insects. *American Naturalist,* 167: 568–577.

Vander Meer, RK, Jouvenaz, DP, Wojcik, DP. 1989. Chemical mimicry in a parasitoid (Hymenoptera: Eucharitidae) of fire ants (Hymenoptera: Formicidae). *Journal of Chemical Ecology,* 15: 2247–2261.

Vander Meer, RK, Wojcik, DP. 1982. Chemical mimicry in the myrmecophilous beetle *Myrmecaphodius excavaticollis. Science,* 218: 806–808.

Van Pelt, AF. 1958. The occurrence of a *Cordyceps* on the ant *Camponotus pennsylvanicus* (De Geer) in the Highlands, NC region. *Journal of the Tennessee Academy of Sciences,* 33: 120–122.

Van Pelt, AF, Van Pelt, SA. 1972. Microdon (Diptera: Syrphidae) in nests of *Monomorium* (Hymenoptera: Formicidae) in Texas. *Annals of the Entomological Society of America,* 65: 977–979.

van Zweden, J, Brask, JB, Christensen, JH, Boomsma, JJ, Linksvayer, TA, d'Ettorre, P. 2010. Blending of heritable recognition cues among ant nestmates creates distinct colony gestalt odours but prevents within-colony nepotism. *Journal of Evolutionary Biology,* 23: 1498–1508.

van Zweden, JS, d'Ettorre, P. 2010. Nestmate recognition in social insects and the role of hydrocarbons. In *Insect Hydrocarbons: Biology, Biochemistry, and Chemical Ecology,* ed. GJ Blomquist, A-G Bagnères, 222–243. Cambridge, UK: Cambridge University Press.

Varone, L, Briano, J. 2009. Bionomics of *Orasema simplex* (Hymenoptera: Eucharitidae), a parasitoid of *Solenopsis* fire ants (Hymenoptera: Formicidae) in Argentina. *Biological Control,* 48: 204–209.

Vazquez, RJ, Porter, SD, Briano, JA. 2006. Field release and establishment of the decapitating fly *Pseudacteon curvatus* on red imported fire ants in Florida. *BioControl,* 51: 207–216.

Vencl, FV, Morton, TC, Mumma, RO, Schultz, JC. 1999. Shield defense of a larval tortoise beetle. *Journal of Chemical Ecology,* 25: 549–566.

Vieira-Neto, E, Mundim, F, Vasconcelos, H. 2006. Hitchhiking behaviour in leaf-cutter ants: An experimental evaluation of three hypotheses. *Insectes Sociaux, 53:* 326–332.

von Beeren, C, Blüthgen, N, Hoenle, PO, Pohl, S, Brückner, A, Tishechkin, AK, Maruyama, M, Brown, BV, Hash, JM, Hall, WE, Kronauer, DJ. 2021a. A remarkable legion of guests: Diversity and host specificity of army ant symbionts. *Molecular Ecology,* https://doi.org/10.1111/mec.16101.

von Beeren, C, Brückner, A, Hoenle, PO, Ospina-Jara, B, Kronauer, DJ, Blüthgen, N. 2021b. Multiple phenotypic traits as triggers of host attacks towards ant symbionts: Body size, morphological gestalt, and chemical mimicry accuracy. *Frontiers in Zoology,* 18: 1–18.

von Beeren, C, Brückner, A, Maruyama, M, Burke, G, Wieschollek, J, Kronauer, DJ. 2018. Chemical and behavioral integration of army ant-associated rove beetles: A comparison between specialists and generalists. *Frontiers in Zoology,* 15: 1–16.

von Beeren, C, Hashim, R, Witte, V. 2012a. The social integration of a myrmecophilous spider does not depend exclusively on chemical mimicry. *Journal of Chemical Ecology,* 38: 262–271.

von Beeren, C, Maruyama, M, Hashim, R, Witte, V. 2011a. Differential host defense against multiple parasites in ants. *Evolutionary Ecology,* 25: 259–276.

von Beeren, C, Maruyama, M, Kronauer, DJC. 2016. Community sampling and integrative taxonomy reveal new species and host specificity in the army ant-associated beetle genus *Tetradonia* (Coleoptera, Staphylinidae, Aleocharinae). *PLOS One,* 11: e0165056.

von Beeren, C, Pohl, S, Witte, V. 2012b. On the use of adaptive resemblance terms in chemical ecology. *Psyche:* 635761.

von Beeren, C, Schulz, S, Hashim, R, Witte, V. 2011b. Acquisition of chemical recognition cues facilitates integration into ant societies. *BMC Ecology,* 11: 1–13.

von Beeren, C, Tishechkin, AK. 2017. *Nymphister kronaueri* von Beeren & Tishechkin sp. nov., an army ant-associated beetle species (Coleoptera: Histeridae: Haeteriinae) with an exceptional mechanism of phoresy. *BMC Zoology,* 2: 1–16.

Wagner, D. 1993. Species-specific effects of tending ants on the development of lycaenid butterfly larvae. *Oecologia,* 96: 276–281.

Wagner, D. 1995. Pupation site choice of a North American lycaenid butterfly: The benefits of entering ant nests. *Ecological Entomology,* 20: 384–392.

Wagner, D, Brown, MJ, Broun, P, Cuevas, W, Moses, LE, Chao, DL, Gordon, DM. 1998. Task-related differences in the cuticular hydrocarbon composition of harvester ants, *Pogonomyrmex barbatus. Journal of Chemical Ecology,* 24: 2021–2037.

Wagner, D, Del Rio, CM. 1997. Experimental tests of the mechanism for ant-enhanced growth in an ant-tended lycaenid butterfly. *Oecologia,* 112: 424–429.

Wallace, J. 1970. Defensive function of a case on a chrysomelid larva. *Journal of the Georgia Entomological Society,* 5: 19–24.

Waller, D. 1980. Leaf-cutting ants and leaf-riding flies. *Ecological Entomology,* 5: 305–306.

Walsh, JP, Tschinkel, W. 1974. Brood recognition by contact pheromone in the red imported fire ant, *Solenopsis invicta. Animal Behaviour,* 22: 695–704.

Walter, DE, Moser, JC. 2010. *Gaeolaelaps invictianus,* a new and unusual species of Hypoaspidine mite (Acari: Mesostigmata: Laelapidae) phoretic on the red imported fire ant *Solenopsis invicta* Buren (Hymenoptera: Formicidae) in Louisiana, USA. *International Journal of Acarology,* 36: 399–407.

Wang, X-L, Yao, Y-J. 2011. Host insect species of *Ophiocordyceps sinensis:* A review. *ZooKeys,* 127: 43.

Wang, Z, Zhuang, H, Wang, M, Pierce, NE. 2019. *Thitarodes shambalaensis* sp. nov. (Lepidoptera, Hepialidae): A new host of the caterpillar fungus *Ophiocordyceps sinensis* supported by genome-wide SNP data. *ZooKeys,* 885: 89.

Ward, PS, Blaimer, BB, Fisher, BL. 2016. A revised phylogenetic classification of the ant subfamily Formicinae (Hymenoptera: Formicidae), with resurrection of the genera *Colobopsis* and *Dinomyrmex. Zootaxa,* 4072: 343–357.

Wardlaw, J, Elmes, G, Thomas, J. 1998. Techniques for studying *Maculinea* butterflies, I: Rearing *Maculinea* caterpillars with *Myrmica* ants in the laboratory. *Journal of Insect Conservation,* 2: 79–84.

Wasmann, E. 1889a. Nachträgliche Bemerkungen zu *Ecitochara* und *Ecitomorpha. Deutsche Entomologische Zeitschrift,* 83: 414.

Wasmann, E. 1889b. Zur Lebens-und Entwicklungsgeschichte von *Dinarda. Wiener Entomologische Zeitung,* 8: 153–162.

Wasmann, E. 1891. *Die zusammengesetzten Nester und gemischten Kolonien der Ameisen: ein Beitrag zur Biologie, Psychologie und Entwicklungsgeschichte der Ameisengesellschaften.* Münster in Westphalen: Verlag der Aschendorffsche Buchdruckerei.

Wasmann, E. 1894. *Kritisches Verzeichniss der myrmekophilen und termitophilen Arthropoden: Mit Angabe der Lebensweise und mit Beschreibung neuer Arten.* Berlin: Verlag Felix L. Dames.

Wasmann, E. 1898a. Lebensweise von *Thorictus foreli.* Mit einem anatomischen Anhang und einer Tafel. *Natur und Offenbarung,* 8: 466–478.

Wasmann, E. 1898b. *Thorictus foreli* als Ectoparasit der Ameisenfühler. *Zoologischer Anzeiger,* 21: 536–546.

Wasmann, E. 1899. Die psychischen Fähigkeiten der Ameisen. *Zoologica,* 11: 1–133.

Wasmann, E. 1901. Zur Lebensweise der Ameisengrillen (*Myrmecophila*). *Natur und Offenbarung,* 47: 129–157.

Wasmann, E. 1902. Zur Kenntnis der myrmecophilen *Antennophorus* und anderer auf Ameisen und Termiten reitender Acarinen. *Zoologischer Anzeiger,* 25: 66–76.

Wasmann, E. 1903. Zur näheren Kenntnis des echten Gastverhältnisses (Symphilie) bei den Ameisen-und Termitengästen. *Biologisches Zentralblatt,* 23: 63–72, 195–207, 232–248, 261–276, 298–310.

Wasmann, E. 1905. Zur Lebensweise einiger in-und ausländischen Arneisengäste. *Zeitschrift für wissenschaftliche Insektenbiologie,* 10: 329.

Wasmann, E. 1915. Neue Beiträge zur Biologie von *Lomechusa* und *Atemeles. Zeitschrift für Wissenschaftliche Zoologie,* 114: 233–402.

Wasmann, E. 1918. Zur Lebensweise und Fortpflanzung von *Pseudacteon formicarum* Verr. (Diptera, Phoridae). *Biologisches Zentralblatt,* 38: 317–329.

Wasmann, E. 1920. *Die Gastpflege der Ameisen.* Berlin: Verlag Gebrüder.

Wasmann, E. 1925. *Die Ameisenmimikry.* Berlin: Verlag Gebrüder Borntraeger.

Watanabe, C. 1984. Notes on Paxylommatinae with review of Japanese species (Hymenoptera, Braconidae). *Kontyū,* 52: 553–556.

Watkins, JF, Gehlbach, FR, Baldridge, RS. 1967. Ability of the blind snake, *Leptotyphlops dulcis,* to follow pheromone trails of army ants, *Neivamyrmex nigrescens* and *N. opacithorax. Southwestern Naturalist,* 12: 455–462.

Watkins, JF, Gehlbach, FR, Kroll, JC. 1969. Attractant-repellent secretions of blind snakes (*Leptotyphlops dulcis*) and their army ant prey (*Neivamyrmex nigrescens*). *Ecology,* 50: 1098–1102.

Webb, JK, Branch, WR, Shine, R. 2001. Dietary habits and reproductive biology of typhlopid snakes from southern Africa. *Journal of Herpetology,* 35: 558–567.

Webb, JK, Shine, R. 1992. To find an ant: Trail-following in Australian blindsnakes (Typhlopidae). *Animal Behaviour,* 43: 941–948.

Weissflog, A, Maschwitz, U, Disney, RHL, Rościszewski, K. 1995. A fly's ultimate con. *Nature,* 378: 137.

Weissflog, A, Maschwitz, U, Seebauer, S, Disney, RHL, Seifert, B, Witte, V. 2008. Studies on European ant decapitating flies (Diptera: Phoridae), II: Observations that contradict the reported catholicity of host choice *Pseudaction formicarum. Sociobiology,* 51: 87.

Wernegreen, JJ, Degnan, PH, Lazarus, AB, Palacios, C, Bordenstein, SR. 2003. Genome evolution in an insect cell: Distinct features of an ant-bacterial partnership. *Biological Bulletin,* 204: 221–231.

Wernegreen, JJ, Kauppinen, SN, Brady, SG, Ward, PS. 2009. One nutritional symbiosis begat another: Phylogenetic evidence that the ant tribe Camponotini acquired *Blochmannia* by tending sap-feeding insects. *BMC Evolutionary Biology,* 9: 292.

Werren, JH, Baldo, L, Clark, ME. 2008. *Wolbachia:* Master manipulators of invertebrate biology. *Nature Reviews Microbiology,* 6: 741–751.

Wesołowska, W. 2006. A new genus of ant-mimicking salticid spider from Africa (Araneae: Salticidae: Leptorchestinae). *Annales Zoologici,* 56: 435–439.

Wesołowska, W, Salm, K. 2002. A new species of *Myrmarachne* from Kenya (Araneae: Salticidae). *Wroclaw,* 13: 409–415.

Wetterer, J, Hugel, S. 2008. Worldwide spread of the ant cricket *Myrmecophilus americanus,* a symbiont of the longhorn crazy ant, *Paratrechina longicornis. Sociobiology,* 52: 157–165.

Wheeler, WM. 1900. The habits of *Myrmecophila nebrascensis* Bruner. *Psyche,* 9: 111–115.

Wheeler, WM. 1907. The polymorphism of ants, with an account of some singular abnormalities due to parasitism. *Bulletin of the American Museum of Natural History,* 23: 1–93.

Wheeler, WM. 1908a. Studies on myrmecophiles, II: *Hetaerius*. *Journal of the New York Entomological Society,* 16: 135–143.

Wheeler, WM. 1908b. Studies on myrmecophiles, I: *Cremastocheilus*. *Journal of the New York Entomological Society,* 16: 68–79.

Wheeler, WM. 1910. *Ants: Their Structure, Development and Behavior.* New York: Columbia University Press.

Whitaker, LM. 1957. A résumé of anting, with particular reference to a captive orchard oriole. *Wilson Bulletin,* 69: 194–262.

Whitford, WG, Bryant, M. 1979. Behavior of a predator and its prey: The horned lizard (*Phrynosoma Cornutum*) and harvester ants (*Pogonomyrmex* spp.). *Ecology,* 60: 686–694.

Whitman, DW. 2008. The significance of body size in the Orthoptera: A review. *Journal of Orthoptera Research,* 17: 117–134.

Wild, AL, Brake, I. 2009. Field observations on *Milichia patrizii* ant-mugging flies (Diptera: Milichiidae: Milichiinae) in KwaZulu-Natal, South Africa. *African Invertebrates,* 50: 205–212.

Williams, R, Whitcomb, W. 1974. Parasites of fire ants in South America. *Proceedings of Tall Timbers Conference on Ecological Animal Control by Habitat Management,* 5: 49–59.

Willis, EO. 1972. The behavior of spotted antbirds. *Ornithological Monographs,* 10: 1–162.

Willis, EO, Oniki, Y. 1978. Birds and army ants. *Annual Review of Ecology and Systematics,* 9: 243–263.

Willson, SK. 2004. Obligate army-ant-following birds: A study of ecology, spatial movement patterns, and behavior in Amazonian Peru. *Ornithological Monographs,* 55: 1–67.

Wilson, EO. 1971. *The Insect Societies.* Cambridge, MA: Belknap Press of Harvard University Press.

Wilson, EO. 1976. The organization of colony defense in the ant *Pheidole dentata* Mayr (Hymenoptera: Formicidae). *Behavioral Ecology and Sociobiology,* 1: 63–81.

Winch, JM, Riley, J. 1985. Experimental studies on the life-cycle of *Raillietiella gigliolii* (Pentastomida: Cephalobaenida) in the South American worm-lizard *Amphisbaena alba:* A unique interaction involving two insects. *Parasitology,* 91: 471–481.

Wing, M. 1951. A new genus and species of myrmecophilous Diapriidae with taxonomic and biological notes on related forms. *Transactions of the Royal Entomological Society of London,* 102: 195–210.

Wirth, S, Moser, JC. 2008. Interactions of histiostomatid mites (Astigmata) and leafcutting ants. In *Integrative Acarology: Proceedings of the 6th European Congress,* ed. M Bertrand, S Kreiter, K McCoy, A Migeon, 378–384. Monpellier, France: European Association of Acarologists.

Wirth, S, Moser, JC. 2010. *Histiostoma blomquisti* n. sp. (Acari: Histiostomatidae): A phoretic mite of the red imported ant, *Solenopsis Invicta* Buren (Hymenoptera: Formicidae). *Acarologia,* 50: 357–371.

Wirth, W, Robinson, W, Kempf, W. 1978. The Rev. Thomas Borgmeier, O. f. m. 1892–1975. *Proceedings-Entomological Society of Washington (USA),* 80: 141–144.

Wisniewski, J, Hirschmann, W. 1992. Gangsystematische Studie von 3 neuen Antennophorus-arten aus Polen (Mesostigmata, Antennophorina). *Acarologia,* 33: 233–244.

Witek, M, Barbero, F, Markó, B. 2014. *Myrmica* ants host highly diverse parasitic communities: From social parasites to microbes. *Insectes Sociaux,* 61: 307–323.

Witek, M, Casacci, LP, Barbero, F, Patricelli, D, Sala, M, Bossi, S, Maffei, M, Woyciechowski, M, Balletto, E, Bonelli, S. 2013. Interspecific relationships in co-occurring populations of social parasites and their host ants. *Biological Journal of the Linnean Society,* 109: 699–709.

Witte, V. 2001. *Organisation und Steuerung des Treiberameisen-verhaltens bei südostasiatischen Ponerinen der Gattung Leptogenys.* PhD dissertation, J. W. Goethe-Universität, Frankfurt / Main.

Witte, V, Hänel, H, Weissflog, A, Rosli, H, Maschwitz, U. 1999. Social integration of the myrmecophilic spider *Gamasomorpha maschwitzi* (Araneae: Oonopidae) in colonies of the South East Asian army ant, *Leptogenys distinguenda* (Formicidae: Ponerinae). *Sociobiology,* 34: 145–159.

Witte, V, Lehmann, L, Lustig, A, Maschwitz, U. 2009. *Polyrhachis lama,* a parasitic ant with an exceptional mode of social integration. *Insectes Sociaux,* 56: 301–307.

Witte, V, Leingärtner, A, Sabaß, L, Hashim, R, Foitzik, S. 2008. Symbiont microcosm in an ant society and the diversity of interspecific interactions. *Animal Behaviour,* 76: 1477–1486.

Witte, V, Schliessmann, D, Hashim, R. 2010. Attack or call for help? Rapid individual decisions in a group-hunting ant. *Behavioral Ecology,* 21: 1040–1047.

Wojcik, D, Smittle, B, Cromroy, H. 1991. Fire ant myrmecophiles: Feeding relationships of *Martinezia dutertrei* and *Euparia castanea* (Coleoptera: Scarabaeidae) with their host ants, *Solenopsis* spp. (Hymenoptera: Formicidae). *Insectes Sociaux,* 38: 273–281.

Wolschin, F, Hölldobler, B, Gross, R, Zientz, E. 2004. Replication of the endosymbiotic bacterium *Blochmannia floridanus* is correlated with the developmental and reproductive stages of its ant host. *Applied and Environmental Microbiology,* 70: 4096–4102.

Wrege, PH, Wikelski, M, Mandel, JT, Rassweiler, T, Couzin, ID. 2005. Antbirds parasitize foraging army ants. *Ecology,* 86: 555–559.

Wright, D. 1983. Life history and morphology of the immature stages of the bog copper butterfly *Lycaena epixanthe* (Bsd. and Le C.) (Lepidoptera: Lycaenidae). *Journal of Research on the Lepidoptera,* 22: 47–100.

Wunderlich, J. 1994. Beschreibung bisher unbekannter Spinnenarten und-Gattungen aus Malaysia und Indonesien (Arachnida: Araneae: Oonopidae, Tetrablemmidae, Telemidae, Pholcidae, Linyphiidae, Nesticidae, Theridiidae und Dictynidae). *Beiträge zur Araneologie,* 4: 559–579.

Yamamoto, S, Maruyama, M, Parker, J. 2016. Evidence for social parasitism of early insect societies by Cretaceous rove beetles. *Nature Communications,* 7: 1–9.

Yan, Y, Li, Y, Wang, W-J, He, J-S, Yang, R-H, Wu, H-J, Wang, X-L, Jiao, L, Tang, Z, Yao, Y-J. 2017. Range shifts in response to climate change of *Ophiocordyceps sinensis,* a fungus endemic to the Tibetan Plateau. *Biological Conservation,* 206: 143–150.

Yanoviak, SP, Kaspari, M, Dudley, R, Poinar, G Jr. 2008. Parasite-induced fruit mimicry in a tropical canopy ant. *American Naturalist,* 171: 536–544.

Yao, I, Akimoto, SI. 2002. Flexibility in the composition and concentration of amino acids in honeydew of the drepanosiphid aphid *Tuberculatus quercicola. Ecological Entomology,* 27: 745–752.

Yélamos, T. 1995. Revision of the genus *Sternocoelis* Lewis, 1888 (Coleoptera: Histeridae), with a proposed phylogeny. *Revue Suisse de Zoologie,* 102: 113–174.

Yoder, JA, Domingus, JL. 2003. Identification of hydrocarbons that protect ticks (Acari: Ixodidae) against fire ants (Hymenoptera: Formicidae), but not lizards (Squamata: Polychrotidae), in an allomonal defense secretion. *International Journal of Acarology,* 29: 87–91.

Yu, DS, Hortsmann, K. 1997. A Catalogue of world Ichneumonidae (Hymenoptera). *Memoirs of the American Entomological Institute,* 58: 1558.

Yu, DS, van Achterberg, K, Horstmann, K. 2007. Biological and taxonomic information of world Ichneumonoidea, 2006. Electronic Compact Disk. *Taxapad,* Vancouver, Canada. http://www.taxapad.com.

Yu, DW, Davidson, DW. 1997. Experimental studies of species-specificity in *Cecropia*-ant relationships. *Ecological Monographs,* 67: 273–294.

Yu, DW, Pierce, NE. 1998. A castration parasite of an ant-plant mutualism. *Proceedings of the Royal Society of London. Series B: Biological Sciences,* 265: 375–382.

Yu, DW, Quicke, D. 1997. *Compsobraconoides* (Braconidae: Braconinae), the first hymenopteran ectoparasitoid of adult *Azteca* ants (Hymenoptera: Formicidae). *Journal of Hymenoptera Research,* 6: 419–421.

Yu, DW, Wilson, HB, Pierce, NE. 2001. An empirical model of species coexistence in a spatially structured environment. *Ecology,* 82: 1761–1771.

Yung, CM. 1938. Morphologische und histologische Studien über Paussidendrüsen. *Zoologische Jahrbücher / Abteilung für Anatomie und Ontogenie der Tiere,* 64: 287–346.

Zagaja, M, Staniec, B, Pietrykowska-Tudruj, E, Trytek, M. 2017. Biology and defensive secretion of myrmecophilous *Thiasophila* spp. (Coleoptera: Staphylinidae: Aleocharinae) associated with the *Formica rufa* species group. *Journal of Natural History,* 51: 2759–2777.

Zakharov, A, Zakharov, R. 2010. The phenomenon of mixed families in red wood ants. *Zoologicheskii Zhurnal,* 89: 1421–1431.

Zhou, Y-L, Ślipiński, A, Ren, D, Parker, J. 2019. A Mesozoic clown beetle myrmecophile (Coleoptera: Histeridae). *eLife,* 8: e44985.

Zientz, E, Beyaert, I, Gross, R, Feldhaar, H. 2006. Relevance of the endosymbiosis of *Blochmannia floridanus* and carpenter ants at different stages of the life cycle of the host. *Applied and Environmental Microbiology,* 72: 6027–6033.

Zimmer, KJ, Isler, ML. 2003. Family Thamnophilidae (typical antbirds). In *Handbook of the Birds of the World,* vol. 8, ed. JD Hoyo, A Elliott, DA Christie. Barcelona, Spain: Lynx Edicions.

Acknowledgments

We gratefully acknowledge the following persons who provided photographic images and illustrations: Gary Alpert, Aleksandrs Balodis, Nicky Bay, Paul Bertner, Christian Brede, Gaspar Bruner, Roberto da Silva Camargo, Sue Chaplin, Zhanqi Chen, Roy Cohutta, John Dawson, Izabela Dziekańska, Peter Eeles, Maria Eisner, Tom Eisner, Konrad Fiedler, Jen Fogarty, Luiz Carlos Forti, Paul Freed, Marion Friedrich, Roy Gross, Amanda Hale, Kwan Han, Bharat Hegde, Gregg Henderson, John Heraty, Hubert Herz, Turid Hölldobler-Forsyth, David Hughes, Robert Jackson, Jeevan Jose, E. Kaiser, Lek Khauv, Kyoichi Kinomura, Roger Kitching, Takashi Komatsu, Pavel Krásenský, Stanislav Krejčík, Daniel Kronauer, Jean-Paul Lachaud, Claude Lebas, Chien C. Lee, Hannu Määttänen, Daniel Martín-Vega, Munetoshi Maruyama, Sean McCann, Konrad Mebert, Mark Moffett, Darlyne A. Murawski, Andry Murray, Margaret Nelson, Ximena Nelson, Joseph Parker, Thomas Parmentier, Stano Pekár, Gabriela Pérez-Lachaud, Martin Pfeiffer, Naomi Pierce, Simon Pollard, Sanford Porter, Diogo B. Provete, Christophe Quentin, Pavan Ramachandra, Wolfgang Sauber, Peter Seufert, Taku Shimada, Marcin Sielezniew, Thomas Stalling, Marie-Lan Taÿ Pamart, Wolfgang Thaler, Michael Thomas, Walter Tschinkel, Benjamin Twist, Iwan van Hoogmoed, Leonard Vincent, Nikolai Vladimirov, Christoph von Beeren, Andreas Weißflog, Tobias Westmeier, Alex Wild, Gil Wizen, Brandon Woo, Steve Yanoviak, Melvyn Yeo.

Konrad Fiedler repeatedly gave us expert advice and clarified several systematic problems we encountered when reviewing the literature on lycaenid myrmecophiles. Al Newton and Margaret Thayer provided multifaceted expert advice throughout our work with myrmecophilous staphylinid beetles. Whenever we consulted Christoph von Beeren, he readily responded and resolved our questions. Jesse E. Taylor gave us valuable advice concerning the systematics of mites. We are grateful to Nobuaki Mizumoto, Mari Muto, and Munetoshi Maruyama, who assisted us in communicating with some of our Japanese colleagues. We are grateful to Friederike Hölldobler for her assistance in preparing the book manuscript. Special thanks go to our friend and colleague Kevin Haight who not only was an important partner in our recent research on myrmecophiles, but also read the entire book manuscript and made many very helpful suggestions. We thank one anonymous reviewer for valuable advice and especially Naomi Pierce for her very detailed, constructive critique and substantial suggestions for chapter 4. C .L. K. gratefully acknowledges the mentorship of Walter R. Tschinkel and his immeasurable contributions to her pursuit of science.

We wish to extend our thanks to our editor Janice Audet, her colleagues Emeralde Jensen-Roberts, Stephanie Vyce, Annamarie Why, and Eric Mulder of Harvard University Press for providing valuable advice that in many ways facilitated the completion of this book and improved its text and visual documentation. Janice Audet made many thoughtful editorial suggestions, which are greatly appreciated. We thankfully acknowledge the superb copyediting by Susan Campbell, the very meticulous proofreading by Shana Jones, and the excellent advice we received from Melody Negron from Westchester Publishing Services who most effectively supervised the final stages of the production process of this book.

Finally, we want to acknowledge the financial support Bert Hölldobler received for the research on myrmecophiles from the German Science Foundation, the National Geographic Society, and Arizona State University.

Index

Note: Page numbers in *italics* refer to illustrations.

and predation, 231–249; staphylinid, trail following by, 260, 266–267. *See also specific genus and species; specific species*

Belytinae, 87

Bengalia: ant species targeted by, 224; brood and booty snatching by, 223–224, *225*

Bengalia emarginata, 223–224

Bengalia jejuna, 223

Bengalia latro, 223

Berghoff, Stefanie, 95–96

Beros, Sara, 34–35

Beugnon, Guy, 265

bicolored antbird, *451, 452–453*

birds, ant-following, 450–456, *451;* ants as prey "beaters" for, 452; butterflies pursuing, 454; calls and eavesdropping of, 455–456; cost exacted on ants, 453; dominance hierarchies of, 454; host specificity of, 454; mixed flocks of, 453–454; nomadic and statary schedule of ants and, 455; obligate followers, 453; occasional followers, 453; regular followers, 453

birds, anting behavior of, 456–460, *458–459, 460*

Birulatus israelensis, 298–299

bivouac, army ant, 64, 249–252

Blattodea, 21

blind snakes, *469, 470–471;* eggs in ant nests, 465, *466;* predation on ants, 467–473

Blochmann, Friedrich, 24

Blochmannia, 24–32, *26–27;* age polyethism and, 30–31; and ant development, 28, 31–32; and ant nutrition, 28–32; and brood-tending ants, 28, 30–31; co-speciation or parallel evolution with *Camponotus,* 25, 31–32; named according to hosts, 25; and nitrogen metabolism, 28–29

Blomquist, Gar, 106, 113

blue jays, 459–460, *460*

Blüthgen, Nico, 436

Bolívar, I., 262

bombardier beetles, 397

bombardier beetles, flanged, 397. *See also* paussine beetles

Boomsma, Koos, 184

booty snatcher beetles *(Homoeusa),* 266–267

booty snatcher flies *(Bengalia),* 223–224, *225*

Borgmeier, Thomas, 48, 65, 68

Bothroponera tesseronoda, 224

Brachymeria, 165

braconid wasps (Braconidae), 70–75; lycaenid butterflies attacked by, 210–214

brain, ant: fungi and, 44–48; trematodes and, 41–43, *42*

"brain worm," 41–43, *42*

Brandstaetter, Andreas, 110–113

Brant, Miriam, 287

Brian, Murray V., 187

brood snatcher flies *(Bengalia),* 223–224, *225*

Brown, Brian V., 49, 54, 62–65

Bruchopria hexatoma, 90

Bruchopria pentatoma, 89–90

Brückner, Adrian, 96

Bruesopria, 90

Bruesopria americana, 90, *90–91*

Bruesopria severi, 90

Brumfield, Robb T., 453

bullet ants: phorid (scuttle) flies and, 62–64; ritual use by humans, *476–477, 477;* scorpions as facultative guest of, 295–297, *298*

butterflies: ant-following birds pursued by, 454; identified as myrmecophiles, 21; mutualistic lycaenid, 149–171; predatory lycaenid, 149, 171, 195–209. *See also specific genus and species*

Caenocholax, 440

Calliteara pudibunda, 218

Calomyrmex, 25, *274–275*

Camargo, Roberto da Silva, 303

Cammaerts, Roger, 266, 391, 393–397

camouflage: chrysomelid beetle, 134–140; tineid and other moth larvae, 140–145. *See also* mimicry

Campbell, Dana, 153

Camponophilus, 314

Camponophilus irmi, 330, *330*

Camponotophilus delvarei, 440

Camponotus: anting behavior and, 457, 459; brood and booty snatcher flies *(Bengalia)* and, 223–224; brood pheromones of, 186–187; census / network analysis of myrmecophiles with, 436, 438–441, *440;* crickets and, 317, 333–335, 336, 337; fungi and, 44–48; *Hyalymenus* mimics of, 276; *Lomechusa* beetles and, 369; mites and, 97, 438; mutualistic symbiosis with *Blochmannia,* 24–32, *26–27;* nestmate recognition among, 111–112; parasitoid wasps and, 72–73, 438–439; phorid (scuttle) flies and, 54, 63–64; polymorphic workers of, 335; reptile eggs in nests of, 464; riodinid butterflies and, 195; snakes and, 469–470; spider mimics of, 277; syrphid flies and, 131–132, 145, 438–439;

trematodes and, 41–43; *Xenodusa* beetles and, 378. *See also specific species*

Camponotus acvapimensis, 277

Camponotus americanus, 46

Camponotus castaneus, 46

Camponotus consobrinus, 469–470

Camponotus floridanus, 28–31, 111–112

Camponotus herculeanus, 26–27, 30, 97

Camponotus japonicus, 111

Camponotus latangulus, 276

Camponotus ligniperdus, 24, 229, 369

Camponotus maculatus, 187

Camponotus modoc, 131, 145, 317

Camponotus nearcticus, 457

Camponotus niveosetosus, 187

Camponotus pennsylvanicus, 46, 54, 459

Camponotus pittieri, 276

Camponotus rufoglaucus, 223, 224

Camponotus samius, 333

Camponotus sansabeanus, 336, 337

Camponotus sericeiventris, 30, 44, 457

Camponotus sericeus, 224

Camponotus textor, 436, 438–441, *440*

Camponotus troglodytes, 221

Camponotus vicinus, 317

Camptosomata, 134–140

capuchin monkeys, 457

carabid beetles (Carabidae), 267, 397–408. *See also* paussine beetles

Carebara urichi, 7

Carlin, Norman, 186

carpenter ants: mutualistic symbiosis with *Blochmannia,* 24–32, *26–27. See also Camponotus*

carrion crows, 456, *458–459*

Cataglyphis: braconid wasps and, 72, 73; histerid beetles and, 118; hydrocarbons of, 108–109; mites and, 103

Cataglyphis bicolor, 103

Cataglyphis bombycina, 103

Cataglyphis ibericus, 73

Cataglyphis niger, 108–109

Cataglyphis savignyi, 103

Cataulacus, 277

Cataulacus brevisetosus, 221

caterpillars. *See* butterflies

cats, anting behavior of, 457

censuses, 436–441

Cephalotes, 37–41, *38–41;* corpse-carrying spiders and, 287–288, *288–289;* membracid bugs mimicking, *283;* nematodes and, 37–41, *38–39*

Cephalotes atratus, 37–40, *38–41*

Ceratoconus setipennis, 65

cestodes, 33–35

of *Haeterius* beetles, 121; of *Pella* beetles, 243–244, *245*

Deeleman-Reinhold, Christa, 290

Dejean, Alain, 171, 265

Dendrocincla fuliginosa (plain-brown woodcreepers), 452–453

Dendrocincla merula, 454

Dendrocolaptes sanctithomae, 452–453

Dendrophillus pygmaeus, 427–428, *429*

dermanyssoid mites, 97–98

Dettner, Konrad, 242, 245

DeVries, Philip J., 155, 158, 190

Deyrup, Mark, 89

Di Giulio, Andrea, 402–408

Diacamma: crickets and, 322; spider mimics of, 290–291, *292*

Diacamma rugosum, 290–291, *292*

diapriid wasps (Diapriidae), 87–92

diapriinine wasps (Diapriinae), 87–90

Diartiger fossulatus, 391–392, *392*

Dicrocoelium dendritcum, 41

Dilocantha lachaudii, 83, *84–85*

Dinarda, 350–358; abnormal *vs.* normal hosts of, 350; appeasement glands of, 356–358, *357, 358;* defensive response of, 352, 356–358, *357,* 358; evolutionary process of, 351–352, 358; in *Formica* ecosystem, 422, 429; host preferences of, 350–351; integration of, 348, 352; location in nest, 352, *352–353,* 356, 358; trail following by, 262; trophallaxis by, 350, 352–356, *354, 355,* 358, 426–427. *See also specific species*

Dinarda dentata, 350–358, *351–352;* location in nest, 352, *352–353,* 356, 358; trail following by, 262; trophallaxis by, 350, 352–356, *354, 355,* 358

Dinarda hagensii, 350, 354

Dinarda lompei, 350

Dinarda maerkelii, 350, 352, 354, 356, 358, 422, 426–427

Dinarda pygmaea, 350

Dinomyrmex (Camponotus) gigas, 63–64, 330, *330*

Dinter, Karin, 267

diplophoresy, 302

Diptera. *See* flies

Disney, Ronald Henry L., 48–49, 66, 68

division of labor: age polyethism and, 12; consequences of, 10; ecological success from, 1; in reproduction, 2, 13–20; among worker ants, 2, 5–13

Dodd, F. P., 76–77, 144

Dolichoderus: beetles and, 115–117; miletine butterflies and, 206; tineid moths and, 142–144

Dolichoderus bidens, 115–117

Dolichoderus bituberculatus, 142–144

Donisthorpe, Horace, 21; on *Claviger* beetles, 389–393; on *Clytra* beetles, 135–138; on *Dinarda* beetles, 351; on ichneumonid wasps, 75–76; on *Lomechusoides* beetles, 379; on *Pella* beetles, 234, 349–350; on phorid flies, 57–60; on syrphid flies, 443, 445

dopamine, lycaenid butterfly secretions and, 155–156

dorsal nectary organ, of lycaenid butterflies, *150,* 150–159, *151, 152–153;* amino acids in secretions, 154–156, 159; behavior-changing effect of secretions, 155–156; cost / benefit ratio of secretions, 170–171; developmental cost of secretions, 166; miletine, 209; neuromodulating component of secretions, 158–159; nitrogenous compounds in, 154–155; nutritive role of, 156–157, 159, 166; parasitoid wasps and, 210–214, *213–215;* in predatory *Phengaris,* 195–196; sugars in secretions, 153–154, 156, 170

Dorylinae, 432. *See also* army ants

Dorylus, 260–261, 409

Dracula beetles, 398–399, 402. *See also* paussine beetles

Drusilla inflatae, 258–260, *260–261*

Drusilla (Santhota) sparsa, 236, *238–239*

Duffield, Richard, 129

DuPont, Steen, 201

Durán, José-Maria, 72–76

Durey, Maëlle, 445

Dziekańska, Izabela, 215

eavesdropping birds, 455–456

Ecclisister, 102–103

Ecitocryptus, 255

Ecitomorpha, 255, 255–256, 260

Eciton: beetles and, 102–103; beetles and, mimetic (myrmecoid), 252–260, *254, 255, 272;* beetles and, staphylinid, 251–262; birds and, ant-following, *451,* 451–456; census / network analysis of myrmecophiles with, 437–438; colony size and myrmecophile diversity with, 409; diapriid wasps and, 87, 89; mites and, 92–98, *94;* myrmecophile body size and, 332; myrmecophiles on trails of, 249–262; phorid commensals with, 64; raiding columns of, 249–252, *252;* static and migratory periods of, 250. *See also specific species*

Eciton burchellii: beetles and, 102–103, 251, *254, 255, 272;* birds and, *451,* 451–455; diversity of myrmecophiles with, 437;

mites and, 95–96; nomadic and statary schedule of, 455

Eciton burchellii foreli, 255

Eciton dulcium, 89

Eciton dulcius, 93–95, *94*

Eciton hamatum, 98, 251–252, *252, 253,* 255

Eciton lucanoides, 255

Eciton mexicanus, 102

Eciton quadriglume, 87

Ecitophya, 255, 255–256, 260

ecological network analysis, 436–441, 449

ecosystem, of ant nests, 22–23, 409–449; army ants, 421, 432–441; colony traits and resistance to infestation in, 441–449; network analysis and colony-level censuses of, 436–441; *Pogonomyrmex badius,* 411–420; wood ant *(Formica rufa),* 420–432

Ectatomma: diversity of myrmecophiles with, 410–411, 436; eucharitid wasps and, 82–83, *84–85;* fungi and, 47; *Hyalymenus* mimics of, 276; mites and, 96–97

Ectatomma quadridens, 276

Ectatomma ruidum, 83, 96–97

Ectatomma tuberculatum, 83, *84–85,* 276, 410–411, 436

ectoparasites / ectoparasitoids: beetles, 103–104, 398–399; braconid wasps, 74; effects at superorganism level, 104; eucharitid wasps, 82, 83; mites, 92–93, 92–95, 96; orasemine wasps, 82; paussine beetles, 398–399, 402

eels, 460–461

eelworms, 35–41

egg mimicry, by crickets, 341–343

Eibl-Eibesfeldt, Eleonore, 49–50

Eibl-Eibesfeldt, Irenaeus, 49–50

Eidmann, Herman, 347

Eisner, Tom, 459–460

Elasmosoma, 72–73

Elasmosoma luxemburgense, 72–73

Elgar, Mark A., 300–301

Elizalde, Luciana, 52

Elmes, Graham, 129–130, 172, 215–218, 443

Elton, Charles Sutherland, 409

Elzinga, Richard, 97–98

Endler, Annett, 110–112

endoparasites / endoparasitoids. *See* ant bodies, parasites on or in

enemy-specific alarm recruitment, 53–54, *54–55*

Eriksson, Ti, 310

Escherich, Karl, 103, 137–138, 379, 397, 399–400

legless worm lizards, 464

Lenoir, Alan, 117–119

Lepidochrysops ignota, 187

Lepidopria pedstris, 90–92

Lepidoptera. *See* butterflies

Lepisiota (Acantholepis) capensis, 157–158, 187

Lepisiota frauenfeldi, 333

Leposternon microcephalum, 464

Leptacinus formicetorum, 422

Leptodeira annulata, 465

Leptogenonia roslii, 433–435

Leptogenys: census / network analysis of myrmecophiles with, 436; diversity of myrmecophiles with, 251; nest ecosystem of, 421, 432–441; nonintegrated *vs.* integrated species with, 433–436. *See also specific species*

Leptogenys borneensis, 433–436

Leptogenys chinensis, 224

Leptogenys distinguenda: mites and, 96; nest ecosystem of, 421, 433–436; silverfish and, 145–146; spiders as integrated guests of, 295, 297; spiders following trails of, 261–262, 262–263, 295; swarm raids of, 433

Leptothorax, 410

Leptotyphlopidae, 467. *See also* blind snakes

lesser liver fluke, 41

limulodid beetles (Limulodidae), trail following by, 260–261

Linepithema humile, 65, 325, 472

Linepithema oblongum, 126

Linksvayer, Timothy, 51

Linsenmair, Eduard, 85–86

Liometopum: crickets and, 338–339, 341; polymorphic workers of, 335

Liometopum apiculatum, 338–339, 341

Liophrurillus flavitarsis, 291

Liotyphlops albirostris, 465, 466

Liphyra: geographic distribution of, 201; predatory, 201–204; protective structures (carapace) of, 201–203, 202, 203, 271; secretory pores of, 202; slug-like larvae of, 201

Liphyra brassolis, 201–204, 205, 271

liquid exchange. *See* trophallaxis

Lithodytes lineatus, 461–463, 462–463

lizards: eggs in ant nests, 464–465, 473; horned, 450, 466–467, 468; legless, 471, 473–476, 474, 475

locomotory mimicry, 291–294, 293; spider, 272–274

Logania malayica, 204–206

Loiácono, Marita S., 87, 89–90

Lomechusa, 358–389; adoption process of, 370–374, 370–376, 385–386; ant contact and handling, 361, 362, 435; ant grooming and tending of, 365–367, 377; appeasement (adoption) glands of, 370–373, 372–373, 374, 385–386; brood pheromones and, 186, 187–190, 385–386; brood predation by, 364, 364, 377; cannibalism of, 364, 422; chemical communication by, 365–376, 385–386; cost of social food flow to, 410; known species of, 359; nomenclature for, 358–359; photographic documentation of, 376, 376–377; pupal cradle of, 366–367, 367; seasonal changes of hosts, 359–361, 369; sensitivity to host odor, 369–370; social integration of, 257, 348; trichromes of, 372–373, 373, 385–386; trophallaxis by, 362–369, 363, 368–369, 371, 376, 377, 384–385. *See also specific species*

Lomechusa atlantica, 360

Lomechusa emarginata, 360, 370–372, 373, 375, 376

Lomechusa myops, 360

Lomechusa paradoxa, 360

Lomechusa pubicollis, 358–359, 360–373, 380; adoption glands / process of, 370, 370–372, 372–373, 374; food solicitation / trophallaxis by, 363, 368–369, 376, 385; pupal structure / cradle of, 366–367, 367; seasonal hosts of, 360, 369

Lomechusa sinuate, 373–376, 375–377

Lomechusina, 359

Lomechusini, 359

Lomechusoides, 358–389; aberrant worker morphology with, 379, 410; adoption process of, 379–382, 382–383, 385–386; ant contact and handling, 361, 362, 363, 382, 435; ant grooming and tending of, 365, 365–367; appeasement (adoption) glands of, 372–373, 373, 374, 378–382, 382–383, 385–386; brood pheromones and, 186, 187–190, 385–386; brood predation by, 364, 364; cannibalism of, 364, 422; chemical communication by, 365–367, 385–386; cost of social food flow to, 410; nomenclature for, 358–359; social integration of, 257; specialized adaptations of, 386–389; trichromes of, 372–373, 373, 378–379, 380, 381, 385–386; trophallaxis by, 382–385, 383, 384–385

Lomechusoides strumosus, 358–359, 361, 378–382; adoption glands / process of, 372–373, 374, 380–383; ant contact and handling, 361, 362; brood predation by, 364; chemical code broken by, 385; cost of social food flow to, 410

Longino, Jack, 457

Lubbock, John, 57

lungworm, 473–474, 475

Lupinus, 160–161

Lycaena tityrus, 156–157

lycaenid butterflies (Lycaenidae), 149–219; benefits from association with ants, 159–171; brood pheromones / pheromone mimics and, 185–190; coloration and markings of, 149; cuckoo strategy (vibrational communication) of, 190–195, 198–200; food solicitation / regurgitation and, 149, 177, 177; hydrocarbons and, 145, 157; indirect parasitism (tripartite symbiosis) of, 207–209; morphology and ant association, 149–153, 150, 151, 152–153; mutualistic, 149–171; nectary organ / secretions of, 150, 150–159, 151, 152–153, 166, 170–171, 195–196, 210–214; nests provided by ants, 166, 167, 168; neuromodulating effect of, 158–159; nutritive contribution to ants, 156–157, 159, 166; oviposition cues for, ants as, 164, 167–169; parasitoids attacking, 161, 165, 210–219; pore cupolas of, 149–150, 150, 151, 196, 206, 210–212; predatory, miletine, 200–209; predatory *Phengaris*, 171, 195–200; protection provided by ants, 160–161, 161, 164–165, 165, 210–212; socially parasitic *Phengaris*, 171, 173–195; tentacle (lateral) organ of, 150, 151, 152–153, 157–159; trail following by, 265

Lyprocorrhe anceps, 422, 425

Lysandra coridon, 151

Lysandra hispana, 153–154

Macaranga plant, 207–209

Macrocheles rettenmeyeri, 93–95, 94

Macrodinychus, 96–97

Macrodinychus derbyensis, 96

Macrodinychus hilpertae, 96

Macrodinychus sellnicki, 96–97

Maculinea, 172–173. See also *Phengaris*

Mahsberg, Dieter, 287

majors (worker ant subcaste), 5, 7

Malayatelura ponerophila, 145–146

mammals, anting behavior of, 457

Margolies, David, 136, 139

Martineziana dutertrei, 113–115, 114–115, 145

Martín-Vega, Daniel, 41–42

Maruyama, Munetoshi, 22, 66, 247–248, 251, 257–258, 359, 369, 388

Maschwitz, Ulrich: on army ants, 251, 432; on *Bengalia* flies, 223–224; on lycaenid butterflies, 153–154, 156–158, 166–167, 170, 175–176, 204–208, 211; on milichiid flies, 269; on phorid flies, 58–60, 65–66; on staphylinid beetles *(Pella)*, 231, 241–242

Neotropical arborrheal myrmicine ant. See
    Cephalotes atratus
Neotypus, 214–215
nest(s): barracks, 11, *11*; bivouac, of army
    ants, 64, 249–252; colony founding in,
    3–5; construction of, 3, 17; disturbance,
    and phorid flies, 60, 62; extension and
    maintenance of, 5, 8; inquilines exploiting,
    347–349; and lycaenid butterflies, 166,
    167, 168; polydomous, 16, 117, 265–266,
    280, 442, 445; species-specific structural
    features of, 17, *18–19*; superorganism
    concept and, 17; three-dimensional casts
    of, 17, *18*; wood ant mounds, 8–9, *10*
nestmate recognition, 105–113; adjustment
    and complexity in, 112–113; dummy
    experiments on, 109–110; hydrocarbons
    and, 33–34, 106–113; neurobiological
    mechanisms in, 110–112; ritualization
    and, 108
network analysis, 436–441, 449
Newcomer, E. J., 150, 153
Newton, Alfred, 359, 369
niches, 409–411; ant trails as, 409; colony size
    and, 409–411; definition of, 409; spatial
    concept of, 409–410; species, 409. See also
    ecosystem
Niphanda fusca, 195
Niphopyralis aurivilli, 144
Niphopyralis myrmecophila, 144
nitidulid beetles, as highwaymen on *Lasius*
    trails, 225–230
nitrogen metabolism, *Blochmannia* and,
    28–29
nitrogenous compounds, in lycaenid
    butterfly secretions, 154–155
noctuid moths, 171
Notothecta flavipes, 422, *423*
Notoxoides pronotalis, 89
Novomessor: crickets and, 318, 333–335, *336,*
    337, 341; new spider family and, 309–310,
    312–313; rescue behavior of, 306
Novomessor albisetosus, 309–310, *312–313,*
    336, 337, 341
Novomessor cockerelli, 309
Nylanderia (Paratrechina), eucharitid wasps
    and, 82
Nylanderia fulva, 96
Nymphister kronaueri, 102–103, *103–104*

Obeza, 439, *440*
ocellated antbirds, 452–453, *454*
Odontomachus: hydrocarbons of, 248–249;
    nematodes and, 36, *36*; reptile eggs in
    nests of, 464–465

Odontomachus brunneus, 248–249, 464–465
Odontomachus haemotodus, 36, *36*
Odontomachus ruginodis, 248–249
odor: nestmate recognition, 105–113. *See also*
    chemical mimicry
odor cloaks, 294–301. *See also* chemical
    mimicry
Oecophylla: brood and booty snatcher flies
    (Bengalia) and, 224; fungi and, 47; lycaenid
    butterflies and, mutualistic, *152–153,*
    166–171, *168, 169*; lycaenid butterflies and,
    parasitoids attacking, 212, *213*; lycaenid
    butterflies and, predatory, 201–206, *202,*
    205; lycaenid butterflies and, trail
    following by, 265; nests of, 167, *168, 169*;
    noctuid moths and, 171; rescue behavior
    of, 306; *Riptortus* bugs mimicking, 278,
    278; spider mimics of, 276–282, 299–301;
    taxonomy of, 167; territorial defense by,
    11, *11*; tineid moths and, 144. *See also*
    specific species
Oecophylla longinoda, 167; lycaenid butter-
    flies and, 204, 265; nests of, 167, *168, 169*;
    noctuid moth and, 171; spiders and, 277;
    territorial defense by, 11, *11*
Oecophylla smaragdina, 167; brood and
    booty snatcher flies *(Bengalia)* and, 224;
    colony founding by, *4–5*; death-feigning
    beetles and, 243; lycaenid butterflies and,
    mutualistic, *152–153,* 166–171, *169*; lycaenid
    butterflies and, parasitoids attacking,
    212, *213*; lycaenid butterflies and, predatory,
    202, 205; rescue behavior of, 306; *Riptortus*
    bugs mimicking, 278, *278*; spider mimics
    of *(Cosmophasis bitaeniata),* 299, 299–301;
    spider mimics of *(Myrmaplata plataleoides),*
    276–282; tineid moths and, 144; wasp
    parasitoid of, 77, *78*
Oliveira, Paulo, 32, 288
Oniki, Yoshika, 451–452
Oodinychus, 97
Oodinychus spatulifera, 97
Ophiocordyceps, 43–48
Ophiocordyceps sinensis, 43–48
Ophiocordyceps unilateralis, 44–46
Orasema minuta, 82
Orasema simplex, 79
Orasema simulatrix, 79, 85
orasemine wasps (Oraseminae), 77–82; ant
    hosts of, 80–82; in ant nest (brood), 80,
    80–81; behavioral impact of, 86; chemical
    mimicry by, 115; ecological impact of,
    85–86; life cycle of, 77–79, *79*; oviposition
    by, 77–80, *78, 79,* 85; plant hosts of, 79–80
Orectognathus, 410

Orivel, Jérôme, 115–117
Orthoptera, 21
Oxypoda, 255
Oxypodini, 350
Oxypria, 88
Ozaenini, 397

Pachycondyla harpax, 68–69
Pachycondyla villosa, 133
Pachythone xanthe, 210
painted ant-nest frog, 461–463, *462–463*
Paltothyreus tarsatus, 463, 464–465
Paralipsis eikoae, 115, *116–117*
Paralucia aurifera, 156
Paraponera clavata: phorid (scuttle) flies
    and, 62–64; ritual use by humans,
    476–477, 477; scorpions as facultative
    guest of, 295–297, *298*
parasite(s): among ant species, 20; fossil
    evidence of, 20–21; niches for, 19–20; of
    other organisms, 20; parasitoids *vs.,* 33;
    social, 20; in superorganism, 19–23.
    See also specific parasites and hosts
Parasites in Social Insects (Schmid-Hempel),
    22, 69
parasitoids: affecting ant appearance and
    behavior, 33–92; *vs.* parasites, 33. *See also*
    specific parasitoids and hosts
paratergite glands, of beetles, *372–373,* 373
Paratrechina, 82
Paratrechina longicornis, 321, 322–324, 340, 345
Parawroughtonilla hirsutus, 433–435
Parker, Joseph, 22, 257–258, 386, 388
Parmentier, Thomas, 22; on *Clytra* beetles,
    138, 265–266; on *Dinarda* beetles, 352,
    354; on hydrocarbon profiles, 147, 248; on
    Pella beetles, 249; on trophic interactions,
    427, *428*; on wood ant ecosystem, 138,
    248, 265–266, 347–349, 420–432, 436–437
Passera, Luc, 87, 92
Pasteels, Jacques M., 236, 238, 266, 370
paths. *See* trails, ant
paussine beetles (Paussinae), 266, 397–408;
    adoption and integration of, 399–401;
    contact with ant queen, 401–402; Dracula-
    like preying of, 398–399, 402; head and
    mandibles of, 398, 402; liquid exuded by,
    403, 403–404; morphologically modified
    larvae of, 402, *402*; rewards from, 400–402;
    stridulation by, 404–408; taxonomy of,
    397–398
Paussini, 397
Paussus arabicus, 400–401
Paussus favieri, 266, 389, 398–408, *399*
Paussus megacephala, 398